Developments in Agricultural and Managed-Forest Ecology, 8

MANAGEMENT AND ANALYSIS OF BIOLOGICAL POPULATIONS

OTHER TITLES IN THIS SERIES

1. Plant Modification for More Efficient Water Use
   by J.F. Stone (Editor)
   1975 xii + 320 pp.

2. Tree Ecology and Preservation
   by A. Bernatzky
   1978 viii + 358 pp.

3. Cycling of Mineral Nutrients in Agricultural Ecosystems
   by M.J. Frissel (Editor)
   1978 viii + 356 pp.

4. Interactions between Non-Pathogenic
   Soil Microorganisms and Plants
   by Y.R. Dommergues and S.V. Krupa (Editors)
   1978 xii + 476 pp.

5. Ecology of Root Pathogens
   by S.V. Krupa and Y.R. Dommergues (Editors)
   1979 x + 282 pp.

6. Soil Disinfestation
   by D. Mulder (Editor)
   1979 xiv + 368 pp.

7. Management of Semi-Arid Ecosystems
   by B.H. Walker (Editor)
   1979 x + 398 pp.

9. Comparison of Forest Water and Energy Exchange Models
   by S. Halldin (Editor)
   1980 xiv + 258 pp.

10. Alternative Methods of Agriculture
   by R. Boeringa (Editor)
   1980 vi + 200 pp.

Developments in Agricultural and Managed-Forest Ecology, 8

MANAGEMENT AND ANALYSIS OF BIOLOGICAL POPULATIONS

by

BEAN-SAN GOH

Department of Mathematics, University of Western Australia, Nedlands, W.A. 6009, Australia

ELSEVIER SCIENTIFIC PUBLISHING COMPANY

Amsterdam — Oxford — New York 1980

ELSEVIER SCIENTIFIC PUBLISHING COMPANY
335 Jan van Galenstraat
P.O. Box 211, 1000 AE Amsterdam, The Netherlands

Distributors for the United States and Canada:

ELSEVIER NORTH-HOLLAND INC.
52, Vanderbilt Avenue
New York, N.Y. 10017

Library of Congress Cataloging in Publication Data

Goh, Bean-San.
Management and analysis of biological populations.

(Developments in agricultural and managed-forest ecology ; 8)
Bibliography: p.
Includes indexes.
1. Population biology--Mathematical models.
2. Wildlife management--Mathematical models. I. Title.
II. Series.
QH352.G63 574.5'24 79-17239
ISBN 0-444-41793-1

ISBN 0-444-41793-1 (Vol. 8)
ISBN 0-444-41515-7 (Series)

Printed in The Netherlands

To

AI-TUAN, STEVEN WERN-YI, *and* MICHAEL CHIEN-YI

CONTENTS

PREFACE

In the real world, an ecosystem is subjected to large perturbations of its initial state and continual disturbances on its dynamics. Therefore it is important to establish that an ecosystem model is stable relative to finite perturbations of its initial state and its dynamics. One object of this book is to demonstrate that Liapunov and Liapunov-like functions can be successfully used to establish, in a nonlinear population model, stability relative to realistic perturbations.

It is shown that a complex ecosystem is stable relative to realistic perturbations if it is a collection of self-regulating species such that the sum total of the stabilizing intraspecific interactions dominates the interspecific interactions. A nontrivial criterion for a complex ecosystem model to have this property is given. For a complex ecosystem model to satisfy this criterion, it is desirable that it has many null interactions between pairs of species.

Another object of this book is to demonstrate the usefulness of optimal control theory in the management of biological populations. Management policies should always be implemented in a *feedback* manner. This would reduce some of the adverse effects of uncertainties in an ecosystem in the real world. After all, Nature itself uses feedback to cope with uncertainties in many processes.

Chapter 2 contains a brief survey of static optimization techniques and optimal control theory for systems which are modelled by differential and difference equations. Chapter 3 contains methods which use Liapunov and Liapunov-like functions to establish that a given population model is stable relative to finite perturbations of its initial state and that it is nonvulnerable relative to large continual disturbances. These methods are applied to single species and two-species populations which are modelled by differential and difference equations. Fisheries provide some of the best examples of the usefulness of mathematics in the management of biological populations. A few management problems in fisheries are examined in Chapter 4. They utilize stability and optimal control theories. Chapter 5 considers stability in an ecosystem model with complexities due to species richness, nonlinearities, time delays and spatial heterogeneity. The very interesting and important problem of pest management is considered in Chapter 6.

I have greatly benefited from discussions with T.T. Agnew, C.W. Clark, M.E. Fisher, G.W. Harrison, A. Hastings, C.S. Hsu, L.S. Jennings, G. Leitmann, D. Ludwig, T.L. Vincent and C.J. Walters. Many of the results in this book were obtained in collaboration with Tom Vincent. This book was completed during my sabbatical leave which was supported by NRC grant No. 3990. I thank Colin Clark and members of the Mathematics Department of the University of British Columbia for their hospitality.

Chapter 1

INTRODUCTION

1.1. MODELS IN ECOLOGY

As in other areas of science and engineering, there exists a wide spectrum of models in ecology. At one end of the spectrum are relatively simple but general models which are used to analyse theoretical questions in ecology. These models are called strategic models by theoretical ecologists. At the other end of the spectrum are detailed computer simulation models; theoretical ecologists call these tactical models. However, some system analysts prefer to refer to simulation models as realistic models. Disclosed in this difference in terminology is the difference in the emphasis that each group places on the models.

It is desirable to have a wide range of models in order to achieve different purposes. For example, consider the flight of a plane between two cities. For the average passenger on the plane, it would be adequate to model the flight of the plane by a particle moving at a constant speed in a straight line over a flat earth from one city to another. The pilot, however, needs a cluster of models which includes strategic models of the dynamics of the plane, and relatively detailed models of the control systems which operate the wings and tail of the plane, the engines, the wheels, the brakes and other subsystems. The pilot's understanding of these models, which are not necessarily specified in explicit mathematical terms, is obtained from introductory courses on aerodynamics and experiences on simulators and planes. Finally, the team of engineers who design the plane requires a deep understanding of the detailed (tactical) models of the aerodynamics of the plane, the engines, the control systems and other subsystems. This understanding of the behavior of the detailed conceptual and mathematical models is obtained from mathematical analyses, simulations on computers, simulations of physical models of the plane in wind tunnels and test flights of the prototype plane.

In the field of ecology, a broad range of models serves different objectives. Strategic models are needed for isolating theoretical questions and studying them one at a time. They also serve as a means of communicating with colleagues, students of ecology and scientists in other fields, in a precise and objective manner the latest advances achieved in the understanding of the dynamical behavior of ecosystems.

On the other hand, tactical models are required for formulating the appropriate control policies in the management of a particular renewable resource or a specific pest population. In this case, only the user of the model needs to have a good understanding of the behavior of the model. The solution of these applied problems is very important in order to maintain strong support for ecological studies. Moreover, simulation models pose many interesting and pertinent questions for the theoretical ecologist to study.

The analyses of strategic models of an ecosystem may provide useful guidelines in the construction of a detailed simulation model. They can also provide suggestions on how to exercise a computer model of a community. For instance, the analysis of a simplified pest—predator system suggests that in the control of pest population, it is sometimes desirable to release pests. Without this result from the simplified model, it would sound absurd to consider the release of pests in a simulation model for managing a pest population. Ideally we should use a combination of strategic and tactical models in studying ecological problems.

1.2. STABILITY CONCEPTS IN ECOLOGY

The word "stability" is used for various purposes in ecology. It is used to describe a lack of change in population levels or population parameters; and to describe the persistence of an ecosystem. In mathematics there are many distinctly different concepts of stability (Kalman and Bertram, 1960 a; Willems, 1970). Recently, several authors (Lewontin, 1969; Holling, 1973; May, 1974; Maynard Smith, 1974; Goh, 1975) have attempted to reconcile the various concepts of stability which are used in ecology and mathematics.

Each concept of stability in ecology could be made more precise by applying it to a mathematical model of an ecosystem. For a precise definition of a concept of stability it is necessary to specify explicitly: (i) the class of admissible perturbations; (ii) the set of admissible initial states; (iii) a set of system responses which characterize desirable or undesirable behavior; and (iv) a time interval. Usually the infinite time horizon $(0, \infty)$ is used. Each set of specifications of these components of stability leads to a particular concept of stability.

Four types of perturbations on an ecosystem model are: (i) impulsive and infrequent perturbations of the initial states; (ii) continual disturbances on the system dynamics; (iii) impulsive and infrequent changes in the system parameters; and (iv) slow and continual changes in the system parameters. The effects of types (i) and (iii) perturbations can be examined together; and those of types (ii) and (iv) can be studied together.

The most common method of studying stability in an ecosystem model is by an examination of the eigenvalues of a matrix at an equilibrium. This method establishes stability only relative to small perturbations of the initial state. Hence it is called *local* stability. It implies that if the initial state is displaced a small distance from an equilibrium and the system is thereafter left alone, then the natural dynamics will drive the state into a decreasing neighbourhood of the equilibrium. Clearly an eigenvalue analysis is only a small initial step in understanding the dynamical behavior of an ecosystem model.

In the real world ecosystems are subjected to large perturbations of the initial state and system dynamics. The most powerful analytical method for studying stability relative to finite perturbations of the initial state of an ecosystem model is the direct method of Liapunov. This method was discovered in 1892 by the Russian mathematician A.M. Liapunov (see Liapunov, 1966). But the method was ignored by mathematicians for a long time. In the West, it was not till the nineteen-fifties that it became a popular tool for the analysis of nonlinear systems.

The direct method of Liapunov requires the construction of certain functions called Liapunov functions. For physical systems the direct method of Liapunov generalizes the principle that a system, which continuously dissipates energy until it attains an equilibrium, is stable.

The population of each species in a spatially homogenous and well mixed ecosystem must be nonnegative. This requires that the concept of global stability for engineering systems must be modified in ecological models. By definition an ecosystem model is globally stable if every trajectory of the model which begins at a positive state remains in the positive orthant for all finite values of the time variable t, and converges to a positive equilibrium as $t \to \infty$. The fact that the density of a viable population in a spatially homogeneous ecosystem must be positive requires that a two-sided energy principle should be used for constructing Liapunov functions for this class of models (Goh, 1977 a). A viable spatially homogeneous single-species population must have net energy absorption when its population level is low; and it must have net energy dissipation when its population level is high. In a spatially homogenous ecosystem the population of each species at extreme densities should have the behavior of a viable single species population.

In an ecosystem model, let $S(0)$ be a set of desirable initial states, $Z(T)$ be a set of undesirable initial states, U be a set of admissible disturbances, and $[0, T]$ be a given time interval. By definition, the ecosystem model is *nonvulnerable* relative to the sets $S(0)$, $Z(T)$ and U during the time interval $[0, T]$, if there is no admissible disturbance which drives the system from a state in $S(0)$ to the set $Z(T)$ during the time interval $[0, T]$. Liapunov-like functions can be used in an effective manner to study nonvulnerability in the class of generalized Lotka–Volterra models, and in a class of nonlinear models.

1.3. ECOLOGICAL ENGINEERING

Ecological engineering problems may be divided into two classes, namely, *design* problems and *control* problems. The question as to whether or not it is desirable to import and introduce a species into an ecosystem is a design problem. On the other hand the formulation of spraying programs for applying an insecticide in the management of a pest is a control problem.

To obtain good and reliable solutions to a design problem, it is necessary to have a relatively accurate model (mathematical or conceptual) of the ecosystem which is being manipulated or created. A reason for this stringent requirement is that it may not be possible to reverse the effects of a decision in a design problem.

In a control problem, there are usually some variables or parameters that can be easily manipulated between certain bounds. These variables may be used in a control program in two ways: (i) in an *open-loop* manner, e.g. the harvesting of a population under a constant quota (yield) policy; and (ii) in a *feedback* (closed-loop) manner, e.g. the harvesting of a population using a fixed effort policy. In the second example, the rate of harvesting is proportional to the product of the applied effort and the population. It is interesting to note that in this example, the state of the system need not be monitored at all, the reason being that the feedback nature of the impact of the control variable on the population dynamics is built-in.

Let x denote the state vector $(x_1, x_2, \ldots, x_n)$ of a system, and u denote the control vector $(u_1, u_2, \ldots, u_m)$. By definition, each function $u(x)$ determines a control policy. A feedback control policy $u(x)$ is usually determined from an analysis of a deterministic model of the system. In feedback control we must specify a target to which the state of the system must be driven. The target could be a point or a region in the state space. A control policy must be able to drive the system from every admissible (operating) initial state to the target. If otherwise, the system is not fully controllable.

The way in which a feedback control policy is *implemented* is very important. In feedback control the state of the system is continuously monitored, and the control variables are adjusted according to a prescribed control policy $u(x)$. It is the updating of the control variables which enables a feedback control system to neutralize some of the adverse effects of uncertainties.

In practice the state of the system is sampled only at discrete times $t, t + \Delta t, t + 2\Delta t, \ldots$ and the control variables are adjusted accordingly. The effectiveness of feedback control to neutralize the adverse effects of a given class of uncertainties decreases as Δt increases. It is usually very difficult to make good estimates of the population levels in an ecosystem. In many cases the interactions between species are not well understood. In spite of these difficulties, management control policies should always be implemented in a feedback manner using the best estimates of the state of the system. A management policy which is not in a feedback form is an improperly designed policy.

Walters and Hilborn (1978) have suggested that generally stochastic optimal control policies which take into account random disturbances on the system dynamics and random observation errors are similar to feedback control policies which are formulated using deterministic models. This problem on the relative merits of using a stochastic optimal control policy and a feedback optimal control based on a deterministic model of a system is very important in the management of ecosystems. For the general linear regulator problem, Fleming and Rishel (1975, p.166) have rigorously proved that the optimal feedback control law is the *same* for the stochastic model as for the deterministic model. For nonlinear problems this question requires further studies (Reed, 1974; Walters, 1975; Walters and Hilborn, 1976).

It is possible to design an efficient feedback control policy using only a crude model of a system if we have available an independent control variable for each state equation. In this case the control variables can be used to enhance the desirable trends in the dynamics of a system and drive it to a prescribed target. This will be demonstrated in Section 6.3.

1.4. LIAPUNOV FUNCTIONS

We shall discuss briefly the properties of a Liapunov function and a general method for constructing Liapunov functions. The applications of Liapunov functions are considered in subsequent chapters.

Let us consider the properties of a Liapunov function for a model in which there are no sign restrictions on the state variables. Suppose the system

$$\dot{x}_i = f_i(x_1, x_2, \ldots, x_m), \quad i = 1, 2, \ldots, m \tag{1.4.1}$$

has an equilibrium at $x = (0, 0, \ldots, 0)$. We have

$$f_i(0) = 0, \quad i = 1, 2, \ldots, m. \tag{1.4.2}$$

Let $\|x\|$ be a norm of x. For example,

$$\|x\| = \min \{|x_1|, |x_2|, \ldots, |x_m|\}, \tag{1.4.3}$$

where $|x_1|$ denotes the absolute value of x_1. Let $r = \|x\|$. Let $G(r)$ be a strictly increasing function of r such that $G(0) = 0$, and $G(r) \to \infty$ as $r \to \infty$. By definition, a continuous scalar function $V(x)$ is radially unbounded if there exists a G function with the above properties such that

$$V(x) \geqslant G(\|x\|) \tag{1.4.4}$$

for all x in the state space $\mathbf{R}^m = \{x \mid (x_1, x_2, \ldots, x_m)\}$,

Let $\mathbf{R}$ be an open region of the state space. Let $V(x)$ be a continuously differentiable scalar function in $\mathbf{R}$. By definition $V(x)$ is a Liapunov function of model (1.4.1) if it has the following properties:

(i) $V(0) = 0$; (1.4.5)

(ii) $V(x)$ is radially unbounded; (1.4.6)

and

(iii) $$\dot{V}(x) = \sum_{i=1}^{m} \frac{\partial V}{\partial x_i} f_i(x) \leqslant 0 \quad \text{for all } x \in \mathbf{R}. \tag{1.4.7}$$

Condition (i) is trivial. It can be replaced by the condition, $V(0) = a$, where a is a positive or a negative constant. Condition (ii) can be weakened considerably. For example, LaSalle (1976, p.30) uses a definition of a Liapunov function which only requires that $V(x)$ is a continuous function and that it satisfies condition (iii). However, if condition (ii) is weakened it is necessary to give additional conditions which ensure that every solution of (1.4.1) which begins in a given closed bounded subset of **R** remains in it for all positive values of t. This requirement will be demonstrated in Example 1.4.1. Condition (ii) ensures that the equations, $V(x) = K_p$, $K_1 < K_2 < K_3 < \cdots$, represent a set of nested closed hypersurfaces (Hahn, 1976, p.99). Condition (iii) can be weakened to handle the case when $V(x)$ is not continuously differentiable, and when $f(x)$ is not a continuous vector function (LaSalle, 1976, p.29).

For population models, the $V(x)$ function is often a separable function. This means that

$$V(x) = c_1 V_1(x_1) + c_2 V_2(x_2) + \cdots + c_m V_m(x_m) \tag{1.4.8}$$

where $c_1, c_2, \ldots, c_m$ are positive constants. In this case $V(x)$ is radially unbounded if, for $i = 1, 2, \ldots, m$, $\mathrm{d}V_i/\mathrm{d}x_i < 0$ for all $x_i \in (-\infty, 0)$, $\mathrm{d}V_i/\mathrm{d}x_i > 0$ for all $x_i \in (0,\infty)$, and $V(x_i) \to \infty$ as $|x_i| \to \infty$. This property of a separable function can be established by using the norm given in (1.4.3).

Example 1.4.1. The following counter-example was constructed by Barbashin and Krasovskii (1952) (see also Hahn, 1967, p.109) to demonstrate that a Liapunov function for global asymptotic stability must be radially unbounded.

Let $u = 1 + x^2$ and

$$\dot{x} = -6x/u^2 + 2y,$$

$$\dot{y} = -2(x + y)/u^2. \tag{1.4.9}$$

Consider the function

$$V(x, y) = (x^2/u) + y^2. \tag{1.4.10}$$

We have $V(x, 0) \to 1$ as $x \to \infty$. Hence for $K > 1$, the equation $V(x, y) = K$ does not define a closed surface. It follows that $V(x, y)$ is not radially unbounded.

Computing $\dot{V}$ along solutions (1.4.9) we get

$$\dot{V} = -4(3x^2 + y^2 u^2)/u^4. \tag{1.4.11}$$

Clearly $\dot{V}(x, y) < 0$ for all $(x, y) \neq (0, 0)$. But we shall show that $(0, 0)$ is not globally asymptotically stable.

The hyperbola

$$y = 2/(x - \sqrt{2}), \quad x > \sqrt{2}, \tag{1.4.12}$$

is a barrier to the trajectories of (1.4.9) which begin in the set $\{(x, y) \mid x > \sqrt{2}, y > 2/(x - \sqrt{2})\}$. This is established by comparing the slope of the hyperbola of (1.4.12) and the slopes of the trajectories of (1.4.8). The slope of the hyperbola is

$$p = dy/dx = -2/(\sqrt{2} - x)^2. \tag{1.4.13}$$

The slope of the trajectory of (1.4.9) which passes through the same point on the hyperbola is

$$q = \dot{y}/\dot{x} = -1/(1 + 2\sqrt{2}x + 2x^2). \tag{1.4.14}$$

Eqs. (1.4.13) and (1.4.14) imply that $q > p$ for all points on the hyperbola given in (1.4.12). Hence the solutions of (1.4.9) which begin in the set $\{(x, y) \mid x > \sqrt{2}, y > 2/(x - \sqrt{2})\}$ do not intersect the hyperbola in (1.4.12), i.e. they do not converge to (0, 0) as $t \to \infty$. Therefore (0, 0) is not globally asymptotically stable.

The next theorem gives a simple set of conditions for global stability.

Theorem 1.4.1. *The equilibrium $x = 0$ of (1.4.1) is globally asymptotically stable if there exists a Liapunov function in R^m such that*

$$\dot{V}(x) < 0 \tag{1.4.15}$$

for all $x \in \mathrm{R}^m$ and $x \neq 0$.

Proof. Let the solution of (1.4.1) which begins at the point $x(0) = a$ converge to the point $x(\infty) = b \neq 0$. It follows that $V(b) = L$ where L is a positive constant.

The region $A = \{x \mid L \leqslant V(x) \leqslant V(a)\}$ is a closed bounded region. The function $V(x)$ is a continuous function. It follows that the minimum of $\dot{V}(x)$ for all $x \in A$ exists. Let this be $-M$. The equations, $V(x) = K_p, K_1, < K_2 < K_3 < \cdots$, represent a set of nested closed hypersurfaces. The conditions, $\dot{V} < 0$ and $\lim V[x(t)] = V(b)$ as $t \to \infty$, imply that $x(t)$ remains in A for all $t > 0$. Along the solution which begins at $x(0) = a$, we have

$$V[x(t)] - V[x(0)] = \int_0^t \dot{V}\,dt < \int_0^t -M\,dt = -Mt. \tag{1.4.16}$$

It follows that $V[x(t)] \to -\infty$. As this is impossible, we conclude that the equilibrium $x = 0$ is globally asymptotically stable.

We shall now consider the modifications which are necessary before a Liapunov function, as defined above, can be applied to a population model.

Suppose a model of a spatially homogenous community is

$$\dot{N}_i = N_i F_i(N_1, N_2, \ldots, N_m), \quad i = 1, 2, \ldots, m \tag{1.4.17}$$

where N_i is the population of the ith species. In this model the state variables must be nonnegative. For all the species to persist it is necessary for model (1.4.17) to have a positive equilibrium or a limit cycle. It is also necessary for all the species to be present initially. Thus a natural concept of global stability for a positive equilibrium of (1.4.17) at $N^\star$ is that every solution of the model which begins in the positive orthant $\mathbf{R}_+^m = \{N \mid N_i > 0, i = 1, 2, \ldots, m\}$ must remain in $\mathbf{R}_+^m$ for all finite values of t and converge to $N^\star$ as $t \to \infty$.

We can establish that $N^\star$ is globally stable by transforming the positive orthant $\mathbf{R}_+^m$ into $\mathbf{R}^m = \{x \mid (x_1, x_2, \ldots, x_m)\}$, and then using Theorem 1.4.1. For example, we could employ the transformation

$$x_i = \ln(N_i/N_i^\star), \quad i = 1, 2, \ldots, m. \tag{1.4.18}$$

However, this preliminary transformation is unnecessary. It may also be undesirable because the transformed model could be more complex.

The alternative and better approach is to translate the stability conditions so that they can be used directly for establishing global stability in the positive orthant. This is achieved by translating condition (ii) of (1.4.6) into an appropriate condition.

Using (1.4.18) we deduce that a scalar function $V(N)$ is "radially unbounded" in $\mathbf{R}_+^m$, if the function

$$W(x) = V(N_1^\star \exp x_1, N_2^\star \exp x_2, \ldots, N_m^\star \exp x_m) \tag{1.4.19}$$

is radially unbounded in $\mathbf{R}^m$. This implies that the equations, $V(N) = K_p$, $K_1 < K_2 < K_3 < \cdots$, represent a set of nested closed hypersurfaces, and that $V(N) \to \infty$ as $N_i \to \infty$ and as $N_i \to 0+$ for $i = 1, 2, \ldots, m$.

In brief, a continuously differentiable function $V(N)$ is a Liapunov function of (1.4.17) in $\mathbf{R}_+^m$ if it has the following properties:

$$\text{(i)} \qquad V(N^\star) = 0; \tag{1.4.20}$$

(ii) The equations, $V(N) = K_p$, $K_1 < K_2 < K_3 < \cdots$, represent a set of nested closed hypersurfaces, and $V(N) \to \infty$ as $N_i \to \infty$ and as $N_i \to 0+$ for $i = 1, 2, \ldots, m$; and

$$\text{(iii)} \qquad \dot{V}(N) = \sum_{i=1}^{m} \frac{\partial V}{\partial N_i} N_i F_i(N) \leqslant 0 \quad \text{for all } N \in \mathbf{R}_+^m. \tag{1.4.21}$$

Theorem 1.4.2. *The positive equilibrium of model (1.4.17) at $N^\star$ is globally stable in the positive orthant, if there exists a Liapunov function $V(N)$ in $\mathbf{R}_+^m$, and*

$$\dot{V}(N = \sum^{m} \frac{\partial V}{\partial N_i} N_i F_i(N) < 0 \quad \textit{for all } N \in \mathbf{R}_+^m \textit{ and } N \neq N^\star. \tag{1.4.22}$$

The proof of this theorem follows directly from that for Theorem 1.4.1, and the use of the properties of the Liapunov function $V(N)$ in the positive orthant $\mathbf{R}_+^m$.

Example 1.4.2. The function

$$V(N) = (\ln N)^2 \tag{1.4.23}$$

is a Liapunov function for the logistic model

$$\dot{N} = N(1 - N). \tag{1.4.24}$$

It can be used to establish that $N = 1$ is globally stable.

The function $(\ln N)^2$ is none other than x^2 under the transformation in (1.4.18). Clearly the change of variables given in (1.4.18) may be used to convert a Liapunov function of model (1.4.1) into a Liapunov function for model (1.4.17).

But for establishing global stability in the positive orthant, the quadratic function

$$V(N) = (N - N^\star)^\mathrm{T}\ P(N - N^\star) \tag{1.4.25}$$

where P is a positive definite matrix, is not a Liapunov function according to the above definition. This is because $V(N)$ does not tend to infinity as $N_i \to 0+$ for $i = 1, 2, \ldots, m$. Hence it cannot be used with Theorem 1.4.2 to establish global stability in the positive orthant without using additional conditions. This is a common shortcoming in proofs of global stability of a competitive equilibrium in economics (Arrow et al., 1959; Quirk and Saposnik, 1968; Arrow and Hahn, 1971; Takayama, 1974).

There is a large amount of literature on the construction of Liapunov functions (Schultz, 1965; Hahn, 1967; Gurel and Lapidus, 1968, 1969; Willems, 1970; Gilpin, 1974; Gatto and Rinaldi, 1977; Goh, 1977 a; Hsu, 1978 a; Harrison, 1979 b). However, most of the Liapunov functions that have been used successfully in the analyses of biological populations can be generated by the variable gradient method.

In the variable gradient method, we guess the gradient of $V(N)$ instead of the function $V(N)$. We choose a set of continuous functions, $G_1(N)$, $G_2(N)$, $\ldots$, $G_m(N)$, such that

$$\frac{\partial G_i}{\partial N_j} = \frac{\partial G_j}{\partial N_i}, \quad i, j = 1, 2, ;\ldots, m \tag{1.4.26}$$

for all $N \in \mathbf{R}_+^m$, and such that the function

$$V(N) = \sum_{i=1}^{m} \int_{N_i^\star}^{N_i} G_i(N_1, N_2, \ldots, S_i, \ldots, N_m)\, \mathrm{d}S_i, \tag{1.4.27}$$

is a Liapunov function for (1.4.17). In other words, the set of functions

$\{G_i(N)\}$ are chosen so that they satisfy the conditions: (i) the integrability conditions in (1.4.26); (ii) the function $V(N)$ in (1.4.27) is "radially unbounded" in $\mathbf{R}_+^m$; and (iii)

$$\dot{V}(N) = \sum_{i=1}^{m} G_i(N)\, N_i F_i\,(N) \leqslant 0 \quad \text{for all } N \in \mathbf{R}_+^m. \tag{1.4.28}$$

Example 1.4.3. Let $c_1, c_2, \ldots, c_m$ be a set of positive constants, and $G_i = c_i\,(S_i - N_i^\star)/S_i$ for $i = 1, 2, \ldots, m$. Clearly condition (1.4.26) is satisfied. Using (1.4.27) we get

$$V(N) = \sum_{i=1}^{m} c_i\,[N_i - N_i^\star - N_i^\star \ln(N_i/N_i^\star)]. \tag{1.4.29}$$

This is a separable function, and $V(N) \to \infty$ as $N_i \to 0+$ and as $N_i \to \infty$ for $i = 1, 2, \ldots, m$. It is "radially unbounded" in $\mathbf{R}_+^m$.

If the constants $c_1, c_2, \ldots, c_m$ are chosen so that

$$\dot{V}(N) = \sum_{i=1}^{m} c_i(N_i - N_i^\star)F_i(N) \leqslant 0 \tag{1.4.30}$$

for all $N \in \mathbf{R}_+^m$, then $V(N)$ is a Liapunov function for (1.4.17).

Example 1.4.4. The following general prey—predator model was considered by Harrison (1979 b):

$$\dot{X} = b(X) - f(X)a(Y),$$
$$\dot{Y} = g(X)h(Y) + d(Y). \tag{1.4.31}$$

We assume that it has a positive equilibrium at $(X^\star, Y^\star)$.

Consider the function

$$V(X, Y) = \int_{X^\star}^{X} \frac{g(X) - g(X^\star)}{f(X)}\, \mathrm{d}X + \int_{Y^\star}^{Y} \frac{a(Y) - a(Y^\star)}{h(Y)}\, \mathrm{d}Y. \tag{1.4.32}$$

Let the equilibrium $(X^\star, Y^\star)$ belong to an open region $\mathbf{R}$. Suppose $V(X, Y) > 0$ for all $(X, Y) \in \mathbf{R}$ and $(X, Y) \neq (X^\star, Y^\star)$. We can show that the equations $V(X, Y) = K_p$, $K_1 < K_2 < K_3 < \cdots$, represent a set of nested closed hypersurfaces in a subregion of $\mathbf{R}$.

Computing $V(X, Y)$ along the solutions of (1.4.31), we get

$$\dot{V} = [g(X) - g(X^\star)]\,[b(X)/f(X) - b(X^\star)/f(X^\star)]$$
$$+ [a(Y) - a(Y^\star)]\,[d(Y)/h(Y) - d(Y^\star)/h(Y^\star)]. \tag{1.4.33}$$

Hence $V(X, Y)$ is a Liapunov function in a neighbourhood of $(X^\star, Y^\star)$ if: (i) $V(X, Y) > 0$ for all $(X, Y) \in R$ and $(X, Y) \neq (X^\star, Y^\star)$; and (ii) $\dot{V}(X, Y) \leqslant 0$ for all $(X, Y) \in R$. For a function to be a Liapunov function in a finite neighbourhood of an equilibrium, it need not be radially unbounded.

Hsu (1978 a) and Harrison (1979 b) have shown that the Liapunov function (1.4.32) can be used to establish stability in a wide range of prey—predator models.

1.5. DIFFERENTIAL EQUATIONS AND DIFFERENCE EQUATIONS IN ECOSYSTEM MODELLING

The application of nonlinear difference equations in ecosystem modelling is becoming popular. One reason for this is that nonlinear difference equations can be solved directly on a digital computer. However, in theoretical ecology a difference equation model is often constructed initially, and a limiting process is used to convert it into a differential equation model. Since it is usually impossible to solve analytically a system of nonlinear differential equations, it is solved on a digital computer. In doing so, it is reconverted into a system of difference equations. Depending on the numerical algorithm for solving differential equations, the resulting system of difference equations may not be the same as in the initial model.

This process is unnecessary. Moreover, sometimes incorrect conclusions are obtained when a difference equation is converted into a differential equation. An example of this occurs in the modelling of the adult population of a single-species population with nonoverlapping generations whose dynamics has a two-point limit cycle. The analogous differential equation would not have an oscillatory behavior. Thus it may be more natural to use difference equations to model ecosystems (Van der Vaart, 1973; Innis, 1974 a).

The present extensive use of differential equations in theoretical ecology may be attributed to the fact that analytical methods for studying them are better developed and are more widely taught than those for difference equations. For instance, optimal control theory for systems of differential equations is better developed than that for difference equations. Moreover differential equation models of two species interactions can be analysed graphically.

For the management of a population with a discrete time delay, there is good reason to use a difference equation model rather than a delay-differential equation model. This is because the optimal control of a delay difference equation produces no more difficulties than that of a model without a delay. In contrast, the optimal control of a delay-differential equation generally leads to considerable mathematical difficulties.

There are two other interesting differences between a differential equation model. It may be difficult to establish the existence of a solution to a nonlinear differential equation during a long period of time. On the other hand, the existence problem for a difference equation model can be established quite simply. Suppose a nonlinear differential equation has a solution which joins a pair of points $(0, x(0))$ and $(T, x(T))$. The solution can be computed in forward time as t increases from $t = 0$ to $t = T$, or in reverse time as t decreases from $t = T$ to $t = 0$. In general a nonlinear difference equation has a unique solution only as t increases. We shall elaborate on these properties of a difference equation model.

Consider the difference equation population model

$$x(t+1) = f[x(t)], \quad t = 0, 1, 2, \ldots \tag{1.5.1}$$

where x is a vector with m components. Let the set of admissible initial states be denoted by X. In population biology the admissible set X is usually

$$X = \{x \mid x_i \geqslant 0, i = 1, 2, \ldots, m\}. \tag{1.5.2}$$

If $f(X) \subseteq X$, then the forward time solution of (1.5.1) from any initial state $x(0) \in X$ exists. Similar comments apply if f is a function of t and x. Therefore the existence problem for an initial value problem in a difference equation model is trivial if model (1.5.1) describes the biology correctly. However, it is sometimes convenient to use simplified expressions for the function $f(x)$ when the set of admissible initial states is restricted to a subset X of the positive orthant. In this case the existence problem is not trivial. The modeller then should verify that $f(X) \subseteq X$. For example, a stock recruitment function that is often used in the modelling of whale populations is $R = aS - bS^2$, where R denotes a recruitment, S denotes the breeding stock which produced the recruitment, and a and b are positive parameters. For large values of S, $R < 0$. This implies that the formula is valid only in a restricted range of values of S.

Generally a nonlinear difference equation has a unique solution from a given point only in the forward time. Consider the scalar equation

$$x(t+1) = 2[x(t) - x^2(t)], \quad t = 0, 1, 2, \ldots . \tag{1.5.3}$$

For $x(0) \in X = \{x \mid 0 < x < 1\}$, this equation has a unique forward solution. But for a given admissible value of $x(t+1)$, there are two possible values for $x(t)$. In particular, if $x(t+1) = 1/4$, we have $x(t) = (1/2) \pm (1/2\sqrt{2})$. This implies that it may not be possible to reconstruct the historical sequence of events which produced the present state of a system by using the current state if a dynamical system is adequately modelled by a set of nonlinear difference equations. Random disturbances make a determination of the past even more problematical. For such a nonlinear system the history of the system can be specified only in terms of a set of equilibria that may have existed in the past. This requires an examination of the range of values of the parameters of the model which would generate stable equilibria.

Recently, important advances have been made in the theory of nonlinear difference equations. The phenomena of chaotic behavior for a single nonlinear difference equation have been unravelled (see May, 1976 a). The applications of Liapunov functions to establish stability, region of ultimate confinement and nonvulnerability are examined in this book. In particular, fairly comprehensive conditions for a scalar nonlinear difference equation to have a global attractor will be established.

CHAPTER 1: SELECTED REFERENCES

Models in ecology

Lotka (1925); Volterra (1931); Gause (1934); Levins (1966, 1968); Watt (1968); Patten (1971); Williamson (1972); Ludwig (1974); May (1974, 1976 b); Maynard Smith (1974); Pianka (1974); Banks (1975); Gilpin (1975 a); Wiegert (1975); Clark (1976 a); Hassell (1976); Christiansen and Fenchel (1977); Pielou (1977).

Stability concepts in ecology

Lewontin (1969); Holling (1973); Botkin and Sobel (1974); Innis (1974 b); May (1974); Maynard Smith (1974); Orians (1974); Wu (1974); Goh (1975, 1976 a, 1977 a).

Ecological engineering

Goh et al. (1974 a, b); Vincent et al. (1975, 1977); Clark (1976 a); Conway (1977); Shoemaker (1977); Holling (1978); Reed (1979); Walters and Hilborn (1978).

Liapunov functions

Kalman and Bertram (1960 a); Schultz (1965); Liapunov (1966); Hahn (1967); Gurel and Lapidus (1968, 1969); Willems (1970); Kilmer (1972); Gilpin (1974); Gatto and Rinaldi (1977); Goh (1977 a, c); Hastings (1978); Hsu (1978 a, b); Takeuchi et al. (1978); Case and Casten (1979).

Differential equations versus difference equations

Kalman (1956); Van der Vaart (1973); Innis (1974 a); May (1976 a); Goh (1978 c); LaSalle (1979).

Chapter 2

OPTIMIZATION TECHNIQUES

2.1. INTRODUCTION

The essence of management is to make optimal decisions subject to realistic constraints. This is none other than an optimization problem. However, in practice there are tremendous difficulties in quantifying the variables, the objectives and the constraints in a given decision problem. Here we shall focus our attention on the task of formulating an optimal policy when a decision problem has already been defined in a mathematical form.

At present, one difficulty encountered in the application of an optimization technique in a decision problem is the numerical problem of computing the optimal solution. This difficulty is especially acute in problems with many state variables. For this reason, in this book we shall only apply optimization techniques to relatively simple problems.

Ecosystems are dynamical systems. Hence the appropriate tool for formulating optimal policies in the management of ecosystems is optimal control theory. However, the understanding of static optimization techniques is useful because a discrete time optimal control problem can be considered as a static optimization problem with a special structure.

Most decision problems in the management of ecosystems involve multiple objectives. There are many difficulties in applying the theory of decision problems with multiple objectives. For this reason, we shall examine only very simple ideas from this theory.

The need for optimization techniques goes back to the ancient Greeks: for example, they were interested in the largest area of land which could be enclosed in a boundary of given length. The solution of optimization problems was one of the main motivations for the creation of calculus. However, before the Second World War, optimization techniques were applied mainly to geometrical problems and problems in physical sciences. The use of optimization techniques as a management tool is a recent development which took place only after the Second World War (Dantzig, 1963; Lawden, 1963; Leitmann, 1966; Bryson and Ho, 1969).

The first modern optimization technique which achieved immediate success in management problems was linear programming. A notable feature of this technique is that it readily handles linear inequality constraints. Even more important is the fact that the simplex method in linear programming provides a very efficient method for computing the optimum solution. The success of linear programming led directly to the discovery and proof of the Kuhn—Tucker Theorem in nonlinear programming. Unfortunately, the Kuhn—Tucker Theorem does not provide a method to compute the optimal solution.

For dynamic problems, Bellman (1957) formulated the principle of optimality which is the core of the theory of dynamic programming. This technique was applied to problems in automatic control. These applications of dynamic programming motivated Pontryagin to develop the Maximum Principle (Pontryagin et al., 1962). Independently of these developments, a small group of scientists was using the classical calculus of variations for finding an optimal thrust program for a rocket (Lawden, 1963; Leitmann, 1966). It soon became clear that the Pontryagin Maximum Principle is simply a more lucid version of some of the necessary conditions of the classical calculus of variations (Berkovitz, 1961). It is unfortunate that most textbooks on optimal theory do not explain that the Pontryagin Maximum Principle does not include the focal point condition of the calculus of variations (Bliss, 1946). In some problems the focal point condition establishes when a sequence of short-term optimal policies fails to be a long-term optimal. Clearly this is of great importance in management.

One important theoretical development in the nineteen-sixties was the derivation of necessary conditions for singular control (Goh, 1966 a, b, c, 1973 a; Kelley et al., 1967; Bell and Jacobson, 1975). Singular control occurs in an optimal control problem in which one or more of the control variables appear linearly in the system dynamics and the objective function. Interesting problems of this kind first appeared in rocket problems. This is because the thrust control variable appears linearly in the system dynamics. Similarly in ecological problems, singular control is very important because control variables often occur linearly in the system dynamics.

The first applications of optimal control theory to ecological problems appeared only in the late nineteen-sixties. It appears that several authors, independently of each other, began to publish their works in this area about the same time (Watt, 1968; Goh, 1969/1970; Becker, 1970; Clark, 1971.)

2.2. STATIC OPTIMIZATION PROBLEMS

In this section we shall examine (i) the minimization of a function of one variable; (ii) the minimization of a function of several variables; (iii) minimization problems with equality constraints; and (iv) linear and nonlinear programming problems.

Consider the problem of minimizing the function $f(x)$ for all x belonging to the admissible set, $S = \{x \mid a \leqslant x \leqslant b\}$ where a and b are constants. The function $f(x)$ is said to have a local minimum at $x^\star \in S$ if $f(x) \geqslant f(x^\star)$ for all x in an admissible neighbourhood of $x^\star$. It has a global minimum at $x^\star$ if $f(x) \geqslant f(x^\star)$ for all $x \in S$.

Theorem 2.2.1. *If $a < x^\star < b$, necessary conditions for $f(x^\star)$ to be a local minimum are $f'(x^\star) = 0$ and $f''(x^\star) \geqslant 0$. A necessary condition for $f(a)$ to be a local minimum is $f'(a) \geqslant 0$. Similarly a necessary condition for $f(b)$ to be a local minimum is $f'(b) \leqslant 0$.*

Proof. By Taylor's Theorem we have

$$f(x^\star + \Delta x) - f(x^\star) = \Delta x f'(x^\star) + (1/2)(\Delta x)^2 f''(x^\star) + (1/3!)(\Delta x)^3 f'''(x^\star + \theta\Delta x) \tag{2.2.1}$$

where $0 < \theta < 1$. Suppose $x^\star$ is an interior point of S and $f'(x^\star) \neq 0$. When Δx is very small the dominant term on the right hand side of (2.2.1) is $\Delta x f'(x^\star)$. We can choose Δx so that $\Delta x\, f'(x^\star) < 0$. This implies $f(x^\star + \Delta x) - f(x^\star) < 0$. Hence $f(x^\star)$ is not a local minimum. It follows that we must have $f'(x^\star) = 0$. If this condition is satisfied, $f(x^\star + \Delta x) - f(x^\star)$ is approximately equal to $(1/2)(\Delta x)^2 f''(x^\star)$ when Δx is small. Clearly we must have $f''(x^\star) \geqslant 0$.

If $x^\star = a$, we have

$$f(a + \Delta x) - f(a) \simeq f'(a)\Delta x. \tag{2.2.2}$$

In this case, $\Delta x \geqslant 0$. Hence for $f'(a)\Delta x$ to be nonnegative we must have $f'(a) \geqslant 0$. Similarly a necessary condition for $f(b)$ to be a local minimum is $f'(b) \leqslant 0$.

Corollary 2.2.1. *If $a < x^\star < b$, sufficient conditions for $f(x^\star)$ to be a local minimum are $f'(x^\star) = 0$ and $f''(x^\star) > 0$. A sufficient condition for $f(a)$ to be a local minimum is $f'(a) > 0$. A sufficient condition for $f(b)$ to be a local minimum is $f'(b) < 0$.*

Corollary 2.2.2. *If $a \leqslant x^\star \leqslant b$, $f(x^\star)$ is a global minimum if $f'(x) < 0$ for $a \leqslant x < x^\star$, and $f'(x) > 0$ for $b \geqslant x > x^\star$. If $f'(x) > 0$ for $a \leqslant x \leqslant b$, then $f(a)$ is the global minimum. If $f'(x) < 0$ for $a \leqslant x \leqslant b$, then $f(b)$ is the global minimum. In all cases, $f(x^\star)$ is a global minimum if $f''(x) > 0$ for all $x \in S$.*

Example 2.2.1. Consider the function

$$f(x) = x - x^\star - x^\star \ln(x/x^\star)$$

where $x^\star$ is a positive constant and the admissible set is $S = \{x \mid x > 0\}$. We have

$$f'(x) = 1 - x^\star/x.$$

Clearly $f'(x^\star) = 0, f'(x) < 0$ for $x^\star > x > 0$, and $f'(x) > 0$ for $x > x^\star$. Hence $f(x)$ has a global minimum at $x = x^\star$.

Alternatively, as

$$f''(x) = 1/x^2 > 0$$

for all $x \in S$, $f(x)$ has a global minimum at $x = x^\star$.

We shall discuss only the theory of minimizing a function. This is because we can find the maximum of $f(x)$ by minimizing $-f(x)$. We have the identity,

$$\max f(x) = -\min\,[-f(x)]. \tag{2.2.3}$$

Let S be a region in the $\mathbf{R}^n$-space. Let $f(x_1, x_2, \ldots, x_n)$ be defined on the set S. If the minimum of $f(x)$ occurs at a boundary point of S, we have a nonlinear programming problem which will be discussed later.

Theorem 2.2.2. *If $x^\star$ is an interior point of the admissible set S, necessary conditions for $f(x^\star)$ to be a local minimum are*

$$\text{(i)} \qquad \partial f(x^\star)/\partial x_i = 0, \quad i = 1, 2, \ldots, n; \tag{2.2.4}$$

and

$$\text{(ii) at } x^\star \textit{ the matrix} \qquad (\partial^2 f(x)/\partial x_i\, \partial x_j) \tag{2.2.5}$$

must be positive semidefinite.

This theorem can be proved by applying Taylor's Theorem. A necessary condition for a matrix to be positive semidefinite is that its leading principal minors are nonnegative.

Let S be an open and convex region. If $\partial f(x^\star)/\partial x_i = 0$, for $i = 1,2, \ldots, n$ and $(\partial^2 f(x)/\partial x_i \partial x_j)$ is positive definite at all points in S, then $f(x^\star)$ is the global minimum of $f(x)$ in the set S.

Example 2.2.2. The following function plays an important role in establishing global stability in an ecosystem model. Let $S = \{x \mid x_i > 0, i = 1, 2, \ldots, n\}$ and the objective function

$$f(x) = \sum_{i=1}^{n} c_i[x_i - x_i^\star - x_i^\star \ln(x_i/x_i^\star)], \tag{2.2.6}$$

where $c_1, c_2, \ldots, c_n$ and $x_1^\star, x_2^\star, \ldots, x_n^\star$ are positive constants. This function has a global minimum at $(x_1^\star, x_2^\star, \ldots, x_n^\star)$.

We have

$$\partial f/\partial x_i = c_i(1 - x_i^\star/x_i), \tag{2.2.7}$$

$$\partial^2 f/\partial x_i^2 = c_i x_i^\star/x_i^2, \tag{2.2.8}$$

$$\partial^2 f/\partial x_i \partial x_j = 0, \quad \text{if } i \neq j. \tag{2.2.9}$$

Clearly $x^\star$ is a solution of (2.2.7) and the matrix $(\partial^2 f/\partial x_i\, \partial x_j)$ is positive definite in S. Hence $f(x)$ has a global minimum at $x^\star$.

Alternatively we note that the function is a linear combination of functions each of which is only a function of one variable. It follows that the optimal value of the function occurs at the point where each of the separate functions is optimal. This completes our analysis of the example.

Theorem 2.2.2 is useful only in a very limited class of problems. This is because it characterizes an optimal solution but does not provide an effective method to compute the optimal solution. It is important because it provides the basis for further theoretical developments. In practice we have to use numerical algorithms to compute an optimal solution. The Davidon—Fletcher—Powell algorithm was a significant advance in numerical methods for solving optimization problems. Since then numerous algorithms have been developed. A new algorithm developed by Fletcher (1975) is one of the best available. Almost all algorithms have been developed for computing a local minimum. From experience, it is found that some of the best algorithms have failed to compute the global minimum in simple problems even when the starting point is near the global minimum. Hence if time and resources permit, it is wise to use at least two different algorithms in a given problem and a large number of starting points for each algorithm.

Consider the constrained minimization problem:

$$\text{Minimize} \quad f(x_1, x_2, \ldots, x_n), \tag{2.2.10}$$

$$\text{subject to} \quad g_i(x_1, x_2, \ldots, x_n) = 0, \quad i = 1, 2, \ldots, m. \tag{2.2.11}$$

By definition, $x^\star$ is an admissible point if it satisfies the constraints (2.2.11).

Theorem 2.2.3. *A necessary condition for* $f(x^\star)$ *to be a local minimum is that there exists a set of constants,* $\lambda_0 \geqslant 0, \lambda_1, \lambda_2, \ldots, \lambda_m$ *such that at* $x^\star$,

$$\partial F(x, \lambda)/\partial x_j = 0, \quad j = 1, 2, \ldots, n, \tag{2.2.12}$$

$$g_i(x) = 0, \quad i = 1, 2, \ldots, m \tag{2.2.13}$$

where

$$F = \lambda_0 f + \sum_{i=1}^{m} \lambda_i g_i. \tag{2.2.14}$$

The constants, $\lambda_0, \lambda_1, \ldots, \lambda_m$, are called Lagrange multipliers. If $\lambda_0 = 0$, there may be more than one set of multipliers. Then $x^\star$ is called an abnormal point (Vincent and Goh, 1972). In static optimization problems, abnormality is a mathematical curiosity. However, if we apply Theorem 2.2.3 to a discrete time optimal control problem, we find that the case when $\lambda_0 = 0$ is quite common. In this case abnormality represents a lack of controllability in a problem. For example, this occurs if the set of available options is sufficient for the control of a system because the time available in a management problem is too short. In truly static optimization problems, it is reasonable to assume normality and let $\lambda_0 = 1$.

The proof of Theorem 2.2.3 can be found in the paper by Vincent and Goh (1972) or the book by Bliss (1946). The proof given in most textbooks on advanced calculus is not valid for the abnormal case.

The usefulness of Theorem 2.2.3 is very limited because the theorem characterizes an optimal solution but does not provide an effective method to compute the optimal solution.

Example 2.2.3. Find the shortest distance from the origin (0,0) to the straight line $x_1 + x_2 = 1$. Graphically it is clear that the shortest distance is given by the straight line joining the points (0,0) and (1/2, 1/2).

In this problem, we minimize $f = (x_1^2 + x_2^2)^{1/2}$ subject to $x_1 + x_2 - 1 = 0$. By definition,

$$F(x,\lambda) = (x_1^2 + x_2^2)^{1/2} + \lambda(x_1 + x_2 - 1) \Rightarrow \partial F/\partial x_1 = x_1/(x_1^2 + x_2^2)^{1/2} + \lambda = 0,$$

$$\partial F/\partial x_2 = x_2/(x_1^2 + x_2^2)^{1/2} + \lambda = 0.$$

We can show that $x = (1/2, 1/2)$ and $\lambda = -1/\sqrt{2}$ satisfy these equations. This analysis illustrates the way Theorem 2.2.3 is often used.

In realistic management problems, we have equality and inequality constraints. For example, the density of a population must be nonnegative. The simplest optimization problem with inequality constraints is a linear programming problem. The standard linear programming problem is as follows:

$$\text{Minimize} \quad [c_1x_1 + c_2x_2 + \cdots + c_nx_n] \tag{2.2.15}$$

$$\text{subject to} \quad \sum_{s=1}^{n} a_{is}x_s = b_i, \quad i = 1, 2, \ldots, m, (m < n), \tag{2.2.16}$$

$$x_i \geqslant 0, \quad i = 1, 2, \ldots, n. \tag{2.2.17}$$

The simplex algorithm provides a very reliable method for computing the optimal solution of a linear programming problem (Dantzig, 1963).

If the objective function or the constraints involve nonlinear functions, we have a nonlinear programming problem. The standard nonlinear programming problem is as follows:

$$\text{Minimize} \quad f(x_1, x_2, \ldots, x_n) \tag{2.2.18}$$

$$\text{subject to} \quad g_s(x_1, x_2, \ldots, x_n) = 0, \quad s = 1, 2, \ldots, p, \tag{2.2.19}$$

$$h_r(x_1, x_2, \ldots, x_n) \geqslant 0, \quad r = 1, 2, \ldots, m. \tag{2.2.20}$$

Let $\lambda_0, \lambda_1, \lambda_2, \ldots, \lambda_p, \mu_1, \mu_2, \ldots, \mu_m$ be a nonzero set of constants. For convenience let

$$L(x,\lambda,\mu) = \lambda_0 f + \sum_{s=1}^{p} \lambda_s g_s - \sum_{r=1}^{m} \mu_r h_r. \tag{2.2.21}$$

By definition, $x^\star$ is an admissible point if it satisfies conditions (2.2.19) and (2.2.20).

Theorem 2.2.4. *A necessary condition for $f(x)$ to be optimal at an admissible point $x^\star$ is that there exists a nonzero set of multipliers, $\lambda_0 \geqslant 0, \lambda_1, \lambda_2, \ldots, \lambda_p$, $\mu_1, \mu_2, \ldots, \mu_m$, such that at $x^\star$,*

$$\mu_r \geqslant 0, \quad \mu_r h_r(x) = 0, \quad r = 1, 2, \ldots, m, \tag{2.2.22}$$

$$\partial L(x,\lambda,\mu)/\partial x_i = 0, \quad i = 1, 2, \ldots, n. \tag{2.2.23}$$

In static problems, we usually assume $x^\star$ is a normal (regular) point and let $\lambda_0 = 1$. This theorem characterizes an optimal solution but it does not provide a constructive method for obtaining it. For a given problem we can use one of the numerous algorithms that are available.

2.3. RELAXED OPTIMIZATION PROBLEM METHOD

By definition, a model is a simplified representation of a real situation. In most decision problems we ignore many factors and constraints so that we have a model of the decision problem which is tractable to analysis. For example in a fishery, we often consider only the problem of harvesting for optimal biomass yield and ignore economic constraints. Similarly many economic decisions are often made without considering their impact on the environment. The act of ignoring constraints enlarges the admissible set of options. This may lead to a worse decision from the overall point of view because the costs in violating the ignored constraints are not taken into account. The purpose of this section is to formalize this process of simplification from the optimization point of view.

Consider the problem of minimizing $f(x)$ for x belonging to an admissible set S. It may be very difficult to compute the global minimum of $f(x)$ when S has a very complicated structure. The relaxed optimization problem method considers tentatively the problem of minimizing $f(x)$ for x belonging to a simpler set Z which is a superset of S. If the global minimum $f(x)$ for $x \in Z$ is at $x^\star$ and $x^\star \in S$, then clearly the global minimum of $f(x)$ for $x \in S$ is at $x^\star$. In a decision problem, if $x^\star$ does not belong to S, it would normally be acceptable if there exists a point $\bar{x} \in S$ such that $f(\bar{x})$ is not too different from $f(x^\star)$. Note that $f(\bar{x})$ need not be an optimal value in the mathematical sense. For future references we shall state these simple and useful results as theorems.

Theorem 2.3.1. *Let $S \subset Z$. If the global minimum of $f(x)$ for $x \in Z$ is at $x^\star$ and $x^\star \in S$, then*

$$\min_{x \in S} f(x) = \min_{x \in Z} f(x). \tag{2.3.1}$$

Theorem 2.3.2. *Let $S \subset Z$ and ϵ be a small positive number. If the global minimum of $f(x)$ for $x \in Z$ is at $x^\star$ and there exists a point $\bar{x} \in S$ such that*

$$f(\bar{x}) = f(x^\star) + \epsilon, \tag{2.3.2}$$

then $\bar{x}$ is a suboptimal solution in the problem of minimizing $f(x)$ for $x \in S$.

The main difficulty in using Theorems 2.4.1 and 2.4.2 is to show that $x^\star \in S$ or that there exists $\bar{x} \in S$ such that $f(\bar{x}) \simeq f(x^\star)$.

Example 2.3.1. Let the set S consist of points which satisfy the equation

$$\begin{aligned} x_2 &= x_1 \sin(1/x_1) && \text{if } x_1 \neq 0, \\ &= 0 && \text{if } x_1 = 0. \end{aligned}$$

Clearly the set S has an intricate structure. Suppose the function to be maintained is

$$f(x) = x_1^2 - x_2^2 .$$

A superset of S with a simpler structure is the set of points, $Z = \{x \mid x_2^2 \leqslant x_1^2\}$. On the set Z, $f(x) \geqslant 0$. We have $f(0) = 0$. Hence the global minimum of $f(x)$ for $x \in Z$ is $f(0) = 0$. The point (0,0) is a member of the set S. Hence $f(0)$ is the global minimum of $f(x)$ for $x \in S$.

Example 2.3.2. Let the set $S = \{x \mid 1 < x \leqslant 2\}$ and $f(x) = x$. Mathematically this problem has no minimum but has an infimum. In decision problems this fine mathematical distinction is often of no consequence. Consider the superset, $Z = \{x \mid 1 \leqslant x \leqslant 2\}$. The function $f(x)$ has a minimum at $x^\star = 1$. Let ϵ be a small positive number. Clearly if $\bar{x} = 1 + \epsilon$, $f(\bar{x})$ is approximately equal to $f(x^\star)$. We say $f(1 + \epsilon)$ is a suboptimal solution.

Example 2.3.3. The following result is needed to establish global stability in some population models. In this case a suboptimal solution is not acceptable. Let the set $S = \{x \mid x_i + e_i > 0, i = 1, 2, \ldots, n\}$ where $e_1, e_2, \ldots, e_n$ is a set of positive constants. Let P be a constant negative definite symmetric matrix such that $P_{ii} < 0$ for $i = 1, 2, \ldots, n$ and $P_{ij} \geqslant 0$ if $i \neq j$. We want to show that the nondifferentiable function

$$Q = \sum_{i=1}^{n} P_{ii}\, x_i^2 + \sum_{j \neq i}^{n} P_{ij} |x_i|\, |x_j| \tag{2.3.3}$$

is negative definite for all $x \in S$.

The problem is solved if we can show that $Q = 0$ at $x = 0$ is the unique global maximum for $x \in S$. We have two difficulties in establishing this result directly. Firstly $Q(x)$ is not a differentiable function at $x = 0$. Secondly the set S is quite complex.

From (2.3.3) we observe that $Q(x)$ is a function of only the absolute values of $x_1, x_2, \ldots, x_n$. This implies that we only have to show that $Q(x)$ is negative definite in the set $W = \{x \mid x_i \geqslant 0, i = 1, 2, \ldots, n\}$. In the set W we can remove the absolute signs from the variables in the function $Q(x)$. Let the function

$$g(x) = \sum_{i=1}^{n} P_{ii}x_i^2 + \sum_{j \neq i}^{n} P_{ij}x_ix_j. \tag{2.3.4}$$

We want to establish that $g(x)$ has a unique global minimum equal to zero on the set W.

Let $\mathbf{R}^n$ be the whole of $(x_1, x_2, \ldots, x_n)$-space. Clearly $\mathbf{R}^n$ is a superset of W. From (2.3.4) $g(x) = x^T Px$. By assumption P is negative definite. Hence $g(x)$ has a unique maximum equal to zero at $x = 0$ for all $x \in \mathbf{R}^n$. But $x = 0$ is an element of the set W. Hence the unique maximum of $g(x)$ for $x \in W$ is equal to zero and occurs at $x = 0$. This implies that $Q(0) = 0$ is the unique global maximum for $x \in S$.

These examples illustrate some technical applications of Theorems 2.3.1 and 2.3.2. These theorems can provide the basis for using simplified models in which some constraints and some processes in a problem are ignored. This is equivalent to minimizing an objective function in a superset of the admissible options. However, the difficult step, which is seldom carried out, is to demonstrate that the solution in the relaxed optimization problem belongs to the original admissible set or that the solution in the relaxed optimization problem can be approximated by a solution which belongs to the original admissible set. In other words, it is unusual to find verification that the simplifying assumptions in a problem are indeed unimportant.

2.4. MULTIPLE OBJECTIVES DECISION PROBLEMS

In a decision problem we usually have several objectives in mind. For example the average driver would like to maximize his safety and minimize the time to get to his destination. The first and major difficulty in such a problem is to set scales for the objective functions so that we can compare the value of one objective with the value of another objective. Even if this basic difficulty has been overcome, there are conceptual difficulties in determining what is meant by one solution being better than another when there is more than one objective.

Suppose we have an admissible set of options which is denoted by S. Let x be an element of S. For simplicity we shall examine the case when there are only two objective functions, $J_1(x)$ and $J_2(x)$. We note that $J = (J_1, J_2)$ is a vector. Let $J(x') = (J_1(x'), J_2(x'))$ and $J(x^\star) = (J_1(x^\star), J_2(x^\star))$. We cannot say that $J(x') < J(x^\star)$ or $J(x') > J(x^\star)$ except in the case when $J(x')$ and $J(x^\star)$ are parallel vectors. It follows that we need other criteria for stating that $J(x')$ is better or worse than $J(x^\star)$.

There are three important ways for establishing that $J(x')$ is better than $J(x^\star)$ or vice-versa. Firstly we can use the concept of non-inferior solution. Secondly we can use a single weighted objective to determine an optimal solution. Thirdly we can temporarily convert one objective into a constraint and minimize the other objective. If this leads to an unsatisfactory solution, we could adjust the constraint on the first objective function and repeat the process until a satisfactory solution is obtained.

By definition, $x^\star$ is a noninferior (Pareto optimal) solution if there does not exist another option x from the admissible set S such that

$$J_1(x) \leqslant J_1(x^\star), \quad J_2(x) \leqslant J_2(x^\star) \quad \text{and} \quad J(x) \neq J(x^\star). \tag{2.4.1}$$

In other words, there does not exist another admissible option which gives a lower value to one objective function without increasing the value of the other. Denote the set of noninferior solutions by A.

Let α_1 and α_2 be positive numbers such that $\alpha_1 + \alpha_2 = 1$. If $J_1(x)$ is more important than $J_2(x)$, then we choose α_1 and α_2 so that $\alpha_1 > \alpha_2$. Thus α_1 and α_2 are weighting factors. In an auxiliary minimization problem, we minimize the objective function,

$$F(x,\alpha) = \alpha_1 J_1(x) + \alpha_2 J_2(x) \tag{2.4.2}$$

for $x \in S$. For each choice of (α_1, α_2), we get one or more solutions in this minimization problem. Denote the set of solutions by this method for all admissible values of α_1 and α_2 by W. We can show that W is a subset of A.

In the third method, we convert $J_1(x)$ or $J_2(x)$ into a constraint. For example we can impose the condition, $J_1(x) \leqslant e_1$, where e_1 is chosen tentatively. We then minimize $J_2(x)$ subject to $x \in S$ *and* $J_1(x) \leqslant e_1$. If the solution of this auxiliary optimization problem is unsatisfactory, we could relax the constraint on $J_1(x)$ by choosing a larger e_1. This process is repeated until satisfactory solutions are obtained. As the value of e_1 is varied, we get a set of solutions which is denoted by E. This method may be the best approach for ecological decision problems with multiple objectives. This is because in ecological problems it is difficult to quantify the objectives so that they are comparable. For example, it is very difficult ot express in dollars and cents the value of a species which has no apparent economic value. And yet there is always a possibility that its extinction could lead to a considerable change in an ecosystem because of subtle and unknown interactions. In such a situation, it is prudent to impose tentatively the constraint that the density of the species should not be allowed to go below a danger level where the probability of extinction is very high.

Example 2.4.1. This example illustrates the difficulties which can be experienced in a very simple decision problem with multiple objectives.

Let the admissible set be

$$S = \{(x,y,z) \mid 1 \leqslant x \leqslant 2, 0 \leqslant y \leqslant 1, 0 \leqslant z \leqslant 1\} .$$

Suppose we wish to "minimize" simultaneously the functions

$$J_1 = x^2 + y, \qquad J_2 = 3 - x + z.$$

After some effort we find that the set S is mapped into the hatched region in Fig. 2.4.1.

The boundary set of points PQ is the set of noninferior solutions. If we minimize the functions, $F = \alpha_1 J_1 + \alpha_2 J_2$ where α_1 and α_2 are positive constants such that $\alpha_1 + \alpha_2 = 1$, we generate the solutions PQ as α_1 and α_2 are varied. If α_1 or α_2 is allowed to be equal to zero, the solution set becomes VPQR. If we minimize $J_2(x)$ subject to the tentative constraint, $J_1(x) \leqslant e_1$, we generate the optimal set PQR as e_1 is varied. Clearly the value of $J_1(x)$ along QR is unsatisfactory but $J_2(x)$ attains its least value.

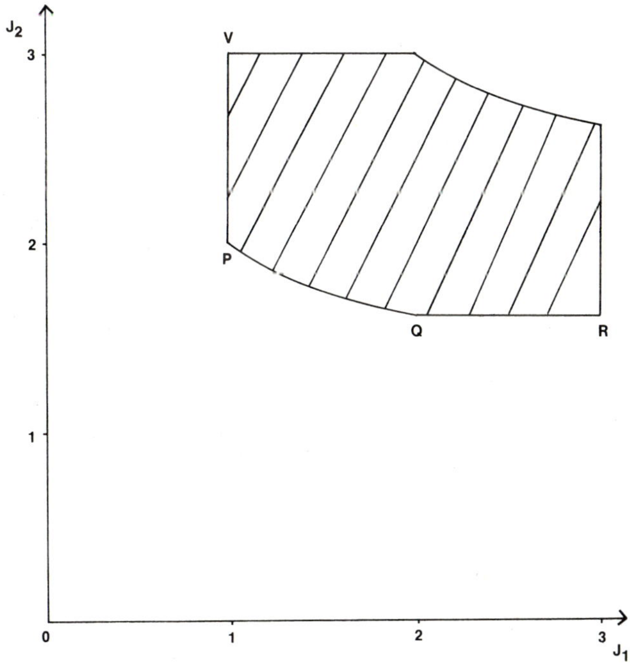

Fig. 2.4.1. The hatched region represents the set of values of the objectives $J_1 = x^2 + y$ and $J_2 = 3 - x + z$ as (x, y, z) is varied in the admissible set.

2.5. CONTINUOUS TIME OPTIMAL CONTROL

Ecosystems are dynamical systems and the most popular way to model an ecosystem is by means of a set of differential equations. It follows that the appropriate tool for formulating optimal policies for ecosystems is continuous time optimal control theory (Goh, 1978 c).

Let x be the state vector. In a managed ecosystem the components of x could be population densities nutrient levels or money invested in the management of the ecosystem. To manage an ecosystem we must be able to manipulate some variables which affect the dynamics of the ecosystem. These variables are called control variables. Let $u = (u_1, u_2, \ldots, u_m)$ be the control vector. In most decision problems the components of u must satisfy certain inequality constraints. Usually we have $u \in U = \{u \mid a_r \leqslant u_r \leqslant b_r, r = 1, 2, \ldots, m\}$.

The standard optimal control problem is as follows:

System: $\dot{x}_i = f_i(t, x, u), \quad i = 1, 2, \ldots, n.$ (2.5.1)

Initially: $x_i(0) = x_{i0}, \quad i = 1, 2, \ldots, n.$ (2.5.2)

Terminally: $\psi^s[x(T), T] = 0, \quad s = 1, 2, \ldots, k.$ (2.5.3)

Constraint: $u \in U.$ (2.5.4)

Objective: $\min g[x(T), T] + \int_0^T L(t, x, u)\, dt.$ (2.5.5)

Condition (2.5.2) simply means that we can measure the state vector at time $t = 0$. Here (x_{i0}) is a constant vector. Condition (2.5.3) represents the target in a decision problem. The specification of the target in an optimal control problem is itself a decision problem. In ecological problems much work remains to be done on how we should specify the target in an optimal control problem which arises in ecology. We shall return to this problem later. Here we assume that the target is specified in a given optimal control problem. The objective function consists of some desirable criterion at the end of the planning period plus a criterion which describes some desirable performance index during the planning period.

In applications of optimal control theory, necessary conditions are more useful than sufficiency conditions. This is because it is extremely difficult to apply sufficiency conditions in a practical problem. There are several sets of necessary conditions in optimal control theory. This reflects the complexity of a dynamical system. A useful set of necessary conditions is summarized in the next theorem.

Necessary conditions can be stated compactly in terms of the Hamiltonian function. By definition, the Hamiltonian function is

$$H(t, x, u, p) = p_0 L(t, x, u) + \sum_{i=1}^{n} p_i f_i(t, x, u). \tag{2.5.6}$$

We have either $p_0 = 0$ or $p_0 = 1$. The functions $p_1(t), p_2(t), \ldots, p_n(t)$ are called costate variables. If $p_0 = 0$, the corresponding optimal set $(x^\star, u^\star)$ is said to be abnormal (Vincent and Goh, 1972). It usually reflects a lack of control in a problem. If $p_0 = 1$, the optimal set $(x^\star, u^\star)$ is said to be normal. Note that many publications use a different Hamiltonian function in which $p_0 = -1$. This leads to a slightly different set of necessary conditions.

Theorem 2.5.1. *A necessary condition for an admissible set* $(x^\star, u^\star)$ *to be optimal is that there exists* $p_0, p_1(t), p_2(t), \ldots, p_n(t)$ *such that*

(i) $$\dot{p}_i = -\partial H(t, x, u, p)/\partial x_i, \quad i = 1, 2, \ldots, n; \qquad (2.5.7)$$

(ii) $$u^\star \text{ minimizes } H(t, x^\star, u, p) \quad \text{for all } u \in U; \qquad (2.5.8)$$

(iii) $$H dT + dg - \sum_{i=1}^{n} p_i(T)\, dx_i(T) = 0 \qquad (2.5.9a)$$

for all dT *and* $dx_i(T)$ *which satisfy*

$$\frac{\partial \psi^s}{\partial T}\, dT + \sum_{i=1}^{n} \frac{\partial \psi^s}{\partial x_i(T)}\, dx_i(T) = 0, \quad s = 1, 2, \ldots, k; \qquad (2.5.9b)$$

and

(iv) *if a control variable switches at time t, then*

$$p_i(t-) = p_i(t+), \quad i = 1, 2, \ldots, n, \qquad (2.5.10a)$$

$$H[t-, x(t), u(t-), p(t-)] = H[t+, x(t), u(t+), p(t+)]. \qquad (2.5.10b)$$

Here $p(t-) = \lim p(t - \epsilon)$ *as* $\epsilon \to 0$.

Corollary 2.5.1. *If* $U = \{u \mid a_r \leqslant u_r \leqslant b_r\}$, *condition (2.5.8) implies*

(i) $$u_r^\star = a_r, \quad \text{only if } \partial H(t, x^\star, u^\star, p)/\partial u_r^\star > 0; \qquad (2.5.11a)$$

(ii) $$u_r^\star = b_r, \quad \text{only if } \partial H(t, x^\star, u^\star, p)/\partial u_r^\star < 0; \qquad (2.5.11b)$$

and

(iii) $a_r < u_r^\star < b_r$ *for* $r = 1, 2, \ldots, m$, *only if at* $(x^\star, u^\star)$

$$\partial H(t, x, u, p)/\partial u_r \equiv 0 \qquad (2.5.11c)$$

and

$$(\partial^2 H(t, x, u, p)/\partial u_r\, \partial u_s) \qquad (2.5.11d)$$

is a positive semidefinite matrix.

If the matrix (2.5.11d) is singular, we have what is known as singular control. This usually occurs when one or more control variables appear linearly in the system dynamics and the objective function. Singular control was first studied seriously because of its appearance in rocket problems (Lawden, 1963). It now appears frequently in ecological problems (Goh, 1969/1970; Cliff and Vincent, 1973; Goh et al., 1974 a; Clark, 1976 a; Wickwire, 1977; Clark et al., 1979). The most useful set of necessary conditions for singular control consists of the generalized Legendre conditions. The most general form of these conditions was obtained by Goh (1966 a,b,c, 1973 b).

Let $u = (v, w)$ and let w appear linearly in the Hamiltonian function $H(x, v, w, p)$. We shall denote partial differentiations by subscripts. The symbols $[\cdot]_w$, $[\cdot]_v$ etc. denote partial differentiations relative to the set $(x, p, v, w, \dot{v}, \ddot{v}, \ldots)$; and $(\cdot)_w$, $(\cdot)_v$ etc. denote partial differentiations relative to the set (x, p, v, w). Total differentiation with respect to time is denoted by a dot or D. Matrix transpose will be denoted by superscript T. The following theorems are not widely known but their importance will increase because of the frequent occurrence of singular control in ecological problems. By definition, the curve in the state space associated with a set $(x^\star, u^\star)$ which satisfies Theorem 2.5.1 is called an extremal.

Theorem 2.5.2. *If along an extremal,* $H_{ww} = 0$ *and* $H_{wv} = 0$, *then necessary conditions for optimality are*

(i) $\quad [DH_w]_w = 0 \quad$ *for all* $t \in (t_1, t_2) \subset [0, T]$; $\qquad$ (2.5.12)

and

(ii) *if* (i) *is satisfied, the matrix*

$$\begin{pmatrix} H_{vv} & -[DH_w]_v^{\mathrm{T}} \\ -[DH_w]_v & -[D^2H_w]_w \end{pmatrix} \tag{2.5.13}$$

must be positive semidefinite for all $t \in (t_1, t_2)$.

Theorem 2.5.2 is called a generalized Legendre condition. This form of the condition is convenient for testing whether or not an extremal is optimal. But it does not lead to conditions for the derivation of a singular extremal. To do so we need another form of the generalized Legendre condition (Goh, 1966 c, 1973 b). Whenever $\dot{v}$ appears it can be eliminated by means of the equation,

$$DH_v = H_{vv}\,\dot{v} + H_{vx}H_p - H_{vp}H_x = 0. \tag{2.5.14}$$

$$\Rightarrow \dot{v} = \phi(x, p, v, w), \tag{2.5.15}$$

where ϕ denotes a function of x, p, v and w. In computing partial derivatives with respect to (x, p, v, w), we must first eliminate $\dot{v}$ whenever it appears before we can compute the partial derivatives.

Theorem 2.5.3. *If along an extremal, $H_{ww} = 0, H_{wv} = 0$, then necessary conditions for optimality are*

(i) $(DH_w)_w = 0$ *for all* $t \in (t_1, t_2) \subset (0, T)$; (2.5.16)

and

(ii) *if* (i) *is satisfied, the matrix*

$$\begin{pmatrix} H_{vv} & \vdots & -(DH_w)_v^{\mathrm{T}} \\ -(DH_w)_v & \vdots & -(D^2H_w)_w + (DH_w)_v\,(H_{vv})^{-1}\,(DH_w)_v^{\mathrm{T}} \end{pmatrix} \tag{2.5.17}$$

must be positive semidefinite for all $t \in (t_1, t_2)$.

Corollary 2.5.3. *If the matrix in (2.5.17) is positive definite, then the matrix*

$$\begin{pmatrix} H_{vv} & H_{vw} \\ -(D^2H_w)_v & -(D^2H_w)_w \end{pmatrix} \tag{2.5.18}$$

is nonsingular

This is so because $H_{vw} = 0$. If the matrix in (2.5.18) is nonsingular it implies that the equations,

$$H_v = 0 \quad \text{and} \quad D^2H_w = 0, \tag{2.5.19}$$

can be solved for v and w in terms of x and p. This then leads to a two-point boundary value problem and the derivation of the extremal as in nonsingular problems.

Theorem 2.5.4. *If along a singular extremal we have*

$$H_{ww} = 0, \qquad H_{wv} = 0, \qquad [D^rH_w]_v = 0, \quad r = 1, 2, \ldots, k-1, \tag{2.5.20}$$

$$[D^sH_w]_w = 0, \quad s = 1, 2, \ldots, 2k-2, \tag{2.5.21}$$

then necessary conditions for optimality are

(i) $[D^{2k-1}H_w]_w = 0$ *for all* $t \in (t_1, t_2) \subset (0, T)$; (2.5.22)

and

(ii) *if* (i) *is satisfied, the matrix*

$$\begin{pmatrix} H_{vv} & (-1)^k[D^kH_w]_v^{\mathrm{T}} \\ (-1)^k[D^kH_w]_v & (-1)^k[D^{2k}H_w]_w \end{pmatrix} \tag{2.5.23}$$

must be positive semidefinite for all $t \in (t_1, t_2)$.

There is a similar theorem to Theorem 2.5.4 if we employ partial derivatives with respect to (x, p, v, w) (see Goh, 1973 b). It can be written down by comparing Theorem 2.5.2 with Theorem 2.5.3.

Example 2.5.1. The exploitation of a renewable resource is in many ways similar to the manner in which a man spends a large sum of money which he has won in a lottery. Suppose he puts all his money into a bank. Let his bank balance at time t be $x(t)$, the interest rate be r and his withdrawal rate be $u(t)$. For simplicity we assume that the interest rate is constant at 10% per annum, i.e. $r = 0.1$. We assume that he does not receive any income from other sources. The dynamics of his bank balance are described by the differential equation $\dot{x} = rx - u$. At time $t = 0$, let $x(0) = 200\,000$. If he is 30 years old at time $t = 0$, it is reasonable for him to choose a planning period of 60 years. This gives $T = 60$. He has to decide from other considerations what he likes to have at the end of the planning period. This will specify the terminal constraints and the terminal part in the objective. He might want to leave a large sum of money to his children. We assume that he wants to leave at least 3 000 dollars for his funeral expenses. Another side decision he has to make is some constraint on his withdrawal rate. Suppose he decides that his withdrawal rate per annum should satisfy the constraint $15\,000 \leqslant u(t) \leqslant 40\,000$. Finally let us suppose that he wants to maximize the total yield which he can get from his money during the 60-year period.

In brief the problem is as follows:

System: $\dot{x} = 0.1x - u$. (2.5.24)

Initially: $x(0) = 200\,000$. (2.5.25)

Terminally: $x(60) \geqslant 3000$. (2.5.26)

Constraint: $15\,000 \leqslant u(t) \leqslant 40\,000$. (2.5.27)

Objective: $\max \int_0^{60} u\, dt$. (2.5.28)

We have described in some detail the many considerations that affect the formulation of a realistic and simple optimal control problem. Clearly subjective factors enter at several points in setting up the problem. Moreover multiple objectives considerations affect the formulation of the problem.

The terminal constraint (2.5.26) is an inequality constraint. It is therefore not in a standard form. We convert it to standard form by considering separately two cases. Firstly we let $x(60) > 3000$; but otherwise, $x(60)$ is unconstrained. In the second case we let $x(60) = 3000$. We can use Theorem 2.5.1 directly if we minimize the objective function,

$$J(u) = \int_0^{60} [-u]dt. \tag{2.5.29}$$

We shall now discuss in some detail the solution of this simple problem in order to demonstrate how an optimal control problem may be solved. By definition the Hamiltonian function

$$H = -u + p(rx - u). \tag{2.5.30}$$

The costate equation is

$$\dot{p} = -\partial H/\partial x = -rp, \tag{2.5.31}$$

$$\Rightarrow p(t) = p(60) \exp[r(60 - t)]. \tag{2.5.32}$$

The transversality condition is

$$H\mathrm{d}T - p(T)\,\mathrm{d}x(T) = 0. \tag{2.5.33}$$

The terminal condition $T - 60 = 0$ implies that $\mathrm{d}T = 0$. The condition, $x(60) > 3000$ (otherwise, $x(60)$ is unconstrained), implies that $\mathrm{d}x(60)$ is arbitrary. In this case condition (2.5.33) implies that $p(60) = 0$. Condition (2.5.32) implies that $p(t) \equiv 0$. If $x(60) = 3000$, we get $\mathrm{d}x(60) = 0$. In this case (2.5.33) implies that $p(60)$ is a constant of integration.

We have

$$\partial H/\partial u = -1 - p(t). \tag{2.5.34}$$

The optimality condition (2.5.8) implies

$$u^\star = 15\,000 \quad \text{only if } \partial H/\partial u > 0 \tag{2.5.35a}$$

and

$$u^\star = 40\,000 \quad \text{only if } \partial H/\partial u < 0. \tag{2.5.35b}$$

In this problem $\partial H/\partial u$ does not vanish identically on a finite time interval and therefore there is no singular control.

If $p(t) \equiv 0$, $\partial H/\partial u$ does not have any change in sign. In this case $u^\star = 40\,000$ for all $t \in (0, 60)$. Condition (2.5.32) implies that $p(t) > 0$ if $p(60) > 0$. In this case $u^\star = 40\,000$ for all $t \in (0, 60)$. If $p(60) < 0$, $p(t)$ is negative for all t. This implies that $\partial H/\partial u = -1 - p(t)$ may have one change in sign during time interval (0, 60).

In the solution of an optimal control problem in ecology, it is desirable to develop an optimal feedback policy. This is obtained by solving the problem for all possible initial states. This is sometimes called the synthesis problem.

The optimal feedback policy in our example is obtained by integrating numerically eq. (2.5.24) and (2.5.31) in reverse time and employing the optimality and transversality conditions. Firstly we integrate these equations in reverse time with $p(60) = 0$ and varying values of $x(60)$ which are greater than 3000. Secondly we integrate these equations in reverse time with $x(60) = 3000$ and varying values of $p(60)$. This process generates a field of solutions which is displayed in Fig. 2.5.1.

If we choose $x(60) = 3000$ and $p(60) = -0.01216$, we generate the optimal trajectory which passes through the point $x(0) = 200\,000$. In this solution $u^\star = 15\,000$ for $0 \leqslant t < 15.9$, and $u^\star = 40\,000$ for $15.9 < t \leqslant 60$. The total yield from this strategy is 2 005 460. If we employ the steady state policy

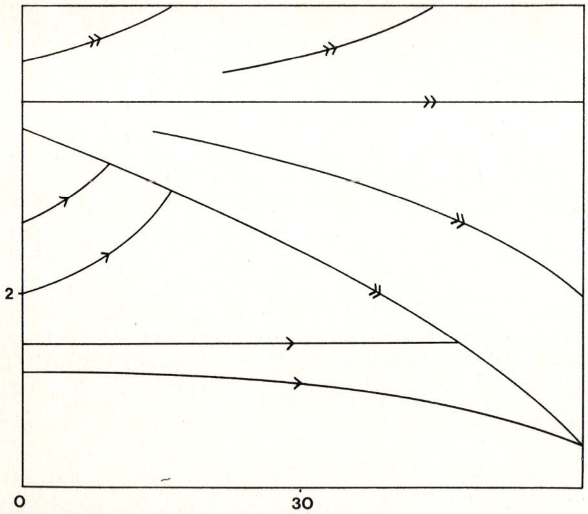

Fig. 2.5.1. Optimal feedback spending policy for a lottery winner. ↠ denotes a u = 40 000 trajectory, and → denotes a u = 15 000 trajectory.

u = 20 000 the total yield is 1 400 000. This completes our analysis of the example.

In an optimal control problem it is possible for a control program to be optimal in the short term and non-optimal in the long term. Example 2.5.1 illustrates this point. In this example $u(t)$ = 40 000 is optimal in the short term, but if $x(0)$ = 200 000, the control $u(t)$ = 40 000 cannot be sustained. For long-term optimality there must be a switch in the control variable. Generally, in an optimal control problem with bounds on the control variables, long-term optimality is achieved by choosing an appropriate switching sequence in the control variables.

If the optimal control variables belong to the interior of the admissible control set, conditions (2.5.10c) and (2.5.11) are tests for short-term optimality. For establishing long-term optimality we have to examine the second variation and show that it is nonnegative (Bliss, 1946; Goh, 1966 a,c). This test is called the focal point condition. It determines when a sequence of short-term optimal policies fails to be a long-term optimal policy. The classic example of this type of dynamical behavior is given by the great circles on a sphere. An arc of a great circle is the shortest path between two points on a sphere provided it is not longer than half a great circle. The focal point condition should become an important condition in future studies of resource management. However, it has not yet been applied to specific problems in resource management. For this reason we shall not discuss it any further.

The above discussion on long-term optimality is within the framework of a standard optimal control problem. But in practice the concept of long-term optimality involves broader issues. In a standard optimal control problem the objective function, the terminal conditions (target), the initial state and the set of admissible control variables are specified. But in the formulation of a long-term optimal policy for a practical problem these very components must be chosen judiciously so as to produce a genuine long-term optimal policy (Vincent et al., 1977).

A major difficulty encountered is the problem of how to specify the terminal conditions. If the planning horizon is infinite some device must be used to prevent an integral objective function from becoming infinite. The standard practice in economics (Clark, 1976 a) is to incorporate a discount factor in the integrand. Another approach is to consider the problem tentatively on a finite time horizon and then shift the terminal time to infinity (Goh, 1969/1970). Recently Vincent et al. (1977) have developed a new approach for formulating long-term optimal policies. In their approach an average cost function is minimized.

We shall now describe the theory of long-term optimal control for autonomous systems which was developed by Vincent et al. (1977). Suppose we have an autonomous system, $\dot{x}_i = f_i(x,u)$, $i = 1, 2, \ldots, n$. Here x is the state vector and u is the control vector. We minimize the average cost which by definition is

$$J[u] = \frac{1}{T}\int_0^T L(x, u)\, dt.$$

Here T is finite and is unconstrained. We impose the boundary condition, $x(0) = x(T)$, so that the average cost policy can be used repeatedly during an infinite time horizon. In this case, $x(0)$ is not specified as in (2.5.2). Let us suppose the control variables must satisfy constraints of the form, $a_r \leqslant u_r \leqslant b_r$, $r = 1, 2, \ldots, m$. Denote the admissible set of control variables by U.

In brief a long-term optimal control problem is as follows:

System: $\dot{x}_i = f_i(x, u), \quad i = 1, 2, \ldots, n.$ (2.5.36)

Constraints: $x_i(0) = x_i(T),$ (2.5.37)

$u \in U.$ (2.5.38)

Objective: $\min \frac{1}{T}\int_0^T L(x, u)\, dt.$ (2.5.39)

By definition, the Hamiltonian function is

$$H(x, u, p) = L(x, u) + \sum_{i=1}^{n} p_i f_i(x, u). \tag{2.5.40}$$

For convenience the necessary conditions for optimality are summarized in the next theorem. A proof of this theorem is given in the paper by Vincent et al. (1977).

Theorem 2.5.4. *A necessary condition for an admissible set* $(x^\star, u^\star, T^\star)$ *to be optimal is that there exists* $p_1, p_2, \ldots, p_n$ *such that*

(i) $\dot{p}_i = -\partial H(x, u, p)/\partial x_i, \quad i = 1, 2, \ldots, n;$ (2.5.41)

(ii) $u^\star$ *minimizes* $H(x^\star, u, p)$ *w.r.t.* $u \in U$; (2.5.42)

and

(iii) $p_i(0) = p_i(T),$ (2.5.43)

$$H[x(T), u(T), p(T)] - \frac{1}{T}\int_0^T L(x, u)dt = 0. \quad (2.5.44)$$

The numerical solution of the optimal control problem displayed in (2.5.36) to (2.5.39) is much harder than that for the standard optimal control problem which is displayed in (2.5.1) to (2.5.5). This is because the problem has mixed end conditions, namely, (2.5.37) and (2.5.43). For this reason we shall seek a special type of solution which satisfies the dynamical equation, the prescribed constraints and the necessary conditions for optimality. The special solution which we shall determine is a steady state solution of (2.5.36) and (2.5.41). It has the added advantage of being easily implemented. To keep the state of the system at such an optimal steady state, we shall, if possible, employ a globally stable control policy whenever the state is displaced from the optimal steady state. We shall adopt the view that in the long run the added costs of such transient control programs are negligible. For this reason we shall not, in any given problem, attempt to evaluate quantitatively the added cost of a globally stable feedback control policy which is used to maintain the state at the optimal steady state solution.

By definition an optimal steady state solution $(x^\star, u^\star, p^\star)$ satisfies the equations,

$$f_i(x, u) = 0, \quad i = 1, 2, \ldots, n, \quad (2.5.45)$$

$$\partial H(x, u, p)/\partial x_i = 0, \quad i = 1, 2, \ldots, n, \quad (2.5.46)$$

and

$$\partial H(x, u, p)/\partial u_r = 0, \quad r = 1, 2, \ldots, m. \quad (2.5.47)$$

If the control variables, $u_1, u_2, \ldots, u_m$, must satisfy the constraints, $a_r \leqslant u_r \leqslant b_r$ for $r = 1, 2, \ldots, m$, condition (2.5.47) must be replaced by the optimality conditions,

$$u_r^\star = a_r \quad \text{only if } \partial H(x^\star, u^\star, p^\star)/\partial u_r^\star > 0. \quad (2.5.48a)$$

$$u_r^\star = b_r \quad \text{only if } \partial H(x^\star, u^\star, p^\star)/\partial u_r^\star < 0. \quad (2.5.48b)$$

and

$$a_r < u_r^\star < b_r \quad \text{only if } \partial H(x^\star, u^\star, p^\star)/\partial u_r^\star = 0. \quad (2.5.48c)$$

These conditions apply for $r = 1, 2, \ldots, m$.

Example 2.5.2. Let N be the density of a fish population and u be the rate of harvest. A model of an exploited fish population is

$$\dot{N} = r\,N(K-N)/K - u \tag{2.5.49}$$

where r and K are positive parameters.

Let $[0, T]$ be a finite but unconstrained time interval. We seek to maximize the average yield from the fishery during the period $[0, T]$. It follows that we minimize

$$J = -\frac{1}{T}\int_0^T u\,\mathrm{d}t. \tag{2.5.50}$$

Let us assume that we can reset the initial state. This is possible because in the long run the yield from an initial transient period is insignificant. This means that $N(0)$ is not specified as in a standard optimal control problem. In order that an average yield policy for a finite time period can be applied repeatedly in an infinite time period, we impose the periodic boundary condition,

$$N(0) = N(T). \tag{2.5.51}$$

From (2.5.40) the Hamiltonian function is

$$H = -u + p[rN(K-N)/K - u]. \tag{2.5.52}$$

The necessary conditions for optimality are

$$\dot{p} = -p[r - 2rN/K], \tag{2.5.53}$$

$$\partial H/\partial u = -1 + p = 0, \tag{2.5.54}$$

$$p(0) = p(T), \tag{2.5.55}$$

$$-u + p[rN(K-N)/K - u] + \frac{1}{T}\int_0^T u\,\mathrm{d}t = 0. \tag{2.5.56}$$

A steady state solution for these conditions is

$$N^\star = K/2, \qquad p^\star = 1, \qquad u^\star = rK/4.$$

We shall show in Section 4.3 that this is the optimal solution if we derive the optimal policy for a finite time horizon and then shift the terminal time to infinity. This provides some assurance that Theorem 2.5.4 provides a valid method for formulating long-term optimal policies.

2.6. DISCRETE TIME OPTIMAL CONTROL

The use of difference equations for modelling interacting populations and ecosystems is becoming popular. One reason for this is the realization that difference equations can provide realistic models of interacting populations when differential equations fail to do so. For example, a single-species population with discrete non-overlapping generations can be realistically described by a difference equation but it cannot be adequately described by a single differential equation.

For the optimal control of a dynamic system with a time delay, there is a very important reason why we should use discrete time optimal control theory rather than continuous time optimal control theory. The optimal control of a difference equation model with a time delay is no more difficult than that for a model without a time delay. On the other hand the optimal control theory for systems described by differential equations with a time delay is extremely difficult to apply; therefore it is best to leave it alone for the time being.

The realization that a discrete time optimal control problem is none other than a mathematical programming problem with a special structure occurred in the early nineteen-sixties. A good account of this approach is given in the books by Tabak and Kuo (1971) and Canon et al. (1970). This means that numerical algorithms developed for nonlinear programming can be used for discrete time optimal control problems.

Let N be a given positive integer, and $t = 0,1,2, \ldots , N-1$. The standard discrete time optimal control problem is as follows:

System: $$x_i(t+1) = f_i[x(t), u(t)], \quad i = 1, 2, \ldots, n. \tag{2.6.1}$$

Initially: $$x_i(0) = x_{i0}, \quad i = 1, 2, \ldots, n. \tag{2.6.2}$$

Terminally: $$\psi^s[x(N), v] = 0, \quad s = 1, 2, \ldots, k. \tag{2.6.3}$$

Constraints: $$a_r \leqslant u_r(t) \leqslant b_r, \quad r = 1, 2, \ldots, m, \tag{2.6.4}$$

Objective: $$\min g[x(N), v] + \sum_{t=0}^{N-1} L[x(t), u(t)]. \tag{2.6.5}$$

In most problems $\{a_r, b_r\}$ is a set of constants. For convenience the set of control vectors which satisfy (2.6.4) is denoted by U. The vector v denotes the control parameters, $v_1, v_2, \ldots, v_h$. By definition the Hamiltonian function at time t is

$$H[x(t), u(t), p(t+1)] = L[x(t), u(t)] + \sum_{i=1}^{n} p_i(t+1) f_i[x(t), u(t)], \tag{2.6.6}$$

where $p(t+1)$ is the costate vector. For convenience we shall use $H\{t\}$ in place of $H[x(t), u(t), p(t+1)]$. The next theorem describes some necessary conditions for optimality.

Theorem 2.6.1. *If* $u(0), u(1), \ldots, u(N-1)$ *is an optimal control sequence,* v *is an optimal vector and* $x(0), x(1), x(2), \ldots, x(N)$ *is an optimal trajectory,* *then there exist costate vectors* $p(0), p(1), p(2), \ldots, p(N)$ *and multipliers* $\lambda_1, \lambda_2, \ldots, \lambda_k$ such that

(i) $$p_i(t) = \partial H\{t\}/\partial x_i(t), \quad i = 1, 2, \ldots, n; \tag{2.6.7}$$

(ii) *for* $r = 1, 2, \ldots, m$

$$\partial H\{t\}/\partial u_r(t) = 0, \quad \text{if } a_r < u_r(t) < b_r, \tag{2.6.8a}$$

$$\partial H\{t\}/\partial u_r(t) > 0, \quad \text{if } u_r(t) = a_r, \tag{2.6.8b}$$

$$\partial H\{t\}/\partial u_r(t) < 0, \quad \text{if } u_r(t) = b_r; \tag{2.6.8c}$$

(iii) $$p_i(N) = \frac{\partial g}{\partial x_i(N)} + \sum_{s=1}^{k} \lambda_s \frac{\partial \psi^s}{\partial x_i(N)}, \quad i = 1, 2, \ldots, n, \tag{2.6.9a}$$

$$0 = \frac{\partial g}{\partial v_j} + \sum_{s=1}^{k} \lambda_s \frac{\partial \psi^s}{\partial v_j}, \quad j = 1, 2, \ldots, h. \tag{2.6.9b}$$

Proof. If $a_r < u_r(t) < b_r$ for $r = 1, 2, \ldots, m$ the above theorem can be established by a simple application of Theorem 2.2.3. We form the augmented function

$$F = g[x(N), v] + \sum_{t=0}^{N-1} [L\{x(t), u(t)\} + \sum_{i=1}^{k} p_i(t+1)\ \{f_i[x(t), u(t)]$$

$$- x_i(t+1)\}] + \sum_{s=1}^{k} \lambda_s \psi^s[x(N), v]. \tag{2.6.10}$$

Conditions (2.6.6) and (2.6.10) give

$$F = g[x(N), v] + \sum_{t=0}^{N-1} H\{t\} - \sum_{t=1}^{N-1} \sum_{i=1}^{n} p_i(t)\, x_i(t)$$

$$- \sum_{i=1}^{n} p_i(N)\, x_i(N) + \sum_{s=1}^{k} \lambda_s \psi^s[x(N), v]. \tag{2.6.11}$$

Necessary conditions for optimality are

$$\frac{\partial F}{\partial x_i(t)} = \frac{\partial H\{t\}}{\partial x_i(t)} - p_i(t) = 0, \quad i = 1, 2, \ldots, n, \tag{2.1.12a}$$

$$\frac{\partial F}{\partial x_i(N)} = \frac{\partial g}{\partial x_i(N)} - p_i(N) + \sum_{s=1}^{k} \lambda_s \frac{\partial \psi^s}{\partial x_i(N)} = 0, \quad i = 1, 2, \ldots, n, \tag{2.6.12b}$$

$$\frac{\partial F}{\partial u_r(t)} = \frac{\partial H\{t\}}{\partial u_r(t)} = 0, \quad r = 1, 2, \ldots, m, \tag{2.6.12c}$$

$$\frac{\partial F}{\partial v_j} = \frac{\partial g}{\partial v_j} + \sum_{s=1}^{k} \lambda_s \frac{\partial \psi^s}{\partial v_j} = 0, \quad j = 1, 2, \ldots, h. \tag{2.6.12d}$$

If $u_r = a_r$ instead of (2.6.12c), we have

$$\frac{\partial F}{\partial u_r(t)} = \frac{\partial H\{t\}}{\partial u_r(t)} > 0.$$

Similarly if $u_r = b_r$, we have $\partial H\{t\}/\partial u_r(t) < 0$.

The equations in (2.6.2) are called costate equations. The conditions in (2.6.8) are called optimality conditions and those in (2.6.9) are called transversality conditions.

Example 2.6.1. Let $S(t)$ be the post-harvest density of a single-species population with discrete and non-overlapping generations. Let $h(t+1)$ be the number of animals or plants which is harvested at time $t+1$. Assume that the harvesting is carried out during a short time interval and takes place before the time for reproduction. A model of this type of population is

$$S(t+1) = S(t)F[S(t)] - h(t+1).$$

It is convenient to relabel $h(t+1)$ and call it $u(t)$. Hence we have

$$S(t+1) = S(t)F[S(t)] - u(t) \tag{2.6.13}$$

where $t = 0, 1, 2, \ldots, N-1$.

If (2.6.13) represents an exploited fish population we usually have the constraint

$$0 \leqslant u(t) \leqslant b \tag{2.6.14}$$

where b is a constant. At time $t = 0$, let us suppose that it is possible to measure the population density before and after the harvesting. The harvesting at time $t = 0$ is also a control variable and has the same status as v in (2.6.5). We could easily adapt Theorem 2.6.1 to take this additional degree of freedom into account but for simplicity we shall ignore it.

The objective in this problem is to maximize the total yield during the period $[1,N]$. This means we want to maximize

$$\sum_{t=1}^{N} h(t) = \sum_{t=0}^{N-1} u(t).$$

In order to utilize Theorem 2.6.1 directly, we shall minimize the function

$$J = -\sum_{t=0}^{N-1} u(t).$$

We shall impose the terminal condition

$$S(N) \geqslant A \qquad (2.6.15)$$

where A is a positive number. This means that the population does become extinct at time $t = N$. Condition (2.6.15) is not in the same form as (2.6.3). We have to decompose it into two cases, namely: (i) $S(N) > A$, but otherwise $S(N)$ is unconstrained; and (ii) $S(N) = A$. This completes the formulation of the problem.

By definition the Hamiltonian function is

$$H\{t\} = -u(t) + p(t+1)\,[S(t)F(S(t)) - u(t)]. \qquad (2.6.16)$$

The costate equation is

$$p(t) = p(t+1)\mathrm{d}[S(t)F(S(t))]/\mathrm{d}S(t). \qquad (2.6.17)$$

We have

$$\partial H\{t\}/\partial u(t) = -1 - p(t+1). \qquad (2.6.18)$$

The transversality condition for the case in which $S(N) > A$ is $p(N) = 0$. When $S(N) = A$, the transversality condition implies $p(N) = c$, a constant of integration.

The conditions, $p(N) = 0$ and (2.6.17), imply that $p(t) \equiv 0$. From (2.6.18), $\partial H\{t\}/\partial u(t) = -1$. It follows that in this case, $u(t) = b$ for $t = 0, 1, 2, \ldots, N-1$. In the case $p(N) = c$, we have to determine the precise switching sequence numerically. In a number of numerical examples in which $F(S)$ are biologically acceptable functions, we find that the dynamic optimal policy is identical to the steady state optimal policy except for the initial and terminal periods.

If b is sufficiently large the steady state optimal policy will satisfy the constraint $0 < u(t) < b$. From (2.6.18) it follows that $p(t+1) + 1 = 0$ for values of t when the system is in equilibrium. Then condition (2.6.17) implies

$$1 = \mathrm{d}[S(t)F(S(t))]/\mathrm{d}S(t). \qquad (2.6.19)$$

This is none other than the familiar condition for an optimal steady state solution. If the optimal steady state has been determined by means of condition (2.6.19), the corresponding optimal control can be obtained from (2.6.13). This completes the analysis of the example.

We have mentioned that a discrete time optimal control with a time delay poses no great difficulty if the problem, as a whole, is considered to be a nonlinear mathematical programming problem. We shall now demonstrate this approach by examining a class of discrete time problems with a time delay.

Let $t = 0, 1, 2, \ldots, N$, and $x(t)$ be the state vector. We want to determine an optimal sequence of control vectors $u(0), u(1), \ldots, u(N-1)$, and an optimal vector v. Briefly the problem is as follows:

System: $\quad x_i(t+1) = f_i[x(t), x(t-1), u(t)], \quad i = 1, 2, \ldots, n.$ (2.6.20)

Initially: $\quad x(-1) = c, \qquad x(0) = d.$ (2.6.21)

Terminally: $\quad \psi^s[x(N), v] = 0, \quad s = 1, 2, \ldots, k.$ (2.6.22)

Constraints: $\quad a_r \leqslant u_r(t) \leqslant b_r, \quad r = 1, 2, \ldots, m.$ (2.6.23)

Objective: $\quad \min g[x(N), v] + \sum_{t=0}^{N-1} L[x(t), x(t-1), u(t)].$ (2.6.24)

By definition the Hamiltonian function

$$H\{t\} = L[x(t), x(t-1), u(t)] + \sum_{i=1}^{n} p_i(t+1) f_i[x(t), x(t-1), u(t)]. \tag{2.6.25}$$

Theorem 2.6.2. *If* $u(0), u(1), \ldots, u(N-1)$ *is an optimal control sequence,* v *is an optimal vector and* $x(-1), x(0), \ldots, x(N)$ *is an optimal trajectory, then there exist costate vectors* $p(0), p(1), \ldots, p(N)$ *and multipliers* $\lambda_1, \lambda_2, \ldots, \lambda_k$ *such that*

(i) $$p_i(t) = \frac{\partial H\{t\}}{\partial x_i(t)} + \frac{\partial H\{t+1\}}{\partial x_i(t)}, \quad i = 1, 2, \ldots, n; \tag{2.6.26}$$

(ii) *for* $r = 1, 2, \ldots, m$

$a_r < u_r < b_r$ *only if* $\partial H\{t\}/\partial u_r = 0,$ (2.6.27a)

$u_r = a_r$ *only if* $\partial H\{t\}/\partial u_r > 0,$ (2.6.27b)

$u_r = b_r$ *only if* $\partial H(t)/\partial u_r < 0;$ (2.6.27c)

(iii) $$p_i(N) = \frac{\partial g}{\partial x_i(N)} + \sum_{s=1}^{k} \lambda_s \frac{\partial \psi^s}{\partial x_i(N)}, \quad i = 1, 2, \ldots, n, \tag{2.6.28a}$$

$$0 = \frac{\partial g}{\partial v_j} + \sum_{s=1}^{k} \lambda_s \frac{\partial \psi^s}{\partial v_j}, \quad i = 1, 2, \ldots, h. \tag{2.6.28b}$$

The proof is similar to that for Theorem 2.6.1. We shall therefore not examine it in detail. Clearly similar theorems apply if there are more complex time delays.

Example 2.6.2. In many fish (animal) populations it takes several years for a fish to mature and reproduce. Clark (1976 b) has developed a difference equation with a time delay for this type of population.

At time t, let $S(t)$ be the post-harvest fish population. Let $E(t + 1)$ be the fraction of the adult population which is impulsively harvested just before spawning takes place for that year. Let λ be the fraction of adults that survive in the period $[t, t + 1]$ in the absence of harvesting. Let $F[S(t - 1)]$ be the recruitment to the adult population at time $t + 1$; and let this be due to the spawning population at time $t - 1$. A model of the harvested population is

$$S(t + 1) = [1 - E(t + 1)] [\lambda S(t) + F(S(t - 1))] . \tag{2.6.29}$$

For convenience let $u(t) = E(t + 1)$ and $G[S(t), S(t - 1] = \lambda S(t) + F[S(t - 1)]$.

Our objective is to maximize the total yield during the planning period $[1,N]$. It follows that we want to minimize

$$J = - \sum_{t=0}^{N-1} u(t)G[S(t), S(t - 1)] . \tag{2.6.30}$$

We shall impose the constraints

$$0 \leqslant u(t) \leqslant b, \tag{2.6.31}$$

$$S(N) \geqslant A \tag{2.6.32}$$

where A is a positive constant. We assume that $S(-1)$ and $S(0)$ are known.

By definition, the Hamiltonian function is

$$H\{t\} = -u(t)G[S(t), S(t - 1)] + p(t + 1)[1 - u(t)] G[S(t), S(t - 1)] . \tag{2.6.33}$$

A necessary condition for optimality is

$$p(t) = [-u(t) + p(t + 1)(1 - u(t))]\lambda$$
$$+ [-u(t + 1) + p(t + 2) (1 - u(t + 1))] \mathrm{d}F[S(t)] / \mathrm{d}S(t). \tag{2.6.34}$$

If $0 < u(t) < b$, a necessary condition for optimality is

$$\partial H\{t\}/\partial u(t) = [-1 - p(t + 1)] \; G[S(t), S(t - 1)] = 0. \tag{2.5.35}$$

Let $S^\star, u^\star, p^\star$ be an optimal steady state solution. Equation (2.6.35) implies that $p^\star + 1 = 0$. This together with equation (2.6.34) imply that

$$1 = \lambda + \mathrm{d}F(S^\star)/\mathrm{d}S^\star . \tag{2.6.36}$$

This equation may be solved for $S^\star$. Eq. (2.6.29) implies that

$$u^\star = 1 - S^\star / [\lambda S^\star + F(S^\star)] . \tag{2.6.37}$$

This completes our analysis of this example.

Singular control theory forms an important part of the continuous time optimal control theory. Singular control usually occurs when one or more control variables appear linearly in the Hamiltonian function. However, a discrete time optimal control problem is none other than a mathematical programming problem with a special structure. From this point of view, the state and control variables in a discrete time optimal control problem have the same status. It follows that there is no special significance to be attached to the fact that control variables only appear linearly in a problem in which the state variables appear nonlinearly.

We shall now examine a simple example to illustrate the above comments on "singular control" in discrete time optimal control problems. Given the system

$$x(t+1) = x(t) - u(t), \quad t = 0, 1, 2, \tag{2.6.38}$$

$$x(0) = 5, \tag{2.6.39}$$

we want to minimize

$$J = x^2(3) + x^2(2) + x^2(1) = \sum_{t=1}^{3} x^2(t). \tag{2.6.40}$$

The Hamiltonian function is

$$H\{t\} = x^2(t) + p(t+1)[x(t) - u(t)]. \tag{2.6.41}$$

A necessary condition for optimality is

$$\partial H\{t\}/\partial u(t) = -p(t+1) = 0. \tag{2.6.42}$$

This equation does not contain $u(t)$ explicitly.

But we can use the system equation to transform the objective function so that the control variable appears nonlinearly in the transformed objective function. Eq. (2.6.38) and (2.6.40) give

$$J = \sum_{t=0}^{2} [x^2(t) - 2x(t)u(t) + u^2(t)].$$

With this new objective function the problem becomes a nonsingular problem.

In the past there has been some confusion between "singular control" and abnormality in discrete time optimal control problem. We shall discuss a simple but effective approach to overcome a common type of abnormality in discrete time optimal control problem. We shall do so by applying it to a simple example. Suppose we want to find $u(0)$, $u(1)$, $u(2)$ which minimize the function,

$$J[u] = \sum_{t=0}^{2} [x^2(t) + u^2(t)]$$

subject to the constraints,

$x(t+1) = x(t) - u(t), \quad t = 0, 1, 2$

$x(0) = 5,$

$x(3) = 0.$

If dynamic programming is used for solving this problem, we first consider the problem in the interval [2,3]. This subproblem is abnormal, the abnormality being due to the fact that there are too many constraints, namely, $x(3) = 0$, $x(3) = x(2) - u(2)$, and $x(2)$ is "given". This means that there is no degree of freedom in the subproblem. This abnormality can be removed by telescoping the two time units in [1, 3] into a single unit of time, and by using (2.6.44) to eliminate condition (2.6.46). These manipulations give the following problem:

$$\min[2x^2(2) + \sum_{t=0}^{1} x^2(t) + u^2(t)]$$

subject to $x(t+1) = x(t) - u(t)$, $t = 0, 1,$

$x(0) = 5.$

In this problem $x(2)$ is unconstrained. It is therefore a normal problem and it can be readily solved by using dynamic programming.

CHAPTER 2: SELECTED REFERENCES

Static optimization problems

Bliss (1946); Dantzig (1963); Fiacco and McCormick (1968); Vincent and Goh (1972); Kuester and Mize (1973); Fletcher (1975).

Multiple objectives decision problems

Starr and Ho (1969); Vincent and Leitmann (1970); Leitmann and Marzollo (1975).

Continuous time optimal control

Bliss (1946); Berkovitz (1961); Pontryagin et al. (1962); Lawden (1963); Goh (1966 a, b, c, 1973 b); Leitmann (1966); Lee and Markus (1967); Bryson and Ho (1969); Intrilligator (1971); Vincent and Goh (1972); Bell and Jacobson (1975); Fleming and Rishel (1975); Vincent et al. (1977).

Discrete time optimal control

Canon et al. (1970); Tabak and Kuo (1971).

Chapter 3

STABILITY AND NONVULNERABILITY

3.1. INTRODUCTION

The simplest way to examine stability in an ecosystem model is by examining the eigenvalues of a matrix which is computed at an equilibrium of the model. But this method can only establish local stability or instability. On the other hand, ecosystems in the real world are subjected to large perturbations of the initial state, and continual disturbances on the system dynamics.

For a given model it is possible to use computer simulations to investigate the behavior of the model for finite perturbations of initial state. But computer simulations cannot guarantee that an equilibrium does indeed have a finite region of attraction. This state of affairs becomes increasingly worse as the number of species in a given ecosystem increases.

A powerful analytical method for establishing that an equilibrium is stable relative to finite perturbations of the initial state is the direct method of Liapunov (LaSalle and Lefschetz, 1961; Hahn, 1967; Gurel and Lapidus, 1968; Willems, 1970). There are many methods for constructing Liapunov functions (Schultz, 1965; Gurel and Lapidus, 1968, 1969). It has often been said that it is very difficult to construct Liapunov functions. This is true. But the statement misses the point that it is equally difficult to use other non-local methods to establish finite stability in a nonlinear model with more than two state variables. Computer simulations are excluded in this comment because they can provide only tentative conclusions on the finite stability of an equilibrium.

In Sections 3.3 to 3.10, we shall demonstrate the usefulness of the direct method of Liapunov by applying it to single species and two species models which are described by differential equations.

Consider the autonomous model

$$\dot{N}_i = N_i F_i(N_1, N_2, \ldots, N_m), \quad i = 1, 2, \ldots, m \tag{3.1.1}$$

where $F_1(N), F_2(N), \ldots, F_m(N)$ are continuous functions in the positive orthant. Suppose it has a positive equilibrium at $N^\star$ where

$$F_i(N^\star) = 0, \quad i = 1, 2, \ldots, m. \tag{3.1.2}$$

Let $x_i = N_i - N_i^\star$ for $i = 1, 2, \ldots, m$. The linearized dynamics is

$$\dot{x}_i = \sum_{j=1}^{m} N_i^\star a_{ij} x_j, \quad i = 1, 2, \ldots, m \tag{3.1.3}$$

where $(a_{ij}) = (\partial F_i/\partial N_j)$ at $N^\star$. We can show that $N^\star$ is locally asymptotically stable if the real parts of all the eigenvalues of $(N_i^\star a_{ij})$ are negative. A better approximation of model (3.1.1) is

$$\dot{N}_i = \sum_{j=1}^{m} N_i a_{ij}(N_j - N_j^\star), \quad i = 1, 2, \ldots, m. \tag{3.1.4}$$

This approximation is better because it contains the second order terms in the Taylor's expansion of (3.1.1). Moreover, model (3.1.4) is valid in a larger neighbourhood than that for (3.1.3). Model (3.1.4) is none other than a Lotka—Volterra model. We conclude that a study of a Lotka—Volterra model in the neighbourhood of an equilibrium can provide more information than an eigenvalue analysis.

Liapunov-like functions can also be used to establish the stability of a limit cycle; and to establish that the solutions of a model are ultimately confined. We shall examine these applications of Liapunov-like functions in Section 3.11. This method for studying a limit cycle and a region of ultimate confinement is underdeveloped.

In Section 3.12, we shall examine how Liapunov-like functions can be used to establish that an ecosystem model is nonvulnerable relative to a given class of continual disturbances. At present, this method of studying nonvulnerability can be readily applied to Lotka—Volterra models, and a class of more general nonlinear models which have qualitatively the same behavior as Lotka—Volterra models. This approach to study the behavior of a system in an uncertain environment is none other than a "worst case" analysis.

The stability theory of difference equations is less developed than that for differential equations. In Sections 3.14 to 3.16, we shall examine how Liapunov and Liapunov-like functions can also be used to establish finite stability, global stability, and nonvulnerability in difference equation population models.

3.2. LOCAL STABILITY

Let $N_1, N_2, \ldots N_m$ be the densities of m interacting populations. Consider a general model of interacting populations which is described by the set of differential equations:

$$\dot{N}_i = N_i F_i(N_1, N_2, \ldots, N_m) \tag{3.2.1}$$

where $i = 1, 2, \ldots, m$. The functions, $F_1, F_2, \ldots, F_m$, are in general nonlinear functions of $N_1, N_2, \ldots, N_m$.

We now state some useful definitions. The set of numbers, $(N_1, N_2, \ldots, N_m) = N$, is called the state vector of the system. The region $\{N \mid N_1 > 0, N_2 > 0, \ldots, N_m > 0\}$ is called the feasible region or the positive orthant. A point $(N_1^\star, N_2^\star, \ldots, N_m^\star)$ is an equilibrium if $\dot{N}_i = 0$ for $i = 1, 2, \ldots, m$. A point $(N_1^\star, N_2^\star, \ldots, N_m^\star)$ is a positive equilibrium if

$$N_1^\star > 0, N_2^\star > 0, \ldots, N_m^\star > 0 \tag{3.2.2}$$

and

$$F_i(N_1^\star, N_2^\star, \ldots, N_m^\star) = 0, \tag{3.2.3}$$

for $i = 1, 2, \ldots, m$. In the absence of migrations we are usually interested only in a positive equilibrium. If an equilibrium is not positive we could consider the model ecosystem as a lower dimension model by deleting the missing species. If necessary we could in this way study a hierarchy of models.

If N is close to $N^\star$, let $x_i = N_i - N_i^\star$ for $i = 1, 2, \ldots, m$. Substituting these into eqs. (3.2.1), and employing Taylor's theorem and ignoring higher order terms in $x_1, x_2, \ldots, x_m$, we get

$$\dot{x}_i = \sum_{j=1}^{m} N_i^\star a_{ij} x_j, \tag{3.2.4}$$

where $i = 1, 2, \ldots, m$ and

$$a_{ij} = \frac{\partial F_i}{\partial N_j}(N^\star), \quad i = 1, 2, \ldots, m. \tag{3.2.5}$$

Theorem 3.2.1. *The equilibrium $(N_1^\star, N_2^\star, \ldots, N_m^\star)$ is locally stable if all the real parts of the eigenvalues of the matrix $(N_i^\star a_{ij})$ are negative.*

Proof. By definition, the equilibrium $N^\star$ is locally stable if all the solutions of (3.2.1) which begin near $N^\star$ will move into decreasing neighbourhoods of $N^\star$. For $t > 0$, the system (3.2.1) must not be disturbed.

By definition, λ_k is an eigenvalue of $(N_i^\star a_{ij})$ if there is a vector X_k such that

$$(N_i^\star a_{ij})X_k = \lambda_k X_k. \tag{3.2.6}$$

In general, λ_k is a complex number and X_k is a complex vector. X_k is called an eigenvector.

Among ecosystem models it is extremely rare to encounter a model with repeated eigenvalues. Let us assume that the eigenvalues of $(N_i^\star a_{ij})$ are distinct. In this case the general solution of (3.2.4) is

$$x = \sum_{i=1}^{m} c_i X_i \exp(\lambda_i t), \tag{3.2.7}$$

where $c_1, c_2, \ldots, c_m$ are arbitrary constants, $\lambda_1, \lambda_2, \ldots, \lambda_m$ are the eigenvalues and $X_1, X_2, \ldots, X_m$ are the corresponding eigenvectors. If the real part of λ_i is negative, $\exp(\lambda_i t) \to 0$ as $t \to \infty$. It follows that if each eigenvalue has a negative real part, x tends to the zero vector as $t \to \infty$.

Example 3.2.1. Consider the prey—predator model:

$$\dot{N}_1 = N_1[31 - 1.1N_1 - 4N_2],$$

$$\dot{N}_2 = N_2[-3 + 0.4N_1 + 0.2N_2].$$

A solution of the system of equations,

$$31 - 1.1N_1 - 4N_2 = 0,$$

$$-3 + 0.4N_1 + 0.2N_2 = 0,$$

is (10, 5). We have $a_{11} = -1.1$, $a_{12} = -4$, $a_{21} = 0.4$, $a_{22} = 0.2$, $N_1^\star = 10$ and $N_2^\star = 5$. It follows that

$$(N_i^\star a_{ij}) = \begin{pmatrix} -11 & -40 \\ 2 & 1 \end{pmatrix}.$$

The eigenvalue equation of this matrix is

$$\lambda^2 + 10\lambda + 69 = 0.$$

It follows that the eigenvalues are $-5 \pm i\sqrt{44}$. Hence the equilibrium (10, 5) is locally stable.

In ecosystem models with a few species, a useful analytical test for the real parts of all the eigenvalues of a matrix to be negative is the Routh—Hurwitz stability criterion. Let A be an $m \times m$ matrix. Its eigenvalue equation is obtained by expanding the equation,

$$\det(A - \lambda I) = 0, \tag{3.2.8}$$

where det denotes the determinant of a matrix and I is an identity matrix. From this operation we get an mth order polynomial equation,

$$\lambda^m + p_1\lambda^{m-1} + p_2\lambda^{m-2} + \cdots + p_{m-1}\lambda + p_m = 0. \tag{3.2.9}$$

The Routh—Hurwitz stability criterion for $m = 2, 3, 4$ is as follows:

$$m = 2, \quad p_1 > 0, p_2 > 0; \tag{3.2.10}$$

$$m = 3, \quad p_1 > 0, p_3 > 0, p_1p_2 > p_3; \tag{3.2.11}$$

$$m = 4, \quad p_1 > 0, p_2 > 0, p_3 > 0, p_4 > 0, (p_1p_2 - p_3)\,p_3 > p_1^2p_4. \tag{3.2.12}$$

These sets of conditions are necessary and sufficient for local asymptotic stability.

Example 3.2.2. This example demonstrates the usefulness of the Routh–Hurwitz stability conditions. A model of three-species competition is as follows:

$$\dot{N}_1 = N_1[10 - 5N_1 - 3N_2 - 2N_3],$$

$$\dot{N}_2 = N_2[9 - 4N_1 - 4N_2 - N_3],$$

$$\dot{N}_3 = N_3[2 - a - N_1 - 2N_2 + aN_3].$$

It has a positive equilibrium at (1, 1, 1).

The positive equilibrium is locally stable if the real parts of all the eigenvalues of the matrix,

$$(N_i^{\star} a_{ij}) = \begin{bmatrix} -5 & -3 & -2 \\ -4 & -4 & -1 \\ -1 & -2 & a \end{bmatrix},$$

are negative. Its eigenvalue equation is

$$\lambda^3 + (9 - a)\lambda^2 + (4 - 9a)\lambda + (1 - 8a) = 0.$$

For stability, we require

$$9 - a > 0, \qquad 1 - 8a > 0 \quad \text{and} \quad (9 - a)(4 - 9a) > 1 - 8a.$$

The last inequality is satisfied if $a < 0.482$ or $a > 8.07$. Clearly for stability, we require $a < 1/8$.

An interesting conclusion from this example is that in a three-species Lotka–Volterra competition model it is not necessary for each species to sustain density dependent mortalities due to intraspecific interactions.

The Routh–Hurwitz stability criterion is most useful for problems in which there is an unknown parameter and in which we are interested in the range of values of the parameter for which the equilibrium is stable.

For models with more than four species, it is very difficult to use the Routh–Hurwitz stability criterion. If a matrix is given numerically, it is quite simple to compute its eigenvalues on a computer. An alternative method to establish that a matrix A is stable is to solve for the unknown symmetric matrix P which satisfies the equation,

$$PA + A^{\mathrm{T}}P = -I. \tag{3.2.13}$$

A necessary and sufficient condition for A to be stable is that P is positive definite (Willems, 1970).

3.3. FINITE AND GLOBAL STABILITY

In this section we shall examine methods which can establish stability relative to realistic perturbations of the initial state. We assume that for $t > 0$, the system is not disturbed.

Consider a model of m interacting populations which is described by the equations,

$$\dot{N}_i = N_i F_i(N_1, N_2, \ldots, N_m), \tag{3.3.1}$$

where $i = 1, 2, \ldots, m$. Assume that it has a positive equilibrium at $N^\star$. This means that

$$N_1^\star > 0, N_2^\star > 0, \ldots, N_m^\star > 0 \tag{3.3.2}$$

and

$$F_i(N_1^\star, N_2^\star, \ldots, N_m^\star) = 0 \tag{3.3.3}$$

for $i = 1, 2, \ldots, m$.

Let Ω be an open region which is completely inside the positive region. Let $N^\star$ belong to Ω. The region Ω is a finite region of stability if Ω is a finite region and all the solutions of (3.3.1) which begin in Ω remain in Ω and tend to $N^\star$ as $t \to \infty$. Note that the solutions must not leave the region Ω. Otherwise, some of the species may become extinct for a temporary period. This concept of stability means that if a disturbance resets the initial state and the system is thereafter left alone, the natural dynamics will drive the state into decreasing neighbourhoods of $N^\star$. It does not directly provide any information on the system's ability to resist continual disturbances which are normally experienced by real ecosystems. Such a limitation in this concept of stability is often overlooked.

If the region Ω is equal to the positive orthant $\{N \mid N_i > 0, i = 1, 2, \ldots, m\}$, the system is globally stable. This concept of global stability is only trivially different from that for physical systems. In the latter, Ω is equal to the whole of the corresponding state space. The two types of global stability are identical because we could map the feasible region into the whole of the state space by means of the transformation:

$$X_i = \ln(N_i/N_i^\star), \tag{3.3.4}$$

where $i = 1, 2, \ldots, m$. However, this change of variables is unnecessary. Moreover, it leads to a loss of insight into the system.

If a disturbance causes one or more species to become extinct, the simplest way to use the concept of finite and global stability is to consider the disturbed ecosystem as a new system of lower dimension. Such catastrophic perturbations of the initial states are excluded unless otherwise stated.

The best method for establishing stability in a finite region or global stability is the direct method of Liapunov (LaSalle and Lefschetz, 1961; Willems, 1970). In a physical system a Liapunov function is a generalization of the free energy of the system. If energy is continuously dissipated, the state of the system will evolve and move towards the state where the energy of the system attains local or global minimum. The corresponding state is none other than an equilibrium of the system.

For ecological systems it is also helpful to employ an energy principle to provide guidelines for the construction of an effective Liapunov function. For ecological systems we require a two-sided energy principle (Goh, 1977 a). The standing biomass in a population is a measure of the free energy embodied in the population. In a viable population, its density must increase at very low densities; and it must decrease at very high densities. It follows that the population must absorb energy from an external source at low densities and it must dissipate energy at very high densities. This two-sided energy principle for a viable population provides a sharp contrast between ecological and physical systems. However, it is plausible that this two-sided energy principle may also be a very useful concept for a physical control system which has to be maintained at an artificial energy level for the proper functioning of the system.

In the region Ω, the function $V(N)$ is a suitable Liapunov function for model (3.3.1) if it has the following properties:

(i) $V(N^\star) = 0$;
(ii) In Ω, $V(N)$ has a global minimum equal to zero at $N^\star$;
(iii) In Ω, the level surface, $V(N) = K$ for each positive value of K, is a closed surface;
(iv) The derivative,

$$\dot{V}(N) = \sum_{i=1}^{m} \frac{\partial V}{\partial N_i} N_i F_i(N), \tag{3.3.5}$$

is nonpositive for all $N \in \Omega$.

Condition (iii) forces us to define Ω in terms of a level surface of $V(N)$. If L is a positive constant, we choose $\Omega = \{N \mid V(N) < L\}$.

Theorem 3.3.1. *Let $N^\star$ be a positive equilibrium of model (3.3.1). The region, $\Omega = \{N \mid V(N) < L\}$, is a region of stability if the function $V(N)$ is a Liapunov function in Ω and $\dot{V}$ is negative definite.*

The proof of this theorem follows in a straightforward manner from the proof of Theorem 1.4.1. Essentially the conditions for a Liapunov function ensure that along the solutions of model (3.3.1), $V[N(t)]$ tends monotonically to zero. The corresponding state vector $N(t)$ tends to $N^\star$. $V[N]$ could be interpreted as a measure of the distance of the state N from $N^\star$.

An effective method for estimating a finite region of stability is to employ an optimization algorithm. Choose tentatively a positive constant K. Maximize $\dot{V}(N)$ subject to $V(N) = K$. If the computed maximum value of $\dot{V}(N)$ is negative, the value of K is increased. This process is carried out until the computed maximum value of $\dot{V}(N)$ is positive. From this process we can determine a positive constant L such that $\dot{V}(N)$ is negative in the region, $\Omega = \{N \mid V(N) < L, N \neq N^\star\}$. This is then a finite region of stability. A robust optimization algorithm which can be used in this method of computing a finite region of stability is Fletcher's algorithm (Fletcher, 1975).

Example 3.3.1. We shall examine a model of a harvested single-species population. Let h be a constant rate of harvest. Consider the model,

$$\dot{N} = 2N(4 - N) - h. \tag{3.3.6}$$

The natural dynamics has a globally stable equilibrium at $N = 4$. A non-trivial equilibrium in the harvested population is a solution of the quadratic equation,

$$2N^2 - 8N + h = 0.$$

This implies two nontrivial equilibria at S_1 and S_2 where

$$S_1 = [4 - \sqrt{(16 - 2h)}]/2, \qquad S_2 = [4 + \sqrt{(16 - 2h)}]/2.$$

These equilibria are positive if $h < 8$.

From (3.3.6) we get

$$\dot{N} = -2(N - S_1)(N - S_2).$$

Graphically in the (t, N)-space, we can show that the lower level equilibrium at S_1 is unstable and the region of stability of the higher level equilibrium is $\{N \mid N > S_1\}$.

A Liapunov function for estimating the region of stability for the equilibrium S_2 is

$$V(N) = |N - S_2|. \tag{3.3.7}$$

We have $\dot{V} = \dot{N}$ if $N > S_2$ and $\dot{V} = -\dot{N}$ if $N < S_2$. It follows that $\dot{V} < 0$ for $N > S_1$ and $N \neq S_2$. The value of $V(N)$ at $N = S_1$ is $|S_1 - S_2|$. Hence an estimate of the region of stability of the equilibrium S_2 is $\{N \mid V(N) < |S_1 - S_2|\}$. With additional arguments which are valid for only single-species models, we cannot conclude from this Liapunov function that the region of stability is $\{N \mid N > S_1\}$.

An interesting conclusion from this example is that a constant quota harvesting policy for a single-species population cannot be globally stable. We can show that this conclusion is not dependent on the assumption that the natural dynamics is described by a logistic equation.

Note that a Liapunov need not be continuously differentiable everywhere in a region Ω.

Example 3.3.2. Let N_1 and N_2 be the densities of a prey and its predator respectively. Let the prey population be harvested at the constant rate of 1.25 units per unit time. We have

$$\dot{N}_1 = N_1(10 - 2.2N_1 - 2N_2) - 1.25,$$

$\dot{N}_2 - N_2(-0.3 + 0.2N_1 - 0.1N_2)$.

This model has an equilibrium at (2.5, 2). It may be rewritten in the form,

$$\dot{N}_1 = N_1[-2.2(N_1 - 2.5) - 2(N_2 - 2) - 1.25/N_1 + 1.25/2.5],$$

$$\dot{N}_2 = N_2[0.2(N_1 - 2.5) - 0.1(N_2 - 2)]. \tag{3.3.8}$$

A suitable Liapunov function is

$$V(N) = N_1 - 2.5 - 2.5 \ln(N_1/2.5) + 10[N_2 - 2 - 2 \ln(N_2/2)].$$

Along solutions of (3.3.8) we have

$$\dot{V} = [(1.25 - 5.5N_1)/(2.5N_1)](N_1 - 2.5)^2 - (N_2 - 2)^2.$$

If $N_1 < 5/22$ and $N_2 = 2$, $\dot{V}$ is positive. It follows that an estimate of the region of stability is $\{N \mid V(N) < 2.5 \ln(22/5) - 2.5 + 5/22\}$. The boundary of this region passes through the point (5/22, 2).

Let Γ denote the positive region $\{N \mid N_1 > 0, N_2 > 0, \ldots, N_m > 0\}$. Assume that the model,

$$\dot{N}_i = N_i F_i(N_1, N_2, \ldots, N_m), \tag{3.3.9}$$

where $i = 1, 2, \ldots, m$, has a positive equilibrium at N^*. A function $V(N)$ is a Liapunov function in Γ if it has the following properties:

(i) $V(N^*) = 0$;

(ii) in Γ, the level surface, $V(N) = K$ for each positive value of K, is a closed hypersurface; and for every $i = 1, 2, \ldots, m$, $V(N) \to \infty$ as $N_i \to 0+$ or as $N_i \to \infty$; and

(iii) along a solution of (3.3.9) which begins inside Γ,

$$\dot{V} = \sum_{i=1}^{m} \frac{\partial V}{\partial N_i} N_i F_i(N) \tag{3.3.10}$$

is nonpositive for all $N \in \Gamma$.

Theorem 3.3.2. *Model (3.3.9) is globally stable if, in Γ, there exists a Liapunov function and $\dot{V}(N)$ is negative definite.*

The proof of this theorem follows directly from that in which the admissible region is the whole of the state space. This is because the transformation, $X_i = \ln(N_i/N_i^*)$, can be used to map the positive region into the complete state space.

We shall outline some applications of this theorem to some general classes of models. More detailed discussions of applications to specific classes of models will be given in subsequent sections.

Consider the general single species model,

$$\dot{N} = NF(N). \tag{3.3.11}$$

Assume that it has a positive equilibrium at $N^\star$. A suitable Liapunov function is

$$V(N) = N - N^\star - N^\star \ln(N/N^\star). \tag{3.3.12}$$

Along solutions of $\dot{N} = NF(N)$, we have

$$\dot{V} = (N - N)F(N). \tag{3.3.13}$$

It follows that $\dot{V} < 0$ for $N > 0$ and $N \neq N^\star$ if (i) for $N^\star > N > 0$, $F(N) > 0$ and (ii) for $N > N^\star$, $F(N) < 0$.

For easy future reference we state this as a theorem.

Theorem 3.3.3. *The single species model, $\dot{N} = NF(N)$, is globally stable if* (i) *it has a positive equilibrium at $N^\star$,* (ii) *for $N^\star > N > 0$, $F(N) > 0$ and* (iii) *for $N > N^\star$, $F(N) < 0$.*

This is a very robust theorem and applies to almost all proposed models of an unexploited single-species population.

When $N/N^\star$ is large the dominant term in the function, $V(N) = N - N^\star - N^\star \ln(N/N^\star)$, is N. Hence for large values of N, $V(N)$ is a measure of the energy embodied in the standing biomass of the population. If $\dot{V} < 0$, it implies that the population is dissipating energy. When $N/N^\star$ is small the dominant term in $V(N)$ is $N^\star \ln(N/N^\star)$. This is again a measure of the energy embodied in the standing biomass of the population. The negative sign in front of this term in $V(N)$ means that the population is absorbing energy if $\dot{V} < 0$ at low densities.

Let s and p be positive parameters. The following functions are good candidates to act as Liapunov functions for global stability in a single-species model:

(a) $$V(N) = (N^s - N^{\star s})/s - N^{\star s} \ln(N/N^\star). \tag{3.3.14}$$

(b) $$V(N) = \frac{(N^p - N^{\star p})}{P} + \frac{N^{\star p+s}}{s\,N^s} - \frac{N^{\star p}}{s}. \tag{3.3.15}$$

(c) $$V(N) = \ln(N/N^\star) + \frac{N^{\star s}}{sN^s} - \frac{1}{s}. \tag{3.3.16}$$

Any function $V(N)$ which has the following properties would be a good candidate to act as a Liapunov function: (i) It has a global minimum at $N^\star$ for all $N > 0$, (ii) $\partial V/\partial N < 0$ for $N^\star > N > 0$ and $\partial V/\partial N > 0$ for $N > N^\star$ and (iii) $V(N) \to \infty$ as $N \to 0+$ or $N \to \infty$.

These Liapunov functions for single species are also useful for multispecies population models. A natural way to construct a Liapunov function for m interacting species is to take a positive linear combination of functions which are good candidates to act as Liapunov functions in single-species models.

Intuitively we can expect this method to succeed if all the interspecific interactions in a complex food web are weak. In this case we can expect the complex food web to behave approximately like a collection of single-species populations. Since we have very good Liapunov functions for single-species models, we can expect that this approach in the construction of a Liapunov function for a multispecies model will be a fruitful approach. In subsequent sections we find that this is indeed the case.

Consider briefly an application of these ideas to the model,

$$\dot{N}_i = N_i F_i\,(N_1, N_2, \ldots, N_m), \tag{3.3.17}$$

where $i = 1, 2, \ldots, m$. Assume that it has a positive equilibrium at $N^\star = (N_1^\star, N_2^\star, \ldots, N_m^\star)$. Let $c_1, c_2, \ldots, c_m$ be positive constants. A good candidate to act as a Liapunov function is

$$V(N) = \sum_{i=1}^{m} c_i\,[N_i - N_i^\star - N_i^\star \ln(N_i/N_i^\star)]\,. \tag{3.3.18}$$

Along solutions of (3.3.17), we have

$$\dot{V} = \sum_{i=1}^{m} c_i(N_i - N_i^\star)F_i(N). \tag{3.3.19}$$

Let Γ be the positive orthant $\{N \mid N_i > 0, i = 1, 2, \ldots,, m\}$.

Theorem 3.3.4. *Model (3.3.17) is globally stable in the positive orthant if it has a positive equilibrium at* $N^\star$ and *there exists a set of positive constants,* $c_1, c_2, \ldots, c_m$, *such that*

$$\sum_{i=1}^{m} c_i(N_i - N_i^\star)F_i(N) < 0 \tag{3.3.20}$$

for all $N \in \Gamma$ *and* $N \neq N^\star$.

This theorem follows directly from Theorem 3.3.2 if the function $V(N)$ (3.3.18) is a suitable Liapunov function. We shall see in a later chapter that this theorem is most effective when each species sustains density-dependent mortalities due to intraspecific interactions, and all the interspecific interactions are relatively weak.

In simple prey—predator models we can have global stability when the predator or prey does not sustain density-dependent mortalities due to intraspecific interactions. To establish such results analytically we need an extension of the direct method of Liapunov.

In the generalization of the direct method of Liapunov, $\dot{V}(N)$ is allowed to vanish at points beside the positive equilibrium. But it is necessary to impose an additional requirement. Apart from the positive equilibrium, the solutions of the model must not remain indefinitely in the set of points where $\dot{V}(N) = 0$.

For a more formal description of this generalization of the direct method of Liapunov, we require the concept of an invariant set (La Salle and Lefschetz, 1961). A set of points denoted by θ is an invariant set if any solution which begins in θ remains in it.

To illustrate this concept, consider the model,

$$\dot{N}_1 = N_1(2 - N_1 - N_2),$$

$$\dot{N}_2 = N_2(-0.5 + 0.5N_1). \tag{3.3.21}$$

The invariant sets in this model are $\{(1,1)\}$, $\{(2,0)\}$, $\{(0,0)\}$, $\{N \mid N_1 = 0\}$ and $\{N \mid N_2 = 0\}$. Clearly the only invariant set in the positive region is $\{(1,1)\}$.

Consider the model,

$$\dot{N}_i = N_i\, F_i\, (N_1, N_2, \ldots, N_m), \tag{3.3.22}$$

where $i = 1, 2, \ldots, m$. Assume it has a positive equilibrium at $N^\star = (N_1^\star, N_2^\star, \ldots, N_m^\star)$. Denote the positive orthant $\{N \mid N_i > 0 \text{ for } i = 1, 2, \ldots, m\}$ by Γ. Suppose there exists a function $V(N)$ which has the following properties: (i) $V(N) > 0$ for all $N \in \Gamma$ and $N \neq N^\star$. (i) In Γ, $V(N)$ has a global minimum equal to zero at $N^\star$. (iii) For every $i = 1, 2, \ldots, m$, $V(N) \to \infty$ as $N_i \to 0+$ or $N_i \to \infty$. Under these assumptions the next theorem describes the crucial conditions which are sufficient for global stability.

Theorem 3.3.5. *Model (3.3.22) is globally stable if*

$$\dot{V}(N) = \sum_{i=1}^{m} \frac{\partial V}{\partial N_i} N_i\, F_i\, (N) \leqslant 0 \tag{3.3.23}$$

for all $N \in \Gamma$ and $\{N^\star\}$ is the only invariant set in the positive orthant.

We shall not prove this theorem (LaSalle and Lefschetz, 1961). Instead we shall apply it to an example to illustrate how it can be used.

Example 3.3.3. Model (3.3.21) has a positive equilibrium at (1,1). The only invariant set in the positive region is $\{(1,1)\}$. A suitable Liapunov function is

$$V(N) = N_1 - 1 - \ln(N_1) + c[N_2 - 1 - \ln(N_2)],$$

where c is positive constant. Along solutions of the model, we have

$$\dot{V} = -(N_1 - 1)^2 - (N_1 - 1)\,(N_2 - 1) + 0.5c(N_2 - 1)\,(N_1 - 1).$$

Choose $c = 2$. We get

$\dot{V} = -(N_1 - 1)^2.$

It follows that $\dot{V} \leqslant 0$ for all $N_1 > 0, N_2 > 0$. By inspection, $\dot{V} = 0$ if $N_1 = 1$. As $\{(1,1)\}$ is the only invariant point in the positive region, the model is globally stable.

Likewise, we can relax the condition, $\dot{V}(N) < 0$ for all $N \in \Omega$ and $N \neq N^{\star}$, in Theorem 3.3.1. Consider the model,

$$\dot{N}_i = N_i F_i (N_1, N_2, \ldots, N_m) \tag{3.3.24}$$

for $i = 1, 2, \ldots, m$. Assume that it has a positive equilibrium at $N^{\star}$. Let Ω be an open subregion of the positive region. Suppose we have a function $V(N)$ which has the following properties: (i) $V(N) > 0$ for all $N \in \Omega$ and $N \neq N^{\star}$; (ii) in Ω, $V(N)$ has a global minimum equal to zero at $N^{\star}$; (iii) in Ω, the level surface, $V(N) = K$ for each value of K, is a closed surface. By necessity we choose $\Omega = \{N \mid V(N) < L\}$ where L is a positive number. Based on these assumptions the next theorem describes the crucial conditions which are sufficient for stability in Ω.

Theorem 3.3.6. *Model (3.3.24) is stable in Ω if $(N^{\star})$ is the only invariant set in Ω and*

$$\dot{V}(N) = \sum_{i=1}^{m} \frac{\partial V}{\partial N_i} N_i F_i (N) \leqslant 0 \tag{3.3.25}$$

for all $N \in \Omega$.

3.4. SINGLE-SPECIES MODELS

A common criticism of the logistic model of a single-species population is that it assumes that the per capita birth rate of the population is a linear function of the density of the population. This criticism is not as serious as some ecologists believe it to be. The crucial assumption in the logistic model is that a single species population can be described by a single nonlinear differential equation. We shall show that the logistic model of a single species has the same qualitative dynamical behavior as all proposed models of a single species, and much more.

From Theorem 3.3.3 we deduce that an effective way to understand the stability behavior of a nonlinear model, $\dot{N} = NF(N)$, is to graph the function $F(N)$ against N. If $F(N)$ has only one change in sign from the positive to the negative as N increases from zero, the model is globally stable.

The standard logistic model is given by the equation,

$$\dot{N} = (r/K)N(K - N). \tag{3.4.1}$$

In this case the function $F(N)$ is a linear function of N with a negative slope, $F(0) = r$ and $F(K) = 0$. For $K > N > 0$, $F(N) > 0$ and for $N > K$, $F(N) < 0$. By Theorem 3.3.3 it is globally stable.

Gilpin and Ayala's (1973) model of a single-species population is described by the equation,

$$\dot{N} = rN[1 - (N/K)^{\theta}], \tag{3.4.2}$$

where r, K and θ are positive constants. This model has an equilibrium at $N = K$. The function, $F(N) = r[1 - (N/K)^{\theta}]$, is monotonic decreasing as N increases from zero. It is positive for $K > N > 0$ and it is negative for $N > K$. By Theorem 3.3.3 it is globally stable.

Gompertz's model is used extensively as a model of cancerous growth (Swann and Vincent, 1977). It is described by the equation,

$$\dot{N} = -\alpha N \ln(N/K), \tag{3.4.3}$$

where α and K are constants. The function, $F(N) = -\alpha \ln(N/K)$, is monotonic decreasing. One unsatisfactory feature of this model is that $F(N) \to \infty$ as $N \to 0+$. The model has a positive equilibrium at $N = K$. $F(N)$ is monotonic decreasing as N increases. For $K > N > 0$, $F(N) > 0$ and for $N > K$, $F(N) < 0$. By Theorem 3.3.3 it is globally stable.

Schoener's (1973) model of a single species population with a constant energy input is

$$\dot{N} = rN[(I/N) - C - bN], \tag{3.4.4}$$

where r, I, C and b are positive constants. $F(N) = r[(I/N) - C - bN]$ is strictly monotonic decreasing as N increases. It has a positive equilibrium at $N^{\star} = [-C + \sqrt{(C^2 + 4bI)}]/(2b)$. For $N^{\star} > N > 0$, $F(N) > 0$ and for $N > N^{\star}$, $F(N) < 0$. It follows that it is globally stable.

The function $F(N)$ in models (3.4.1) to (3.4.4) decreases monotonically as N increases. It follows that these models do not describe a single species with an Allee effect (see Odum, 1971). The Allee effect occurs in a population in which there are difficulties in breeding at low densities. In some populations (schooling fish) the survival of the members is enhanced up to a certain density by aggregation. Beyond the optimum population density, the mortality rate increases as the density increases because of competition for food space and increased stress. Volterra (1938) put forward a model to describe this type of situation. It is

$$\dot{N} = N[-d + bN - aN^2], \tag{3.4.5}$$

where a, b and d are positive constants. The function, $F(N) = -d + bN - aN^2$, has a single hump for $N > 0$. It has positive equilibria at $N_1 = [b - \sqrt{(b^2 - 4ad)}]/(2a)$ and $N_2 = [b + \sqrt{(b^2 - 4ad)}]/(2a)$ if $b^2 > 4ad$. The lower level equilibrium N_1 is unstable. The higher level equilibrium N_2 is stable. Its region of attraction is $\{N \mid N > N_1\}$. This model has received very little attention from ecologists.

In the ecological literature, Pianka (1972) and others have graphically described single-species models of populations which sustain the Allee effect.

An analytical model of a single-species model of a population which experiences an Allee effect is given by the equation,

$$\dot{N} = N[b + N(a - N)/(1 + cN)], \tag{3.4.6}$$

where a, b, c are positive constants. The function, $F(N) = b + N(a - N)/(1 + cN)$, has a maximum at $N^{\star\star} = [-1 + \sqrt{(1 + ac)}]/c$. For $N < N^{\star\star}$, $F(N)$ may be roughly approximated by the function, $b + aN$. For $N > N^{\star\star}$, $F(N)$ may be roughly approximated by $b + a/c - N/c$. It follows that for $N > N^{\star\star}$, model (3.4.6) approximately reduces to the familiar logistic model. For $b > 0$, model (3.4.6) has a single positive equilibrium at

$$N^\star = [(a + bc) + \sqrt{\{(a + bc)^2 + 4b\}}]/2.$$

For $N^\star > N > 0$, $F(N)$ is positive and for $N > N^\star$, $F(N)$ is negative. It follows that model (3.4.6) is globally stable if $b > 0$. For $b < 0$, the model has another equilibrium positive at

$$N_1 = [(a + bc) - \sqrt{\{(a + bc)^2 + 4b\}}]/2.$$

This equilibrium is unstable while the region of stability of the higher level equilibrium $N^\star$ is $\{N \mid N > N_1\}$.

Example 3.4.1. A model of a single-species population which sustains an Allee effect is described by the equation,

$$\dot{N} = N[b + N(4 - N)/(1 + 2N)]. \tag{3.4.7}$$

The function $F(N)$ in this model has a maximum at $N = 1$, $F(0) = b$. If $b > 0$, the model has a positive equilibrium at $N^\star$ where

$$N^\star = 2 + b + \sqrt{(4 + 4b + 2b^2)}.$$

Fig. 3.4.1 displays the function $F(N)$ for $b = -0.5$ and $b = 0.4$. When $b = -0.5$ the model has two equilibria, N_1, N_2. The lower level equilibrium N_1 is unstable. The region of stability of the equilibrium N_2 is $\{N \mid N > N_1\}$. When $b = 0.1$, the model is globally stable.

3.5. EXPLOITED SINGLE-SPECIES MODELS

An important problem in the management of an exploited population is the stability of the population. We can distinguish two types of constant (open loop) harvesting policies. In a constant quota policy, a constant number of animals in a population is harvested per unit time. This type of policy requires that the technology to harvest the animals is very efficient and the capacity to carry out the harvest is in excess. An example of this type of policy is the constant quota policy of the International Whaling Commission prior to 1974. In the other type of constant harvesting policy, the effort applied per unit time to catch the animals in a population is fixed. A good

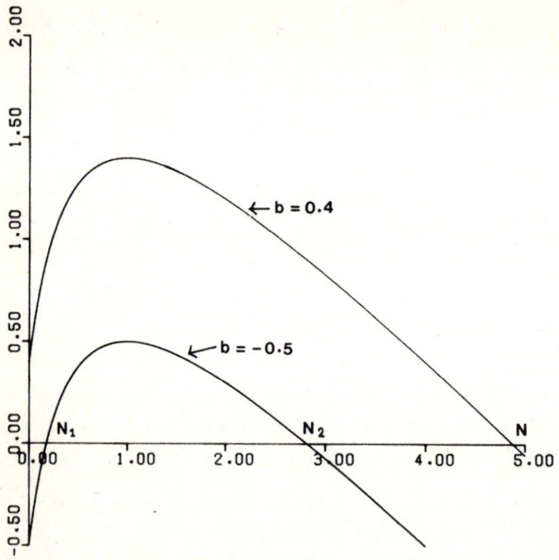

Fig. 3.4.1. Per capita growth rates of populations which sustain the Allee effect.

working example of this policy is the fixed number of boats and pots which are licensed to catch Rock Lobsters in Western Australia.

It has not been sufficiently emphasized that qualitively these two types of constant harvesting policies are very different. Constant quota harvesting policies are usually unstable (Goh, 1969/1970) while constant effort harvesting policies are usually stable (Goh, 1977 c).

Let E be a measure of the effort applied in harvesting a population. Usually, E is set equal to qf, where q is the catchability coefficient and f is the number of boats, men or other units used in the harvesting of the population. The harvesting alters the dynamics of a population. If the natural dynamics is described by $\dot{N} = NF(N)$, the exploited population is given by

$$\dot{N} = NF(N) - EN. \tag{3.5.1}$$

If $F(N)$ is strictly monotonic decreasing as N increases and $\dot{N} = NF(N)$ has a positive equilibrium, the unexploited population is globally stable. In this case, $\partial F/\partial N < 0$ for $N > 0$. Clearly if this is satisfied, $\partial(F-E)/\partial N < 0$ for $N > 0$. We conclude that if $F(N)$ is strictly monotonic decreasing for $N > 0$, the exploited and unexploited populations have the same qualitative behavior, provided in each case there is a positive equilibrium. This conclusion applies to the logistic model, Gilpin and Ayala's model, Gompertz's model and Schoener's model which were examined in the previous section.

If a population sustains an Allee effect, the population under a constant effort harvesting policy could qualitatively have a different behavior from the exploited population. If a constant effort harvesting policy is applied to model (3.4.6), we have

$$\dot{N} = N\{b + N(a - N)/(1 + cN)\} - EN. \tag{3.5.2}$$

If $b > 0$, the unexploited population is globally stable. Suppose $E > b$. Let $d = E - b$. Model (3.5.2) has two positive equilibria at N_1 and N_2, where

$$N_1 = [(a - cd) - \sqrt{\{(a - cd)^2 - 4d\}}]/2, \tag{3.5.3}$$

$$N_2 = [(a - cd) + \sqrt{\{(a - cd) - 4d\}}]/2. \tag{3.5.4}$$

The equilibrium at N_1 is unstable. The region of stability of the equilibrium at N_2 is $\{N \mid N > N_1\}$. This follows because $\dot{N}_1 > 0$ for $N_2 > N > N_1$ and $\dot{N} < 0$ for $N > N_2$. If $E < b$, there is a unique positive equilibrium and the exploited population is globally stable.

Assume that a constant quota policy is prescribed for the harvesting of a single-species population. Let h be the constant rate of harvest. A model of the harvested population is

$$\dot{N} = NF(N) - h. \tag{3.5.5}$$

An effective way to establish the stability of the exploited model is to plot $F(N)$ and h/N on the same graph paper, as in Fig. 3.5.1. Usually, model (3.5.5) has two positive equilibria or none. Suppose it has two positive equilibria at N_1 and N_2 where $N_1 < N_2$. The lower level equilibrium N_1 is unstable because $\dot{N} < 0$ for $N_1 > N > 0$ and $\dot{N} > 0$ for $N_2 > N > N_1$. The region of stability of the equilibrium at N_2 is $\{N \mid N > N_1\}$ if $\dot{N} > 0$ for $N_2 > N > N_1$ and $\dot{N} < 0$ for $N > N_2$.

To illustrate these conclusions, consider the exploited logistic model

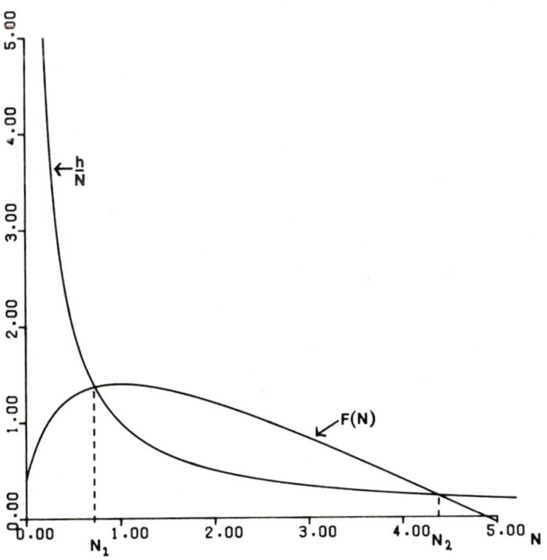

Fig. 3.5.1. A constant quota harvesting policy which creates equilibria at N_1 and N_2.

$$\dot{N} = (r/K)N(K - N) - h. \tag{3.5.6}$$

If $h > rK/4$, there is no positive equilibrium. If $h = rK/4$, we have the so called "maximum sustainable yield" policy. There is a single unstable equilibrium at $N = K/2$. If $h < rK/4$, the model has two positive equilibria at S_1 and S_2, where

$$S_1 = (1/2)[K - \sqrt{(K^2 - 4hK/r)}], \tag{3.5.7}$$

$$S_2 = (1/2)[K + \sqrt{(K^2 - 4hK/r)}]. \tag{3.5.8}$$

In this case, we have

$$\dot{N} = -(r/K)(N - S_1)(N - S_2). \tag{3.5.9}$$

It follows that $\dot{N} < 0$ for $S_1 > N > 0$ and $\dot{N} > 0$ for $S_2 > N > S_1$. Hence the equilibrium at S_1 is unstable. For $S_2 > N > S_1, \dot{N} > 0$ and for $N > S_2, \dot{N} < 0$. Hence the region of stability for the equilibrium at S_2 is $\{N \mid N > S_1\}$.

Recently, Agnew (1979) has introduced a new and more realistic expression for the effects of harvesting on a single-species population. This harvesting expression reduces to a constant quota policy at high population densities; and it also reduces to the classical constant effort policy at low population densities. In this harvesting term, the handling time of the harvest is explicitly included in the expression. We further generalize this harvesting term in the model by adding the effects of competition among the men, boat, or other means used in the harvesting of a population.

We shall use the same type of reasoning which is used for developing an expression for a functional response of a predator (Holling, 1966; Murdoch, 1973). Let Δt be a short time interval during which the density N of an exploited population does not change significantly. Let f be the applied effort per unit time which is used in the harvesting. In the case of a fish population, f is the number of boats used in the harvesting which is standardized in some manner. Let C be the catch per unit effort (boat) during the time interval Δt. During the same time interval, let the time spent in handling the catch be wC where w is a constant. It follows that the time available for searching for the animals in the population is $(\Delta t - wC)$. Let q be the catchability coefficient as used in fishery. It follows that the catch per unit effort (boat) during the time interval Δt is

$$C = q(\Delta t - wC)N. \tag{3.5.10}$$

This implies that the catch per unit effort during a unit time interval is given by the equation

$$C/\Delta t = qN/(1 + qwN). \tag{3.5.11}$$

The total catch per unit time is $fC/\Delta t$. It follows that a model of an exploited population is

$$\dot{N} = NF(N) - qfN/(1 + qwN). \tag{3.5.12}$$

If there is competition among the boats or other types of units used in the population, the expression (3.5.10) may be replaced by

$$C = q(\Delta t - wC)N/(1 + af) \tag{3.5.13}$$

where f is the applied effort and a is a positive constant. From this equation we get

$$\frac{C}{\Delta t} = \frac{qN}{(1 + af + qwN)} . \tag{3.5.14}$$

It follows that a model of an exploited population which incorporates the handling time of the harvest and the competition among the units engaged in the harvest of the animals is given by the equation,

$$\dot{N} = NF(N) - qfN/(1 + af + qwN). \tag{3.5.15}$$

Suppose a constant effort harvesting policy is prescribed. In this case, f is a constant. A good way to determine how many positive equilibria there are in model (3.5.15) is to plot the functions, $F(N)$ and $H(N) = qf/(1 + af + qwN)$, on the same graph paper. The function $H(N)$ is strictly monotonic decreasing as N increases. $H(0) = qf(1 + af)$. Fig. 3.5.2 illustrates what can happen in a population which sustains an Allee effect. In the particular case given in Fig. 3.5.2, the lower level equilibrium at N_1 is unstable. The region of stability of the higher level equilibrium at N_2 is $\{N \mid N > N_1\}$.

Suppose the unexploited model can be adequately described by the logistic model. We have

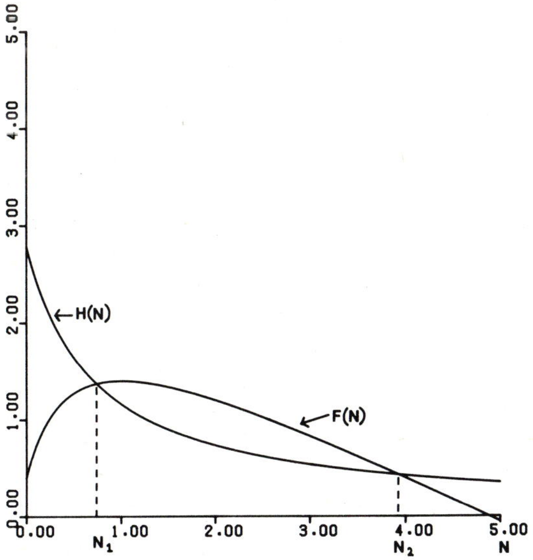

Fig. 3.5.2. N_1 and N_2 are equilibria of model (3.5.15).

$$\dot{N} = (r/K)N(K - N) - qfN/(1 + af + qwN). \tag{3.5.16}$$

If $H(N) = qf/(1 + af + qwN)$, we have $H(0) = qf/(1 + af)$. Graphically, we find that for the model to have a positive equilibrium, we must have $r > qf/(1 + af)$. If this is satisfied, the positive equilibrium is at

$$N^* = (Kqw - 1 - af) + \sqrt{\{(Kqw - 1 - af)^2 + 4qwK(r + raf - qf)/r\}}\,]/(2qw). \tag{3.5.17}$$

Graphically, it is clear that $\dot{N} > 0$ for $N^* > N > 0$, and $\dot{N} < 0$ for $N > N^*$. Hence model (3.5.16) is globally stable. If the competition among the units (boats) engaged in harvesting is weak, $a \ll 1$. In this situation, if f is large it is likely that $r < qf/(1 + af)$, and there is no positive equilibrium.

If $F = r[1 - (N/K)^\theta]$ and $1 > \theta > 0$, model (3.5.15) has a dynamical behavior which is qualitatively similar to that in the logistic model. If $\theta > 1$, it is possible for model (3.5.15) to have two positive equilibria. The lower level equilibrium is unstable; the higher level equilibrium is stable relative to initial states that are larger than the lower level equilibrium.

If the function $F(N)$ is as in the Gompertz's or Schoener's model, the model (3.5.15) always has a globally stable positive equilibrium. This can be established by plotting $F(N)$ and $H(N) = qf/(1 + af + qwN)$ on the same paper. We note that in these models, $N = 0$ is an asymptote of $F(N)$, while $H(N)$ has an asymptote at $N = -(1 + af)/(qw)$. Hence qualitatively, these models of an exploited population can behave differently from the logistic model of an exploited population.

3.6. MODELS WITH UNSPECIFIED PARAMETERS

Let $a_1, a_2, \ldots, a_p$ be nonnegative parameters and $b_1, b_2, \ldots, b_n$ be parameters which can have either sign. In theoretical ecology we are often confronted with models of the form,

$$\dot{N}_i = N_i F_i(N; a, b), \quad i = 1, 2, \ldots, m, \tag{3.6.1}$$

in which the functional form of the functions, $F_1, F_2, \ldots, F_m$, is given but the parameters $\{a, b\}$ are unspecified. Traditionally we study this model by examining in sequence the following questions: (i) for a given set of parameters $\{a, b\}$, does the model have a positive equilibrium? (ii) If a positive equilibrium exists, is it locally stable? (iii) Is the model globally stable?

In theoretical ecology, question (i) is unrealistic because we usually do not have a given set of values for the parameters $\{a, b\}$. Many of us bemoan this state of affairs but we should not let it hold up efforts to develop insight into ecological problems. The more realistic and effective approach is to postulate the existence of a positive equilibrium at $(N_1^*, N_2^*, \ldots, N_m^*)$ and then ask whether the set of equations for N^*,

$$F_i(N^\star; a, b) = 0, \quad i = 1, 2, \ldots, m, \tag{3.6.2}$$

has a positive solution in the set of parameters. In most problems, the number of parameters is usually much larger than that of the species. Hence this problem usually has a continuum of solutions.

For an illustration of these ideas, consider the Lotka—Volterra model:

$$\dot{N}_i = N_i \left[b_i + \sum_{j=1}^{m} a_{ij} N_j \right] \tag{3.6.3}$$

where $i = 1, 2, \ldots, m$. The nontrivial equilibrium satisfies the equation,

$$b_i + \sum_{j=1}^{m} a_{ij} N_j^\star = 0, \quad i = 1, 2, \ldots, m. \tag{3.6.4}$$

If we choose the parameters $\{a_{ij}, b_k\}$ first, then compute the nontrivial equilibrium we may find that it is not positive. In fact, from computer experiments (Goh and Jennings, 1977) the probability of success decreases rapidly with m. It is more practical to assume that an equilibrium exists at $(N_1^\star, N_2^\star, \ldots, N_m^\star)$ which could be any positive point, and then ask what the parameters, $b_1, b_2, \ldots, b_m$, should be in order that the assumption is valid. We achieve this by letting

$$b_i = -\sum_{j=1}^{m} a_{ij} N_j^\star, \tag{3.6.5}$$

for $i = 1, 2, \ldots, m$. Clearly the parameters $\{a_{ij}\}$ can be freely prescribed according to whether we wish to model a competition, a mutualism or a prey—predator interaction.

For more complex situations, we can use an auxiliary nonlinear programming problem. We first assume that a positive equilibrium exists at $(N_1^\star, N_2^\star, \ldots, N_m^\star)$. We consider the parameters, $a_1, a_2, \ldots, a_p, b_1, b_2, \ldots, b_n$, as independent variables. Then we choose a reasonable objective function like

$$L = a_1^2 + a_2^2 + \cdots + a_p^2 + b_1^2 + b_2^2 + \cdots + b_n^2. \tag{3.6.6}$$

We minimize the function $L(a, b)$ subject to the conditions,

$$F_i(N^\star; a, b) = 0, \quad i = 1, 2, \ldots, m, \tag{3.6.7}$$

$$a_s \geqslant 0, \quad s = 1, 2, \ldots, p. \tag{3.6.8}$$

If we employ a numerical optimization algorithm (Fletcher, 1975), we could rapidly generate a set of admissible parameters $\{a, b\}$.

3.7. TWO-SPECIES LOTKA—VOLTERRA MODELS

The Lotka—Volterra model of a two-species interaction is normally studied graphically. The case of two competing species occupies a prominent place in this approach because of widespread interest in the coexistence of competing species. The graphical analysis suggests that for two competing species, local stability implies global stability. We shall prove this result analytically. For a Lotka—Volterra model of three or more competing species, this result is not necessarily true. In a Lotka—Volterra model of a single prey and a single predator system, it is possible to have a positive equilibrium which is only locally stable.

Consider the Lotka—Volterra model for two-species system:

$$\dot{N}_i = N_i \left[b_i + \sum_{j=1}^{m} a_{ij} N_j \right], \quad i = 1, 2. \tag{3.7.1}$$

This type of model may be justified by the fact that it is an approximation of a nonlinear model near an equilibrium. Denote the nontrivial equilibrium by $(N_1^\star, N_2^\star)$. It is a solution of the equations,

$$b_i + \sum_{j=1}^{2} a_{ij} N_j^\star = 0, \quad i = 1, 2. \tag{3.7.2}$$

If the equilibrium is positive, $N_1^\star > 0$ and $N_2^\star > 0$.

The diagonal elements of the matrix, $A = (a_{ij})$, represent intraspecific interactions. If the equilibrium level of the first species $N_1^\star$ is low, it is likely that a_{11} is positive. This follows from Allee's principle, as discussed in Odum (1971). If $N_1^\star$ is large, it is likely that density-dependent mortalities due to intraspecific interactions (like crowding) will occur and a_{11} would be negative. Similar relationships hold between $N_2^\star$ and a_{22}.

Let $x_i = N_i - N_i^\star$ for $i = 1, 2$. Substitute x_1, x_2 into (3.7.1) and ignore higher order terms in x_1 and x_2; we then get

$$\dot{x}_1 = N_1^\star a_{11} x_1 + N_1^\star a_{12} x_2, \tag{3.7.3a}$$

$$\dot{x}_2 = N_2^\star a_{21} x_1 + N_2^\star a_{22} x_2 . \tag{3.7.3b}$$

By the Routh—Hurwitz stability criterion, system (3.7.3) is stable, if and only if,

$$N_1^\star a_{11} + N_2^\star a_{22} < 0, \tag{3.7.4}$$

and

$$N_1^\star N_2^\star (a_{11} a_{22} - a_{12} a_{21}) > 0. \tag{3.7.5}$$

Note that in theoretical ecology, it is accepted that neutral stability is irrelevant. Hence (3.7.4) and (3.7.5) are necessary and sufficient conditions for $(N_1^\star, N_2^\star)$ to be locally (asymptotically) stable.

In a two-species competition, $a_{12} < 0$ and $a_{21} < 0$. In a two-species mutualism, $a_{12} > 0$ and $a_{21} > 0$. For both cases, $a_{12}a_{21} > 0$. It follows that in order for (3.7.4) and (3.7.5) to be satisfied at a positive point $(N_1^\star, N_2^\star)$, it is necessary that $a_{11} < 0$ and $a_{22} < 0$ must both be satisfied. This means that each species must be self-regulating (See Fig. 3.7.1 and 3.7.2). In the case of amensalism, where $a_{12} < 0$ and $a_{21} = 0$, we can similarly show that it is necessary for $a_{11} < 0$ and $a_{22} < 0$. Similarly, for stability to exist in a commensalism in which $a_{12} > 0$ and $a_{21} = 0$, we must have $a_{11} < 0$ and $a_{22} < 0$. In a model of an interaction between a single prey and a single predator, it is not necessary for a_{11} and a_{22} to be both negative.

Example 3.7.1. Let N_1 and N_2 be prey and predator densities respectively. Consider the prey—predator model:

$$\dot{N}_1 = N_1(5 - 2N_1 - 3N_2),$$

$$\dot{N}_2 = N_2(-2 + N_1 + N_2).$$

This system has a positive equilibrium at (1, 1). Conditions (3.7.4) and (3.7.5) are satisfied. Hence the equilibrium is locally stable. However the computer solution shows that the trajectory of this model through the initial state (1.0, 1.6) tends rapidly to $(0, \infty)$ as $t \to \infty$. Hence the system has only a locally stable positive equilibrium. This example is interesting because there is a widespread belief that in a Lotka—Volterra model of a two-species interaction, local stability implies global stability.

This model can help to explain how a pest explosion takes place, Let a predator (pest) feed on a prey which regulates itself. A change in the environmental conditions can displace the state of the system from its domain of attraction for its positive equilibrium. This can lead to an explosion of the predator population and the destruction of the prey population.

We shall now describe a simple set of conditions for global stability in a Lotka—Volterra model of a two-species interaction.

Theorem 3.7.1. *Conditions which are sufficient for global stability in a Lotka—Volterra model of a two-species interaction are* (i) *the nontrivial equilibrium* $(N_1^\star, N_2^\star)$ *is positive,* (ii) *the positive equilibrium is locally stable and* (iii) *each species sustains density-dependent mortalities due to intraspecific interactions.*

Proof. From a graphical analysis of a given Lotka—Volterra model of a two-species interaction, it is clear that for the system to persist, its nontrivial equilibrium must be positive.

Condition (iii) means that $a_{11} < 0$ and $a_{22} < 0$. It follows that (3.7.4) is satisfied because $N_1^\star > 0$ and $N_2^\star > 0$. Condition (ii) implies that $\det(A) > 0$ where $A = (a_{ij})$.

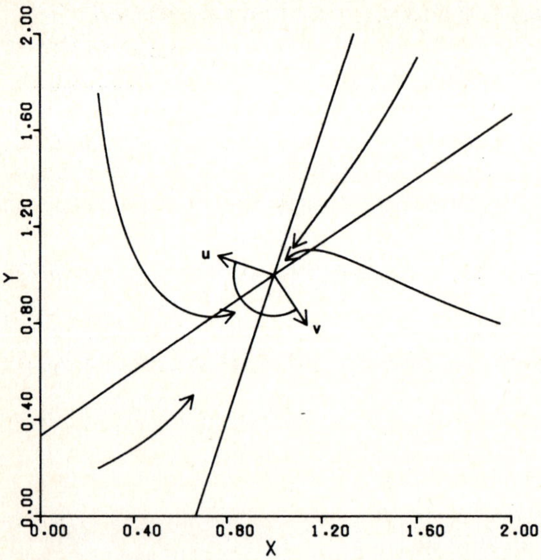

Fig. 3.7.1. Computer solutions of a globally stable Lotka—Volterra model of a two-species mutualism. The model is $\dot{X} = X(2 - 3X + Y)$, and $\dot{Y} = Y(1 + 2X - 3Y)$.

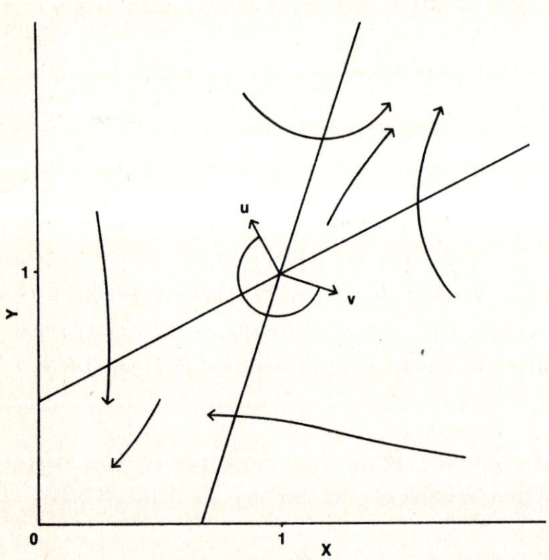

Fig. 3.7.2. Computer solutions of an unstable Lotka—Volterra model of a two-species mutualism. The model is $\dot{X} = X(-1 - X + 2Y)$ and $\dot{Y} = Y(-2 + 3X - Y)$.

By definition, $(N_1^\star, N_2^\star)$ is the nontrivial equilibrium. It follows that we have

$$b_1 + a_{11}N_1^\star + a_{12}N_2^\star = 0, \tag{3.7.6a}$$

$$b_2 + a_{21}N_1^\star + a_{22}N_2^\star = 0. \tag{3.7.6b}$$

These equations can be used to eliminate b_1 and b_2 from model (3.7.1). We then get

$$\dot{N}_1 = N_1[a_{11}(N_1 - N_1^\star) + a_{12}(N_2 - N_2^\star)], \tag{3.7.7a}$$

$$\dot{N}_2 = N_1[a_{21}(N_1 - N_1^\star) + a_{22}(N_2 - N_2^\star)]. \tag{3.7.7b}$$

A suitable Liapunov function for this model is

$$V(N) = c_1[N_1 - N_1^\star - N_1^\star \ln(N_1/N_1^\star)] + c_2[N_2 - N_2^\star - N_2^\star \ln(N_2/N_2^\star)], \tag{3.7.8}$$

where c_1 and c_2 are positive constants which will be prescribed later on. Along solutions of (3.7.7), we have

$$\begin{aligned}\dot{V} &= c_1(N_i - N_1^\star)(\dot{N}_1/N_1) + c_2(N_2 - N_2^\star)(\dot{N}_2/N_2)\\ &= c_1a_{11}(N_1 - N_1^\star)^2 + c_1a_{12}(N_1 - N_1^\star)(N_2 - N_2^\star)\\ &\quad + c_2a_{21}(N_2 - N_2^\star)(N_1 - N_1^\star) + c_2a_{22}(N_2 - N_2^\star)^2\\ &= (½)(N - N^\star)^{\mathrm{T}}(CA + A^{\mathrm{T}}C)(N - N^\star)\end{aligned} \tag{3.7.9}$$

where $C = \mathrm{diag}(c_1, c_2)$, and A^{T} is the matrix transpose of A. It follows that sufficient conditions for model (3.7.1) to be globally stable are (i) there exists a positive equilibrium $(N_1^\star, N_2^\star)$ and (ii) there exists a positive diagonal matrix such that $CA + A^{\mathrm{T}}C$ is negative definite.

We shall now show that the conditions of the theorem imply that there exists a positive diagonal matrix C such that $CA + A^{\mathrm{T}}C$ is negative definite. The matrix, $CA + A^{\mathrm{T}}C$, is definite if

$$c_1a_{11} < 0, \qquad c_2c_{22} < 0 \tag{3.7.10}$$

and

$$4c_1c_2a_{11}a_{22} - (c_1a_{12} + c_2a_{21})^2 > 0. \tag{3.7.11}$$

Condition (iii) means that $a_{11} < 0$ and $a_{22} < 0$. Hence (3.7.10) is satisfied. For condition (3.7.11), there are three cases: (a) $a_{12}a_{21} = 0$, (b) $a_{12}a_{21} > 0$ and $a_{12}a_{21} < 0$.

In case (a), either $a_{12} = 0$ or $a_{21} = 0$ or both are equal to zero. If $a_{12} = 0$, condition (3.7.11) becomes

$$c_2[4c_1a_{11}a_{22} - c_2a_{21}^2] > 0. \tag{3.7.12}$$

Clearly it is possible to choose positive numbers, c_1 and c_2, so that this inequality is satisfied. Similar arguments apply if $a_{21} = 0$ or $a_{12} = a_{21} = 0$.

Biologically, this case contains interactions known as amensalism, commensalism and neutralism (Odum, 1971).

In case (b), $a_{12}a_{21} > 0$. This means that either $a_{12} < 0$ and $a_{21} < 0$ or $a_{12} > 0$ and $a_{21} > 0$. These subcases are known as competition and mutualism respectively. Condition (3.7.11) can be rewritten as

$$4c_1c_2(a_{11}a_{22} - a_{12}a_{21}) - (c_1a_{12} - c_2a_{21})^2 > 0. \tag{3.7.13}$$

Choose positive numbers c_1 and c_2 so that

$$c_1a_{12} - c_2a_{21} = 0. \tag{3.7.14}$$

This is possible as $a_{12}a_{21} > 0$. By assumption, the positive equilibrium (N_1^*, N_2^*) is locally stable. From (3.7.5), this implies that $(a_{11}a_{22} - a_{12}a_{21})$ is positive. It follows that (3.7.13) is satisfied by this choice of c_1 and c_2.

Case (c) is a prey—predator interaction. The parameters, a_{12} and a_{21}, are non-zero and have opposite signs. Hence it is possible to select positive numbers, c_1 and c_2, so that

$$c_1a_{12} + c_2a_{21} = 0. \tag{3.7.15}$$

As $a_{11} < 0$ and $a_{22} < 0$, it follows that condition (3.7.11) is satisfied. This completes the proof.

There is hardly any mathematical analysis of the Lotka—Volterra models of mutualism in the ecological literature. Ironically this class of models, which has been largely ignored, turns out to be mathematically the most tractable. As noted in the proof of Theorem 3.7.1, in a two-species Lotka—Volterra model of mutualism, local stability implies global stability. This result remains true in a Lotka—Volterra model of mutualism between any finite number of species (Goh, 1979).

Similarly, in two-species Lotka—Volterra models of competition, local stability implies global stability. However, in three-species Lotka—Volterra models of competition, local stability need not imply global stability. Example 3.2.2 demonstrates this.

If $a_{11} < 0$, $a_{22} < 0$ and $a_{12}a_{21} \leqslant 0$, the interaction matrix (a_{ij}) is said to be qualitatively stable (May, 1974). This means that for (a_{ij}) to be associated with a locally stable positive equilibrium, it is not necessary to know the exact magnitudes of the parameters so long as they satisfy the above sign constraints. For such a matrix, there exists a positive diagonal matrix C such that $CA + A^TC$ is negative definite. Thus in a two-species interaction, either amensalism, neutralism, commensalism, prey—predator or host—parasite, the conditions for global and qualitative stability are both satisfied if each species is self-regulating. Similarly, in cases involving more than two species, if the interaction matrix A satisfies the strengthened conditions for qualitative stability and the corresponding Lotka—Volterra model has a positive equilibrium, the model is globally stable.

In a Lotka—Volterra model of a prey—predator system, it is possible to have global stability even though only one of the species sustains density-dependent motralities due to intraspecific interactions. We shall establish global stability for a Lotka—Volterra model of a prey—predator system in which only the predator sustains density-dependent mortalities due to intra-specific interactions. This case is very important in pest control. A popular method in biological control is to introduce a predator to suppress the pest (prey). In general a pest population does not sustain significant density-dependent mortalities due to intraspecific interactions when its density is maintained at economically acceptable levels by the introduced predator or by insecticides.

Let N_1, N_2 be prey and predator densities respectively. Let b, d, E, α_{12} and α_{22} be positive constants. Consider the Lotka—Volterra model,

$$\dot{N}_1 = N_1[b - \alpha_{12}N_2], \tag{3.7.16a}$$

$$\dot{N}_2 = N_2[-d + E\alpha_{12}N_1 - \alpha_{22}N_2]. \tag{3.7.16b}$$

The parameter E is the conversion efficiency of prey into predator; b is the per capita birth rate of the prey and d is the per capita death rate of the predator. The model has a unique positive equilibrium at $(N_1^\star, N_2^\star)$ where

$$N_1^\star = (b\alpha_{22} + d\alpha_{12})/(E\alpha_{12}^2), \qquad N_2^\star = b/\alpha_{12}. \tag{3.7.17}$$

We shall prove the following theorem which established that the model (3.7.16) is a very robust system.

Theorem 3.7.2. *The system described by the equations (3.7.16) is globally stable if the parameters, b, d, E, α_{12}, α_{22}, are all positive.*

Proof. The positive equilibrium $(N_1^\star, N_2^\star)$ satisfies the equations,

$$b_1 - \alpha_{12}N_2^\star = 0, \tag{3.7.18a}$$

$$-d + E\alpha_{12}N_1^\star - \alpha_{22}N_2^\star = 0. \tag{3.7.18b}$$

Substitute these into (3.7.16). We get

$$\dot{N}_1 = N_1(-\alpha_{12})(N_2 - N_2^\star), \tag{3.7.19a}$$

$$\dot{N}_2 = N_2[E\alpha_{12}(N_1 - N_1^\star) - \alpha_{22}(N_2 - N_2^\star)]. \tag{3.7.19b}$$

A suitable Liapunov function is

$$V(N) = E[N_1 - N_1^\star - N_1^\star \ln(N_1/N_1^\star)] + [N_2 - N_2^\star - N_2^\star \ln(N_2/N_2^\star)]. \tag{3.7.20}$$

Along solutions of system (3.7.19), we have

$$\dot{V}(N) = -\alpha_{22}(N_2 - N_2^\star)^2. \tag{3.7.21}$$

Clearly $\dot{V}(N) < 0$ for $N_1 > 0$, $N_2 > 0$ and $N_2 \neq N_2^\star$. Inside the set $\{N \mid N_2 = N_2^\star\}$, the only invariant subset is $\{(N_1^\star, N_2^\star)\}$. By Theorem 3.3.5, this model is globally stable.

Suppose that the predator has an alternative prey which provides it with an abundant and steady supply of energy per predator. In this case, a model of the prey—predator system is

$$N_1 = N_1[b_1 - \alpha_{12}N_2], \tag{3.7.22a}$$

$$\dot{N}_2 = N_2[b_2 + E\alpha_{12}N_1 - \alpha_{22}N_2] \tag{3.7.22b}$$

where b_1, b_2, E, α_{12} and α_{22} are positive constants. The nontrivial equilibrium is at $(N_1^\star, N_2^\star)$ where

$$N_1^\star = (b_1\alpha_{22} - b_2\alpha_{12})/(E\alpha_{12}^2), \qquad N_2^\star = b/\alpha_{12}. \tag{3.7.23}$$

If $b_2 > b_1\alpha_{22}/\alpha_{12}$, $N_1^\star$ is negative and the nontrivial equilibrium is not positive. Graphically, we can show that all the solutions which begin in the positive quadrant tend to the point $(0, b_2/\alpha_{22})$. Clearly this is a highly desirable situation if the prey is a pest. In fact there is a well known example of the usefulness of an alternative prey for the predator in the biological control of a pest (prey) (Doutt and Nakata, 1965). If $b_2 < b_1\alpha_{22}/\alpha_{12}$, the nontrivial equilibrium is positive. By arguments which are similar to those in the proof of Theorem 3.7.2 it can be shown that system (3.7.22) is globally stable when it has a positive equilibrium.

Suppose that only the prey sustains density dependent mortalities due to intraspecific interactions. A model of this type of prey—predator system is given by the equations,

$$\dot{N}_1 = N_1[b - \alpha_{11}N_1 - \alpha_{12}N_2], \tag{3.7.24a}$$

$$\dot{N}_2 = N_2[-d + E\alpha_{12}N_1] \tag{3.7.24b}$$

where b, d, E, α_{11} and α_{12} are positive constants. Its nontrivial equilibrium is at $(N_1^\star, N_2^\star)$ where

$$N_1^\star = d/(E\alpha_{12}), \qquad N_2^\star = (bE\alpha_{12} - d\alpha_{11})/(e\alpha_{12}^2) \tag{3.7.25}$$

If $(N_1^\star, N_2^\star)$ is positive, we can use arguments similar to those in the proof of Theorem 3.7.2 to establish that this model is globally stable.

It appears that we have not exhausted all globally stable Lotka—Volterra models of two-species interactions. Computer simulations suggest that there is global stability in the following prey—predator model:

$$\dot{N}_1 = N_1(0.8 + 0.5N_1 - 1.3N_2), \tag{3.7.26a}$$

$$\dot{N}_2 = N_2(-1.0 + 2N_1 - N_2). \tag{3.7.26b}$$

Biologically, this model is very interesting. It suggests that it is possible for a predator N_2, which regulates itself, to provide a robust regulation of a pest (prey N_1) which is not self-regulating. In the absence of the predator, the density of the prey will explode.

3.8. GILPIN AND AYALA'S COMPETITION MODEL

The Lotka—Volterra models have often been severely criticized. One of the criticisms is that in such a model, the per capita rate of change of the density of each species is a linear function of the densities of the interacting species.

In single-species models, we have seen that the linearity assumption in the logistic model is not a serious limitation. The most important assumption of the logistic model of a single-species model is that it is a differential equation model. We note that these observations apply to a lesser extent in Lotka—Volterra models of multispecies systems. If we use Lotka—Volterra models only to describe second order dynamics in a neighbourhood of a positive equilibrium, we will always get some correct and relevant information. The trouble arises when we insist that a Lotka—Volterra model should be valid globally.

Gilpin and Ayala (1973) found that the Lotka—Volterra models do not fit globally to the data from a series of carefully conducted experiments. For this reason they proposed a new model in which the zero isolines are concave. Let N_1 and N_2 be the densities of two competing populations. Gilpin and Ayala's competition model is

$$\dot{N}_1 = r_1 N_1 [1 - (N_1/K_1)^{\theta_1} - \alpha_{12}(N_2/K_1)], \tag{3.8.1a}$$

$$\dot{N}_2 = r_2 N_2 [1 - \alpha_{21}(N_1/K_2) - (N_2/K_2)^{\theta_2}], \tag{3.8.1b}$$

where r_1, r_2, θ_1, θ_2, α_{12} and α_{21} are positive constants.

Let $(N_1^\star, N_2^\star)$ be the nontrivial equilibrium of (3.8.1). We have

$$1 - (N_1^\star/K_1)^{\theta_1} - \alpha_{12}(N_2^\star/K_1) = 0, \tag{3.8.2a}$$

$$1 - \alpha_{21}(N_1^\star/K_2) - (N_2^\star/K_2)^{\theta_2} = 0. \tag{3.8.2b}$$

For a more compact notation, we shall require the parameters,

$$a_{11} = r_1(N_1^\star/K_1)^{\theta_1}, \qquad a_{12} = r_1\alpha_{12}N_2^\star/K_1,$$

$$a_{21} = r_2\alpha_{21}N_2^\star/K_2, \qquad a_{22} = r_2(N_2^\star/K_2)^{\theta_2}. \tag{3.8.3}$$

It is convenient to use the variables, x_1 and x_2, where

$$x_1 = N_1/N_1^\star, \qquad X_2 = N_2/N_2^\star. \tag{3.8.4}$$

Under this change of variables and with the help of (3.8.2), the system (3.8.1) becomes

$$\dot{X}_1 = X_1[-a_{11}(X_1^{\theta_1} - 1) - a_{12}(x_2 - 1)], \tag{3.8.5a}$$

$$\dot{X}_2 = X_2[-a_{21}(X_1 - 1) - a_{22}(X_2^{\theta_2} - 1)]. \tag{3.8.5b}$$

A suitable Liapunov function for (3.8.5) is

$$V = X_1 - 1 - \ln X_1 + X_2 - 1 - \ln X_2. \tag{3.8.6}$$

Along solutions of the system (3.8.5) we have

$$\dot{V} = -a_{11}(X_1 - 1)(X_1^{\theta_1} - 1) - a_{22}(X_2 - 1)(X_2^{\theta_2} - 1)$$
$$-(a_{12} + a_{21})(X_1 - 1)(X_2 - 1). \tag{3.8.7}$$

By inspection $\dot{V}$ is negative in the region $\{X \mid X_1 > 1 \text{ and } X_2 > 1\}$and the region $\{X \mid X_1 < 1 \text{ and } X_2 < 1\}$. If $1 > \theta_1 > 0$ and $1 > \theta_2 > 0$, it is difficult to establish simple analytical conditions that $\dot{V}$ is negative in the positive orthant minus the point (1, 1). If $\theta_1 \geqslant 1$ and $\theta_2 \geqslant 1$, we shall derive some simple analytical conditions for $\dot{V}$ to be negative in the positive orthant minus the point (1, 1).

Gilpin and Ayala applied their model to a competition between *Drosophila pseudoobscura* (N_1) and *Drosophila willistoni* (N_2). The data for the parameters in this competition are

$$r_1 = 4.513, \quad K_1 = 791, \quad \theta_1 = 0.12, \quad \alpha_{12} = 0.0869,$$
$$r_2 = 1.496, \quad K_2 = 1332, \quad p_2 = 0.35, \quad \alpha_{21} = 0.713. \tag{3.8.8}$$

The corresponding model has a positive equilibrium at $(N_1^\star, N_2^\star) = (441, 616)$. From these data, we get

$$a_{11} = 4.207, \quad a_{12} = 0.305,$$
$$a_{21} = 0.493, \quad a_{22} = 1.142. \tag{3.8.9}$$

These parameter values indicate that each species is strongly self-regulating and the interspecific interactions are relatively weak. We establish that $\dot{V}$ is negative for all $X_1 > 0$ and $X_2 > 0$ and $(X_1, X_2) \neq (1,1)$, by plotting a large number of level surfaces, $\dot{V}(X) = C$, where C is a negative number. From this we deduce that model (3.8.1) is globally stable if its parameter values are those given in (3.8.8).

We shall establish simple analytical conditions for global stability in Gilpin and Ayala's competition model when $\theta_i \geqslant 1$. For m competing species, Gilpin and Ayala's competition model is

$$\dot{N}_i = r_i N_i \left[1 - (N_i/K_i)^{\theta_i} - \sum_{j \neq i}^{m} \alpha_{ij}(N_j/K_i) \right] \tag{3.8.10}$$

where $i = 1, 2, \ldots, m$. Assume that it has a positive equilibrium at $(N_1^\star, N_2^\star, \ldots, N_m^\star)$. We have $N_i^\star > 0$ and

$$1 - (N_i^\star/K_i)^{\theta_i} - \sum_{j \neq i}^{m} \alpha_{ij}(N_j^\star/K_i) = 0 \tag{3.8.11}$$

where $i = 1, 2, \ldots, m$. For convenience let us define a matrix β by the equations,

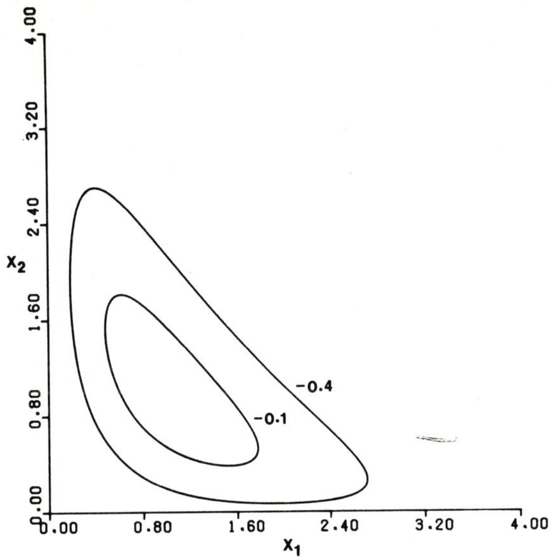

Fig. 3.8.1. Level surfaces of the $\dot{V}(x)$ function in (3.8.7).

$$\beta_{ii} = (N_i^\star/K_i)^{\theta_i - 1} \tag{3.8.12}$$

and

$$\beta_{ij} = \alpha_{ij} \quad \text{if } i \neq j \tag{3.8.13}$$

where $i,j = 1, 2, \ldots, m$.

Theorem 3.8.1. *If $\theta_i \geqslant 1$ for $i = 1, 2, \ldots, m$, model (3.8.10) is globally stable if there exists a positive equilibrium $N^\star$ and there exists a positive diagonal matrix C such that $C\beta + \beta^T C$ is positive definite.*

Proof. Substituting (3.8.11) into (3.8.10), we get

$$\dot{N}_i = r_i N_i [(N_i^\star/K_i)^{\theta_i} - (N_i/K_i)^{\theta_i} - \sum_{j \neq i}^{m} \alpha_{ij}(N_j - N_j^\star)/K_i], \tag{3.8.14}$$

where $i = 1, 2, \ldots, m$.

Let $s_1, s_2, \ldots, s_m$ be positive constants which will be prescribed later. A suitable Liapunov function for model (3.8.14) is

$$V(N) = \sum_{i=1}^{m} s_i [N_i - N_i^\star - N_i^\star \ln(N_i/N_i^\star)]. \tag{3.8.15}$$

Along solutions of (3.8.14) we have

$$\dot{V} = -\sum_{i=1}^{m} (s_i r_i / K_i^{\theta_i}) (N_i - N_i^\star)(N_i^{\theta_i} - N_i^{\star\theta_i})$$

$$- \sum_{i=1}^{m} \sum_{j \neq i}^{m} (s_i r_i \alpha_{ij} / K_i)(N_i - N_i^\star)(N_j - N_j^\star). \tag{3.8.16}$$

For convenience, let $c_i = s_i r_i / K_i$ for $i = 1, 2, \ldots, m$. Clearly $c_1, c_2, \ldots, c_m$ are positive constants.

If $\theta \geqslant 1$ and Z is a real number, we have

$$|Z^\theta - 1| \leqslant |Z - 1| \tag{3.8.17}$$

for all $Z > 0$. Let $Z_i = N_i / N_i^\star$ for $i = 1, 2, \ldots, m$. We then have

$$-(N_i - N_i^\star)(N^{\theta_i} - N^{\theta_i}) = -|N_i - N_i^\star|\,|Z_i^{\theta_i} - 1| N_i^{\star\theta_i} \leqslant -N_i^{\star\theta_i - 1}(N_i - N_i^\star)^2 \tag{3.8.18}$$

for $i = 1, 2, \ldots, m$.

Conditions (3.8.12), (3.8.13), (3.8.16) and (3.8.18) imply that

$$\dot{V} \leqslant -\sum_{i=1}^{m} \sum_{j=1}^{m} c_i \beta_{ij} (N_i - N_i^\star)(N_j - N_j^\star)$$

$$= -(1/2)(N - N^\star)^{\mathrm{T}} (C\beta + \beta^{\mathrm{T}} C)(N - N^\star). \tag{3.8.19}$$

By assumption, there exists a positive diagonal matrix C such that $C\beta + \beta^{\mathrm{T}}C$ is positive definite. It follows that $\dot{V} < 0$ for all N in the positive region minus $N^\star$. Hence (3.8.10) is globally stable.

We shall describe a simple way to establish that a given matrix β has a positive diagonal matrix C such that $C\beta + \beta^{\mathrm{T}}C$ is positive definite. We require a matrix P which is defined by the equations,

$$P_{ii} = \beta_{ii} \quad \text{for } i = 1, 2, \ldots, m, \tag{3.8.20a}$$

and

$$P_{ij} = -\beta_{ij} \quad \text{if } i \neq j. \tag{3.8.20b}$$

If all the leading principal minors of P are positive, then there exists a positive diagonal matrix C such that $C\beta + \beta^{\mathrm{T}}C$ is positive definite (Johnson, 1974).

For completeness, we shall establish global stability for a modification of model (3.8.10). Let E_{ij} for $i, j = 1, 2, \ldots, m$ be non-negative constants and $E_{ii} = 1$ for $i = 1, 2, \ldots, m$. A model of competition among m species is

$$\dot{N}_i = r_i N_i \left[1 - \sum_{j=1}^{m} E_{ij} (N_j / K_j)^{\theta_j} \right] \tag{3.8.21}$$

where $i = 1, 2, \ldots, m$. In this model we only require that $\theta_i > 0$ for $i = 1, 2, \ldots, m$. We shall assume that it has a positive equilibrium at $(N_1^\star, N_2^\star, \ldots, N_m^\star)$.

Let $Y_i = N_i / K_i$ for $i = 1, 1, \ldots, m$. We get

$$\dot{Y}_i = r_i Y_i \left[1 - \sum_{j=1}^{m} E_{ij} Y_j^{\theta_j}\right] \tag{3.8.22}$$

where $i = 1, 2, \ldots, m$. Let $Y_i^\star = N_i^\star / K_i$ for $i = 1, 2, \ldots, m$. We have

$$1 - \sum_{k=1}^{m} E_{ij} Y_j^{\theta_j} = 0 \tag{3.8.23}$$

for $i = 1, 2, \ldots, m$.

Theorem 3.8.2. *Model (3.8.21) is globally stable if it has a positive equilibrium $N^\star$ and there exists a positive diagonal matrix C such that $CE + E^{\mathrm{T}}C$ is positive definite.*

Proof. Substitute (3.8.23) into (3.8.22). We get

$$\dot{Y}_i = -r_i Y_i \sum_{j=1}^{m} E_{ij} (Y_j^{\theta_j} - Y_j^{\star\,\theta_j}), \tag{3.8.24}$$

for $i = 1, 2, \ldots, m$.

Let $s_1, s_2, \ldots, s_m$ be positive constants which will be prescribed later. A suitable Liapunov function for (3.8.24) is

$$V(Y) = \sum_{i=1}^{m} s_i \left[\frac{Y_i^{\theta_i}}{\theta_i} - \frac{Y_i^{\star\theta_i}}{\theta_i} - Y_i^{\theta_i} \ln(Y_i/Y_i^\star)\right]. \tag{3.8.25}$$

Let $c_i = s_i r_i$ for $i = 1, 2, \ldots, m$. Clearly $c_1, c_2, \ldots, c_m$ are positive constants. Along solutions of (3.8.24), we have

$$\dot{V} = -(1/2)\,(y - y)^{\mathrm{T}}(CE + E^{\mathrm{T}}C)\,(y - y^\star) \tag{3.8.26}$$

where $y_i = Y_i^{\theta_i}$ for $i = 1, 2, \ldots, m$ and $C = \mathrm{diag}(c_1, c_2, \ldots, c_m)$. By assumption, there exists a positive diagonal matrix C such that $CE + E^{\mathrm{T}}C$ is positive definite. Hence $\dot{V}$ is negative in the positive region minus the equilibrium $H^\star$. It follows that the model is globally stable.

3.9. PREY—PREDATOR SYSTEM WITH TYPE 2 FUNCTIONAL RESPONSE

By definition, the functional response of a predator is the rate at which prey is harvested by a predator (Holling, 1966). If we take into account the time a predator uses in handling the prey it has captured, we find the predator has a Type 2 functional response. According to Hassell et al. (1976 a, b), Type 2 functional response is the most common type of functional response among arthropod predators.

In the ecological literature, Kolmogoroff's Theorem (1936) has been the most versatile tool for the qualitative analysis of two-species interactions. It has been used very successfully by Rescigno and Richardson (1973) and May (1974) in their analysies of two-species interactions. The discovery of biologically meaningful Liapunov and Liapunov-like functions provides a new tool which is as versatile as Kolmogoroff's Theorem. We shall examine examples in which Kolmogoroff's Theorem does not apply but in which it is possible to use a Liapunov function to establish global stability. Moreover, Kolmogoroff's Theorem applies only to two-species interactions wherereas the direct method of Liapunov and its extensions apply to interactions between any number of species (Goh, 1978 b).

Let X and Y be the densities of a prey and its predator respectively. A model of prey—predator system is given by the equations

$$\dot{X} = X[g(X) - aY/(1 + wX)], \tag{3.9.1a}$$

$$\dot{Y} = Y[-d + eaX/(1 + wX)], \tag{3.9.1b}$$

where $g(X)$ is the per capita rate of change of prey density in the absence of the predator and a, d, e, w are positive parameters. The parameter e is the conversion efficiency of prey into predator and d is the mortality rate of the predator. The expression, $aX/(1 + wX)$, is a Type 2 functional response of a predator. An interesting feature of this model is that the predator does not sustain any density-dependent mortalities due to intraspecific interactions.

Assume that the model has a positive equilibrium at $(X^\star, Y^\star)$. We have $X^\star > 0$ and $Y^\star > 0$ and

$$g(X^\star) - aY^\star/(1 + wX^\star) = 0, \tag{3.9.2a}$$

$$-d + eaX^\star/(1 + wX^\star) = 0. \tag{3.9.2b}$$

It follows that $X^\star = d/(ea - wd)$ and $Y^\star = g(X^\star)\,(1 + wX^\star)/a$. For this equilibrium to be positive, we must have $ea > wd$ and $g(X^\star) > 0$. For convenience, let

$$S = 1/[(1 + wX)\,(1 + wX^\star)]. \tag{3.9.3}$$

Theorem 3.9.1. *Model (3.9.1) is globally stable if it has a unique positive equilibrium at $(X^\star, Y^\star)$, and*

$$(X - X^\star)[g(X) - g(X^\star) + awY^\star S(X - X^\star)] < 0 \tag{3.9.4}$$

for all $X > 0$, $Y > 0$ and $X \neq X^\star$.

Proof. Substitute (3.9.2) into (3.9.1). We get

$$\dot{X} = X[g(X) - g(X^\star) - aY/(1 + wX) + aY^\star/(1 + wX^\star)], \tag{3.9.5a}$$

$$\dot{Y} = Y[eaX/(1 + wX) - eaX^\star/(1 + wX^\star)]. \tag{3.9.5b}$$

A suitable Liapunov function is

$$V = X - X^\star - X^\star \ln(X/X^\star) + c[Y - Y^\star - Y^\star \ln(Y/Y^\star)], \tag{3.9.6}$$

where c is a positive constant which will be prescribed later. Along solutions of (3.9.5), we have

$$\begin{aligned}\dot{V} &= (X - X^\star)[g(X) - g(X^\star)] - a(X - X^\star)S(Y + wX^\star Y - Y^\star - wXY^\star) \\ &\quad + cea(Y - Y^\star)S(X - X^\star) = (X - X^\star)[g(X) - g(X^\star) + awY^\star S(X - X^\star)] \\ &\quad - a(1 + wX^\star)(X - X^\star)S(Y - Y^\star) + cea(X - X^\star)S(Y - Y^\star).\end{aligned}$$

Set $c = (1 + wX^\star)/e$; we get

$$\dot{V} = (X - X^\star)[g(X) - g(X^\star) + awY^\star S(X - X^\star)]. \tag{3.9.7}$$

It follows that $\dot{V} < 0$ for $X > 0$, $Y > 0$ and $X \neq X^\star$.

From an inspection of (3.9.7), we have $\dot{V} = 0$ when $X = X^\star$. On the set $\{(X, Y) \mid X = X^\star\}$, the only invariant set is $\{(X^\star, Y^\star)\}$. This is because by assumption, (3.9.1) has a unique positive equilibrium. By Theorem 3.3.5, model (3.9.1) is globally stable.

We show by a simple example that it is possible to have global stability when the per capita birth rate of the prey is not a monotonic decreasing function of its density. In this case, the prey population exhibits an Allee effect (Odum, 1971). In the same example, it is shown that Kolmogoroff's Theorem cannot be used to establish global stability. This observation is interesting as May (1974) has shown that Kolmogoroff's Theorem applies effectively to most proposed models of prey—predator interactions.

Example 3.9.1. Consider the prey—predator model,

$$\dot{X} = X[g(X) - Y/(1 + 0.05X)], \tag{3.9.8a}$$

$$\dot{Y} = Y[-6/13 + 0.1X/(1 + 0.05X)] \tag{3.9.8b}$$

where $g(X) = 2 + X(4 - X)/(1 + 2X)$. The function $g(X)$ has a maximum at $X = 1$, $g(0) = 2$ and $g(4 + \sqrt{18}) = 0$. For $X > 0$, $g(X)$ has a single hump. For $X < 1$, $g(X)$ can be roughly approximated by $2 + 4X$, and for $X > 1$, it can roughly be approximated by $4 - X/2$.

The nontrivial equilibrium $(X^\star, Y^\star)$ is at (6, 1.4). Condition (3.9.4) implies that

$$\dot{V} = -\frac{(X - 6)^2(0.65X^2 + 11.7X + 1.3)}{13(1 + 2X)(1 + 0.05X)}. \tag{3.9.9}$$

Hence $\dot{V}$ is negative for $X > 0$, $Y > 0$ and $X \neq 6$. We can establish from a graphical analysis that there is only a single equilibrium in the feasible region and the only invariant set on the line is $\{(6, 1.4)\}$. Hence model (3.9.8) is globally stable.

Model (3.9.8) is of the form,

$$\dot{X} = XF(X,Y), \qquad \dot{Y} = YG(X,Y) \tag{3.9.10}$$

One of the conditions in Kolmogoroff's Theorem is the inequality,

$$X(\partial F/\partial X) + Y(\partial F/\partial Y) < 0. \tag{3.9.11}$$

For (3.9.8), this inequality requires that

$$X[\partial g/\partial X + 0.05Y/(1 + 0.05X)^2] - Y/(1 + 0.05X) < 0. \tag{3.9.12}$$

At the point (0.5, 0.3), the expression on the left hand side in (3.9.12) is approximately equal to 0.0269. Hence this inequality is violated.

We shall now establish some simple and robust conditions for global stability in a class of prey—predator models, in which the per capita death rate of the predator is a complex nonlinear function of its density. This model could apply to a prey—predator system in which the survival rate of the predators is increased up to a certain density by aggregation.

Consider the prey—predator model:

$$\dot{X} = X[g(X) - aY/(1 + wX)], \tag{3.9.13a}$$

$$\dot{Y} = Y[-f(Y) + eaY/(1 + wX)]. \tag{3.9.13b}$$

Let this model have a positive equilibrium at (X^*,Y^*). The function S is as displayed in (3.9.3).

Theorem 3.9.2. *Model (3.9.13) is globally stable if it has a positive equilibrium at (X^*,Y^*) and if*

$$(X - X^*)[g(X) - g(X^*) + awY^*S(X - X^*)] + (Y - Y^*)[(1 + wX^*)/e]$$
$$[f(Y^*) - f(Y)] < 0 \tag{3.9.14}$$

for all $X > 0$, $Y > 0$ and $(X,Y) \neq (X^*,Y^*)$.

Proof. The function V displayed in (3.9.6) is a suitable Liapunov function if $c = (1 + wX^*)/e$. The expression (3.9.14) is none other than $\dot{V}$ computed along solutions of (3.9.13). Hence, it is negative for $X > 0$, $Y > 0$ and $(X,Y) \neq (X^*,Y^*)$. It follows that model (3.9.13) is globally stable.

Example 3.9.2. Consider the prey—predator model,

$$\dot{X} = X[g(X) - Y/(1 + 0.05X)], \tag{3.9.15a}$$

$$\dot{Y} = Y[-f(Y) + 0.1Y/(1 + 0.05X)], \tag{3.9.15b}$$

where $-f(Y) = -(6/13 - 56/1220) + Y(1 - y)/(1 + 8Y)$ and $g(X) = 2 + X(4 - X)/(1 + 2X)$. The function $-f(Y)$ has a maximum at $Y = 0.25$. For $Y > 0$, it is a function with a single hump. For $Y < 0.25$, it can be roughly approximated by $-6/13 + 56/1220 + Y$, and for $Y > 0.25$, it can be roughly approximated by $(-6/13 + 56/1200 + 1/8) - Y/8$. For $0.25 > Y > 0$, the function $-f(Y)$ increases monotonically as Y increases. This violates one of the conditions in Kolmogoroff's Theorem.

Model (3.9.15) has a positive equilibrium at (6, 1.4). We have

$$\dot{V} = \frac{-(X-6)^2(0.65X^2 + 11.7X + 1.3)}{13(1+2X)(1+0.05X)} - \frac{1.07(Y-1.4)^2(0.4+12.2Y)}{(1+8Y)} \tag{3.9.16}$$

By inspection, $\dot{V} < 0$ for $X > 0$, $Y > 0$ and $(X, Y) \neq (6, 1.4)$. Hence (3.9.15) is globally stable.

3.10. NONLINEAR TWO-SPECIES MODELS

There is only a very limited set of simple functions which can be used to construct a model of an ecosystem. This imposes a major constraint in the construction of realistic models of ecosystems. A reason why the graphical method of Rosenzweig and MacArthur (1963) is so popular and effective for two-species interactions is that it overcomes this constraint to a large extent.

Some ecologists would be delighted if there were a firm mathematical foundation for a vague verbal statement like the following: If each species of an ecosystem is self-regulating at all densities and all the interspecific interactions are relatively weak, the ecosystem is globally stable. In fact we can show that there is a precise mathematical foundation for making such a verbal statement for two-species interactions. However, the validity of such a verbal statement in multispecies interactions decreases rapidly as the number of species in an ecosystem increases. This is because the sum of all the interspecific interactions acting on a species in an ecosystem eventually overcomes the self-regulating intraspecific interactions.

Let N_1 and N_2 be the densities of two interacting species. A general model of two-species interactions, which could be prey—predator interaction, competition or mutualism, is given by the equations,

$$\dot{N}_1 = N_1F_1(N_1, N_2), \qquad \dot{N}_2 = N_2F_2(N_1, N_2). \tag{3.10.1}$$

We do not require explicit and precise knowledge of the functions, $F_1(N_1, N_2)$, $F_2(N_1, N_2)$.

We need only information similar to that which is used in a graphical analysis of the model. An advantage of this analytical approach is that it is applicable to multispecies interactions. The graphical analysis is basically limited to two-species interactions. A subtle and difficult problem in applying the graphical analysis to three-species interactions is that the angles between the zero isocline planes by themselves may not determine the stability or instability of a positive equilibrium.

Assume that model (3.10.1) has a positive equilibrium at (N_1^*, N_2^*). This means that the zero isoclines, $F_1(N_1, N_2) = 0$ and $F_2(N_1, N_2) = 0$, intersect at a positive point. The conditions which we shall prescribe will ensure that there is only one positive equilibrium.

By Taylor's Theorem, model (3.10.1) may be exactly rewritten in the form,

$$\dot{N}_1 = N_1[(\partial F_1/\partial N_1)(N_1 - N_1^*) + (\partial F_1/\partial N_2)(N_2 - N_2^*)], \tag{3.10.2a}$$

$$\dot{N}_2 = N_2[(\partial F_2/\partial N_1)(N_1 - N_1^*) + (\partial F_2/\partial N_2)(N_2 - N_2^*)], \tag{3.10.2b}$$

where the partial derivatives are computed at a pair of points, N', N'', along the line joining the points N and N^*. Unfortunately Taylor's Theorem does not explicitly give N' and N'' as functions of N. For this reason, it is appropriate to look at these equations as the set of equations of the form,

$$\dot{N}_1 = N_1[A_{11}(N_1, N_2)(N_1 - N_1^*) + A_{12}(N_1, N_2)(N_2 - N_2^*)], \tag{3.10.3a}$$

$$\dot{N}_2 = N_2[A_{21}(N_1, N_2)(N_1 - N_2^*) + A_{22}(N_1, N_2)(N_2 - N_2^*)], \tag{3.10.3b}$$

where $A_{11}, A_{22}, A_{12}, A_{21}$ are certain nonlinear functions of N_1 and N_2.

From an inspection of (3.10.2) we conclude that A_{11}, A_{22} are measures of the per capita intraspecific interactions and A_{12}, A_{21} are measures of the per capita interspecific interactions.

Let H be a constant 2×2 matrix with nonnegative elements. Suppose precise descriptions of the functions, A_{11}, A_{22}, A_{12}, A_{21}, are not available, but we have estimates of the values of these functions for all $N_1 > 0$ and $N_2 > 0$. Let these functions satisfy the inequalities,

$$A_{11}(N_1, N_2) \leqslant -H_{11}, \qquad A_{22}(N_1, N_2) \leqslant -H_{22}, \tag{3.10.4a}$$

$$|A_{12}(N_1, N_2)| \leqslant H_{12}, \qquad |A_{21}(N_1, N_2)| \leqslant H_{21} \tag{3.10.4b}$$

for all $N_1 > 0$ and $N_2 > 0$.

If $H_{11} > 0$ and $H_{22} > 0$, condition (3.10.4a) means that each species is self-regulating at all densities. Condition (3.10.4b) implies that the values of the per capita interspecific interactions are uniformly bounded at all densities. The expression, $H_{11}H_{22} - H_{12}H_{21}$, is a measure of the combined strength of the intraspecific interactions relative to the combined strength of the interspecific interactions. If $H_{11}H_{22} - H_{12}H_{21} > 0$, it means that the intraspecific interactions are stronger than the interspecific interactions.

Theorem 3.10.1. *Model (3.10.3) is globally stable if* (i) *it has a positive equilibrium at* (N_1^*, N_2^*), (ii) $H_{11} > 0$ *and* $H_{22} > 0$ *and* (iii) $H_{11}H_{22} - H_{12}H_{21} > 0$.

Proof. A suitable Liapunov function is

$$V(N) = c_1[N_1 - N_1^* - N_1^* \ln(N_1/N_1^*)] + c_2[N_2 - N_2^* - N_2^* \ln(N_2/N_2^*)], \tag{3.10.5}$$

where c_1 and c_2 are positive constants which will be prescribed later.

Along solutions of (3.10.3), we have

$$\dot{V} = c_1(N_1 - N_1^\star)[A_{11}(N_1, N_2)(N_1 - N_1^\star) + A_{12}(N_1, N_2)(N_2 - N_2^\star)]$$
$$+ c_2(N_2 - N_2^\star)[A_{12}(N_1, N_2)(N_1 - N_1^\star) + A_{22}(N_1, N_2)(N_2 - N_2^\star)]$$
$$\leqslant c_1 A_{11}(N_1 - N_1^\star)^2 + c_2 A_{22}(N_2 - N_2^\star)^2$$
$$+ c_1 |A_{12}|\,|N_1 - N_1^\star| + c_2 |A_{21}|\,|N_1 - N_1^\star|\,|N_2 - N_2^\star|. \tag{3.10.6}$$

Conditions (3.10.4) and (3.10.6) imply that

$$\dot{V} \leqslant -c_1 H_{11}(N_1 - N_1^\star)^2 - c_2 H_{22}(N_2 - N_2^\star)^2$$
$$+ c_1 H_{12}|N_1 - N_1^\star)\,|N_2 - N_2^\star| + c_2 H_{21}|N_1 - N_1^\star|\,|N_2 - N_2^\star|. \tag{3.10.7}$$

Let $X_1 = |N_1 - N_1^\star|$ and $X_2 = |N_2 - N_2^\star|$. We have

$$\dot{V} = -(1/2)X^{\mathrm{T}}(CH + H^{\mathrm{T}}C)X \tag{3.10.8}$$

where $C = \mathrm{diag}(c_1, c_2)$. If there exists a positive diagonal matrix C such that $CH + H^{\mathrm{T}}C$ is positive definite, $\dot{V} < 0$ for all $N \neq N^\star$.

The matrix $CH + H^{\mathrm{T}}C$ is positive definite if

$$2c_1 H_{11} > 0, \qquad 2c_2 H_{22} > 0, \tag{3.10.9}$$

$$4c_1 c_2 H_{11} H_{22} - (c_1 H_{12} + c_2 H_{21})^2 > 0. \tag{3.10.10}$$

Clearly the conditions, $H_{11} > 0$ and $H_{22} > 0$, imply that (3.10.9) is satisfied. Condition (3.10.10) is equivalent to the inequality,

$$4c_1 c_2 (H_{11} H_{22} - H_{12} H_{21}) - (c_1 H_{12} - c_2 H_{21})^2 > 0. \tag{3.10.11}$$

If $H_{12} \neq 0$, we choose $c_2 = 1$ and $c_1 = H_{21}/H_{12}$. This and the condition, $H_{11}H_{12} - H_{12}H_{21} > 0$, imply that condition (3.10.11) is satisfied. If $H_{12} = 0$ and $H_{21} \neq 0$, we employ a similar argument. If $H_{12} = 0$ and $H_{21} = 0$, inequality (3.10.10) is clearly satisfied.

It follows that $\dot{V} < 0$ for all $N \neq N^\star$. Hence model (3.10.3) is globally stable.

Corollary 3.10.1. *If $\dot{V} < 0$ for all $N_1 > 0, N_2 > 0$ and $N \neq N^\star$, the positive equilibrium is unique.*

Proof. Along the solutions of model (3.10.1), we have

$$\dot{V} = c_1(N_1 - N_1^\star)F_1(N_1, N_2) + c_2(N_2 - N_2^\star)F_2(N_1, N_2). \tag{3.10.12}$$

By inspection, $\dot{V} = 0$ when $N = N^\star$. Suppose there exists a positive equilibrium at N' which is not equal to $N^\star$. We have $N_1' > 0, N_2' > 0$ and

$$F_1(N_1', N_2') = 0, \qquad F_2(N_1', N_2') = 0. \tag{3.10.13}$$

From (3.10.12) we have $\dot{V}(N_1', N_2') = 0$. This contradicts the assumption that $\dot{V} < 0$ for all $N_1 > 0, N_2 > 0$ and $N \neq N^\star$. It follows that (3.10.1) has a unique positive equilibrium.

Example 3.10.1. We shall examine a slightly modified version of Schoener's (1974) model for competition between two species with pure interference and fixed energy input to the population. Schoener's model is modified to ensure that $(\dot{N}_i/N_i)$ remains finite as $N_i \to 0+$ for $i = 1$ and 2. Let r_i, I_i, C_i and e_i for $i = 1, 2$ be positive constants. The modified Schoener's model for competition between two species is

$$\dot{N}_1 = r_1 N_1 [I_1/(N_1 + e_1) - \gamma_{11} N_1 - \gamma_{12} N_2 - c_1], \tag{3.10.14a}$$

$$\dot{N}_2 = r_2 N_2 [I_2/(N_2 + e_2) - \gamma_{21} N_1 - \gamma_{22} N_2 - c_2]. \tag{3.10.14b}$$

From (3.10.1) and (3.10.14), we get

$$\partial F_1/\partial N_1 = -I_1/(N_1 + e_1)^2 - r_1\gamma_{11}, \tag{3.10.15a}$$

$$\partial F_2/\partial N_2 = -I_2/(N_2 + e_2)^2 - r_2\gamma_{22}, \tag{3.10.15b}$$

$$|\partial F_1/\partial N_2| = r_1\gamma_{12}, \qquad |\partial F_2/\partial N_1| = r_2\gamma_{21}. \tag{3.10.15c}$$

A suitable H matrix is defined by the equations:

$$H_{11} = r_1\gamma_{11}, \qquad H_{22} = r_2\gamma_{22}, \qquad H_{12} = r_1\gamma_{12}, \qquad H_{21} = r_2\gamma_{21}. \tag{3.10.16}$$

From an inspection of (3.10.15) and (3.10.16), we have

$$\partial F_1/\partial N_1 \leqslant -r_1\gamma_{11}, \qquad \partial F_2/\partial N_2 \leqslant -r_2\gamma_{22}, \tag{3.10.17a}$$

$$|\partial F_1/\partial N_2| \leqslant r_1\gamma_{12}, \qquad |\partial F_2/\partial N_1| \leqslant r_2\gamma_{21}. \tag{3.10.17b}$$

By Theorem 3.10.1, model (3.10.14) is globally stable if (i) it has a positive equilibrium at $(N_1^\star, N_2^\star)$, (ii) $r_1\gamma_{11} > 0$, $r_2\gamma_{22} > 0$ and (iii) $r_1r_2\gamma_{11}\gamma_{22} - r_1r_2\gamma_{12}\gamma_{21} > 0$. As r_1 and r_2 are positive, these inequalities may be replaced by the conditions, $\gamma_{11} > 0$, $\gamma_{22} > 0$ and $\gamma_{11}\gamma_{22} - \gamma_{12}\gamma_{21} > 0$.

It is interesting to note that if these conditions are satisfied, then model (3.10.14) has exactly the same qualitative dynamic behavior as the associated Lotka—Volterra model,

$$\dot{N}_1 = N_1[-\gamma_{11}(N_1 - N_1^\star) - \gamma_{12}(N_2 - N_2^\star)], \tag{3.10.18a}$$

$$\dot{N}_2 = N_2[-\gamma_{21}(N_1 - N_1^\star) - \gamma_{22}(N_2 - N_2^\star)], \tag{3.10.18b}$$

where $(N_1^\star, N_2^\star)$ is the positive equilibrium of (3.10.14). This follows because (3.10.18) is globally stable if $(N_1^\star, N_2^\star)$ is positive, $\gamma_{11} > 0$, $\gamma_{22} > 0$ and $\gamma_{11}\gamma_{22} - \gamma_{12}\gamma_{21} > 0$. This observation shows that Lotka—Volterra models are more robust than some ecologists believe. Basically their static criterion for judging whether a model is correct or not is not necessarily valid from the qualitative and dynamical point of view.

Example 3.10.2. An important problem in the management of an ecosystem is whether the system is stable if it is enriched or harvested. Suppose we have a two-species ecosystem. Let E_1 and E_2 be measures of the constant efforts applied in the harvesting of the populations. A model of the exploited ecosystem is

$$\dot{N}_1 = N_1 [F_1(N_1, N_2) - E_1], \tag{3.10.19a}$$

$$\dot{N}_2 = N_2[F_2(N_1, N_2) - E_2]. \tag{3.10.19b}$$

Let $G_1(N_1, N_2) = F_1(N_1, N_2) - E_1$ and $G_2(N_1, N_2) = F_2(N_1, N_2) - E_2$. If F_1 and F_2 satisfy the inequalities in (3.10.4) so will G_1 and G_2 because E_1 and E_2 are constants. Hence model (3.10.19) is globally stable if (i) the unexploited model (3.10.1) satisfies Theorem 3.10.1 and (ii) model (3.10.19) has a positive equilibrium. It follows that such systems are robust under constant effort harvesting policies.

3.11. REGION OF ULTIMATE CONFINEMENT

May (1972) pointed out that the eigenvalue analysis may lead to an erroneous conclusion that a model ecosystem is unstable. This is because a model ecosystem with an unstable equilibrium could have a globally stable and acceptable limit cycle. The populations in a model ecosystem with a stable limit cycle will eventually have regular oscillations. There are a number of well known examples of regular fluctuations of populations Odum, 1971, (p. 188) especially among animal populations in the Arctic. This suggests that limit cycles are important in theoretical ecology.

The main mathematical tool available for the study of limit cycles is the Poincare—Bendixon theorem. Unfortunately it is limited to models with two state variables. This has prompted May (1974) to state on p. 94 of his well known book that ecology would benefit from a mathematical breakthrough which could extend this type of analysis to problems with three or more state variables. In fact, there is a mathematical method available. However, it is not well developed and has seldom been used even for the analysis of a physical system. The concept of a region of ultimate confinement includes a limit cycle as a special case and it is valid for problems with three or more state variables.

The best method to estimate a region of ultimate confinement is to use a Liapunov-like function. This function has the usual properties of a Liapunov function except that its time derivative along solutions of a model can be positive inside a region of ultimate confinement. A region of ultimate confinement has slightly different properties from a region of ultimate boundedness (Aggarwal, 1972). The trajectories of a model which begin inside a region of ultimate confinement cannot leave it at all. In contrast, the trajectories of a model are permitted to leave a region of ultimate boundedness for some intermediate time periods.

Let $W(N)$ be a Liapunov-like function for estimating a region of ultimate confinement for the model,

$$\dot{N}_i = N_i F_i(N_1, N_2, \ldots, N_m) \tag{3.11.1}$$

where $i = 1, 2, \ldots, m$. Let L_o and L_u be positive constants, where $L_o > L_u$. Let $S = \{N \mid W(N) \leqslant L_u\}$, $Y = \{N \mid L_o > W(N) > L_u\}$ and $\Omega = S \cup Y$ (union of S and Y). Let $N^\star$ be a point in the region of ultimate confinement of (3.11.1). Note that $N^\star$ need not be an equilibrium point. The function $W(N)$ is a Liapunov-like function if it behaves like a Liapunov function in Y. Its behavior in S need not be specified.

However, in practice a Liapunov-like function $W(N)$ has the following properties:

(i) $W(N) > 0$ for all $N \in \Omega$ and $N \neq N^\star$.

(ii) In Ω, $W(N)$ has a unique global minimum equal to zero at $N^\star$.

(iii) In Y, the level surface, $W(N) = K$ for each admissible positive value of K, is a closed surface.

(iv) The time derivative along solutions of the model,

$$\dot{W}(N) = \sum_{i=1}^{m} \frac{\partial W}{\partial N_i} N_i F_i(N), \tag{3.11.2}$$

must be nonpositive for all $N \in Y$.

If Ω is equal to the positive orthant $\Gamma = \{N \mid N_i > 0, i = 1, 2, \ldots, m\}$, we replace (iii) by the following condition:

(iii) $W(N) \to \infty$ as $N_i \to 0+$ or $N_i \to \infty$ for $i = 1, 2, \ldots, m$.

This is an important case in theoretical ecology. It means that if a disturbance does not directly cause the extinction of any species and if the model ecosystem is left alone for $t > 0$, the natural dynamics of the system would drive the state into the region of ultimate confinement. If this region is bounded away from the boundary planes $N_i = 0$, this system property clearly ensures the survival of the ecosystem.

Example 3.11.1. Consider the single variable model:

$$\dot{N} = N[\ln N - (\ln N)^3] = NF(N). \tag{3.11.3}$$

It has nontrivial equilibria at $N^\star = e^{-1}$, $N^\star = 1$ and $N^\star = e$, where e is the exponential constant.

The function $F(N)$ has the following properties:

(i) If $e^{-1} > N > 0$, $F(N) > 0$,
(ii) if $1 > N > e^{-1}$, $F(N) < 0$,
(iii) if $1 > N > e$, $F(N) > 0$, and
(iv) if $N > 1$, $F(N) < 0$.

Hence (a) the equilibrium $N^\star = e^{-1}$ is stable, (b) the equilibrium $N^\star = 1$ is unstable and (c) the equilibrium $N^\star = e$ is stable.

A suitable Liapunov-like function is

$$W(N) = \{\ln(N/2)\}^2. \tag{3.11.4}$$

It has a global minimum at $N' = 2$. It tends to infinity as $N \to 0$ or $N \to \infty$. Along solutions of (3.11.3), we have

$$\dot{W} = 4\,[\ln(N/2)]\,[\ln N - (\ln N)^3\,].$$

Let $Y = \{N \mid W(N) > [\ln(e/2)]^2\}$. Clearly $\dot{W} < 0$ for all $N \in Y$. Hence an estimate of the region of ultimate confinement of model (3.11.3) is $S = \{N \mid W(N) \leqslant [\ln(e/2)]^2\}$. By direct arguments, it can be shown that the smallest region of ultimate confinement is $\{N \mid e \geqslant N \geqslant e^{-1}\}$.

Example 3.11.2. Consider the model:

$$\dot{N}_1 = N_1\,[\ln N_2 + (1 - R^2)\ln N_1], \tag{3.11.5a}$$

$$\dot{N}_2 = N_2[-\ln N_1 + (1 - R^2)\ln N_2], \tag{3.11.5b}$$

where

$$R^2 = (\ln N_1)^2 + (\ln N_2)^2. \tag{3.11.6}$$

This model has an unstable positive equilibrium at (1,1). A suitable Liapunov-like function is $W(N) = R^2$. Along solutions of (3.11.5), we have

$$\dot{W} = 2R^2(1 - R^2). \tag{3.11.7}$$

Clearly $\dot{W} < 0$ for $R^2 > 1$, and $\dot{W} > 0$ for $R^2 < 1$. It follows that any solution which begins in the region, $\{N \mid 1 > R^2 > 0\}$, tends to the limit cycle,

$$R^2 = (\ln N_1)^2 + (\ln N_2^2) = 1. \tag{3.11.8}$$

Similarly any solution which begins in the region, $\{N \mid R^2 > 1, N_1 > 0, N_2 > 0\}$, tends to the same limit cycle. Fig. 3.11.1 displays the limit cycle and some typical trajectories from the interior and exterior regions of it. Except for the initial state (1,1), the limit cycle is globally stable. The region, $S = \{N \mid W(N) \leqslant 1\}$, is a region of ultimate confinement. It is in fact the smallest region of ultimate confinement; we have such a good estimate of it because the limit cycle is a level surface of the Liapunov-like function $W(N)$.

Example 3.11.3. We shall now examine a model with three state variables, in which the region of ultimate confinement has only a finite region of stability. The model is as follows:

$$\dot{N}_1 = N_1\,[\ln N_2 + (1 - R^2)\,(2 - R^2\,)\ln N_1\,], \tag{3.11.9a}$$

$$\dot{N}_2 = N_2[-\ln N_1 + (1 - R^2)\,(2 - R^2\,)\ln N_2\,], \tag{3.11.9b}$$

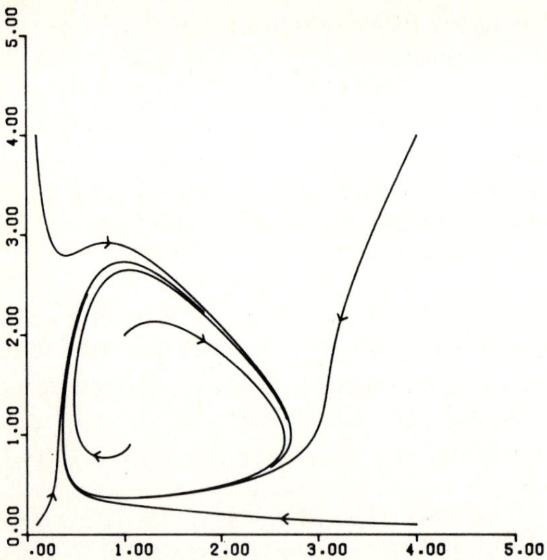

Fig. 3.11.1. Limit cycle and converging trajectories of model (3.11.5).

$$\dot{N}_2 = N_3[(1 - R^2)(2 - R^2)\ln N_3] \tag{3.11.9c}$$

where

$$R^2 = (\ln N_1)^2 + (\ln N_2)^2 + (\ln N_3)^2. \tag{3.11.10}$$

This model has an unstable positive equilibrium at (1,1,1). The function, $W(N_1, N_2, N_3) = R^2$, is a suitable Liapunov-like function. It has all the properties of a Liapunov function in the positive region except that $\dot{W}$ is positive in the region, $W^2 < 1$, without the point (1,1,1). Along solutions of (3.11.9), we have

$$\dot{W} = 2R^2(1 - R^2)(2 - R^2). \tag{3.11.11}$$

In this model we have a continuum of limit cycles. On the hypersurface, $R^2 = 1$, there is one limit cycle for each value of N_3. This follows because on the hypersurface, $R^2 = 1$, $\dot{N}_3 = 0$.

In this example the smallest region of ultimate confinement is $S = \{N \mid W(N) \leqslant 1\}$. All solutions which begin in the exterior region, $\{N \mid 2 > W(N) > 1\}$, will ultimately enter the remain in the region S. Also all solutions which begin in the interior region, $1 > W > 0$, will tend to the hypersurface, $W = 1$. In an analysis of a region of ultimate confinement, we are usually not interested in or able to examine what happens inside the region.

It is interesting and important to note that a definition of a concept of "inside" or "outside" in an m-dimensional space when $m \geqslant 2$, requires a closed, simply connected and bounded hypersurface of $(m - 1)$ dimensions.

A simple way to construct such a hypersurface is to employ a level surface of a Liapunov-like function.

The method of using Liapunov-like functions to estimate regions of ultimate confinement in a dynamic model poses a great challenge to applied mathematicians to use it effectively in the analysis of realistic models of ecosystems. It is clearly more difficult to use than the direct method of Liapunov for establishing global stability. The reason for this is that it demands that the level surfaces of a suitable Liapunov-like function must be related to the exact boundaries of the smallest region of ultimate confinement.

3.12 NONVULNERABILITY

In the real world ecosystems are continually disturbed by external forces like migrating populations, weather changes, pollution and outbreaks of fire and disease. It is desirable that the state of an ecosystem remains in an acceptable region of its state space in spite of large and continual disturbances. The concept of nonvulnerability was developed (Goh, 1971, 1975, 1976 a) to describe a system's ability to withstand unpredictable, large and continual disturbances. The nonvulnerability approach to describe a system's response to unpredictable disturbances is more effective than that based on the theory of stochastic differential equations. The latter leads to the Kolmogoroff or the Fokker—Planck equation which poses very serious computational problems, except in very simple population problems.

Contributions to the concept of nonvulnerability and related concepts of practical stability, resilience and total stability have been made by LaSalle and Lefschetz (1961), Lashiher and Storey (1972), Holling (1973), Harrison (1979 a), Vincent and Anderson (1979), and Vincent and Skowronksi (1979).

Consider the model ecosystem:

$$\dot{N}_i = N_i F_i(N_1, N_2, \ldots, N_m; u_1, u_2, \ldots, u_n], \qquad (3.12.1)$$

where $i = 1, 2, \ldots, m$. The essential elements of a definition of nonvulnerability are, (i) a specified time interval $[0,T]$, (ii) a set $S(0)$ which contains desirable states of the system, (iii) a set $Z(T)$ which contains undesirable states and (iv) a given class of continual disturbances, $\{u_1(t), u_2(t), \ldots, u_n(t)\}$.

It is reasonable to assume that we have good estimates of the bounds for the disturbance functions, $u_1(t), u_2(t), \ldots, u_n(t)$. Let

$$-\zeta_r \leqslant u_r(t) \leqslant \xi_r, \qquad (3.12.2)$$

where ζ_r and ξ_r, for $r = 1, 2, \ldots, n$, are constants. Assume that the set of admissible functions, $\{u_1, u_2, \ldots, u_n\}$, is piecewise continuous. For convenience we shall denote the set of admissible unpredictable disturbances by U.

By definition, system (3.12.1) is vulnerable during the time interval $[0,T]$ if there exists an admissible set of functions, $u_1(t), u_2(t), \ldots, u_n(t)$, which could drive the system from an initial state in $S(0)$ to a state in $Z(T)$ during the time interval $[0,T]$. Otherwise, the system is said to be nonvulnerable. Clearly in this concept of vulnerability we adopt the most pessimistic point of view. For this reason we could have a more balanced view if in practice the parameters ζ_r and ξ_r for $r = 1, 2, \ldots, n$, are underestimated.

The above vulnerability problem can be formulated as an optimal control problem. Unfortunately at present, the computational cost in this accessory optimal control problem is high in problems with more than two state variables. The most effective approach is to make a conservative analysis with the help of Liapunov functions.

Let $V(N)$ denote a suitable Liapunov function for model (3.12.1) in the absence of disturbances, i.e. when $u_1 = 0, u_2 = 0, \ldots, u_n = 0$. It is convenient to define the sets, $S(0)$ and $Z(T)$, in terms of $V(N)$. Let V_S and V_Z be positive constants. Let $S(0) = \{N \mid V(N) \leqslant V_S\}$ and $Z(T) = \{N \mid V(N) \geqslant V_Z\}$. Along solutions of (3.12.1), we have

$$\dot{V}(N;u) = \sum_{i=1}^{m} \frac{\partial V}{\partial N_i} N_i F_i(N;u). \tag{3.12.3}$$

Strictly speaking, when $V(N)$ is used for this type of analysis it is not a Liapunov function. This is because we do not require $\dot{V}$ to be negative definite in the positive orthant. For this reason it is more accurate to call it a Liapunov-like function.

Theorem 3.12.1. *Model (3.12.1) is nonvulnerable relative to the sets, U, $S(0)$ and $Z(T)$, during the time interval $[0,T]$ if the global maximum of $\dot{V}(N;u)$ for all $u \in U$ and for all $N \in \{N \mid V_S \leqslant V(N) + V_Z\}$ is less than $(V_Z - V_S)/T$.*

Proof. By assumption,

$$\dot{V} < (V_Z - V_S)/T \tag{3.12.4}$$

for all $u \in U$ and for all $N \in \{N \mid V_S \leqslant V(N) \leqslant V_Z\}$. This implies that any trajectory which begins on the hypersurface, $\{N \mid V(N) = V_S\}$, cannot be driven to the set $Z(T)$ during the time interval $[0,T]$. Any trajectory of (3.12.1) which begins inside the set, $\{N \mid V(N) < V_S\}$, must first intersect the hypersurface, $V(N) = V_S$, before it is driven towards $Z(T)$. It follows that the system cannot be driven from the set $S(0)$ to the set $Z(T)$ during the time interval $[0,T]$.

Theorem 3.12.2. *Model (3.12.1) is nonvulnerable relative to the sets, U, $S(0)$ and $Z(\infty)$, during the time interval $[0,\infty]$, if there exists a positive number G such that $V_S \leqslant G < V_Z$, and if the global maximum of $\dot{V}(N;u)$ for all $u \in U$ and for all $N \in \{N \mid V(N) = G\}$ is negative.*

Proof. The assumption that $\dot{V}(N;u)$ is negative for all $u \in U$ and for all $N \in \{N \mid V(N) = G\}$ implies that all trajectories of model (3.12.1) which begin on the hypersurface, $V(N) = G$, will enter and remain indefinitely in the region, $\{N \mid V(N) \leqslant G\}$. As $G < V_Z$, it follows that the system cannot be driven from the region, $\{N \mid V(N) \leqslant G\}$, to the set $Z(\infty)$. The set $S(0)$ is a subset of the set, $\{N \mid V(N) \leqslant G\}$. Hence the system is nonvulnerable relative to the sets, U, $S(0)$ and $Z(\infty)$, during the time interval $[0,\infty]$.

Theorems 3.12.1 and 3.12.2 are potentially very effective sufficient conditions for a system to be nonvulnerable. Unfortunately, a major obstacle to the application of these theorems is due to the rudimentary nature of numerical algorithms for computing the global maximum of a function of many variables (Dixon and Szego, 1975). In Chapter 6 we shall discuss an algebraic method which can be applied to model ecosystems with many species. However, it is a very conservative estimate of the set $S(0)$ in an analysis on whether a system is vulnerable or not.

Example 3.12.1. Let $u(t)$ be a description of the effects of a disturbance on the dynamics of a single-species population. When $u(t) = 0$, let the population be described by the logistic model. Assume that the disturbance affects the population on a percentage basis during a unit time interval. Examples of this type of disturbance are the mortalities caused by weather and diseases. It follows that a model of a continually disturbed population is given by the equation,

$$\dot{N} = (r/K)N(K-N) - u(t)N. \tag{3.12.5}$$

If the disturbance acts favorably on the population, e.g. a boost in its food supply, the function $u(t)$ is negative. If the disturbance is adverse, e.g. a disease, $u(t)$ is positive. Assume that we have

$$-\zeta \leqslant u(t) \leqslant \xi \tag{3.12.6}$$

where ζ and ξ are constants.

In this simple model, the worst possible way for the disturbance to act on the population is for $u(t)$ to be equal to ξ for all time. If this occurs, we have

$$\dot{N} = N[(r-\xi) - (r/K)N]. \tag{3.12.7}$$

It follows that a necessary and sufficient condition for the disturbed population to persist in this pessimistic approach is $\xi < r$.

This suggests that a way to study the vulnerability or nonvulnerability of a system is to pose and answer the following question: what is the worst possible way for a prescribed set of disturbances to act on the dynamics of a system? One way to answer this question is to set up and analyse an accessory optimal control problem. This is a practical method for problems with a few species.

Let us suppose that the disturbance $u(t)$ acts in a different manner on the population dynamics. For example, let the disturbance be due to predation by a very efficient and self-regulating predator. In this case, we have

$$\dot{N} = (r/K)N(K - N) - u(t). \qquad (3.12.8)$$

We shall assume that $u(t)$ satisfies the inequality (3.12.6).

We can deduce the system's response to $u(t)$ by comparing model (3.12.8) with a model of a fishery which is subjected to a constant quota harvesting policy. The latter is discussed in Section 3.5. We conclude that system (3.12.8) is nonvulnerable for all time if

$$S(0) = \{N \mid N > (1/2)\,[K - \sqrt{(K^2 - 4\xi K/r)}]\},$$

$$Z(\infty) = \{N \mid N = 0\} \quad \text{and} \quad \xi < rK/4.$$

If $\xi > rK/4$, the population is rapidly driven to extinction.

Example 3.12.2. Let N_1 and N_2 be the densities of a prey and its predator respectively. Suppose a model of the prey—predator system which is subjected to continual disturbances is given by the equations:

$$\dot{N}_1 = N_1\,[22 - 0.1N_1 - 2.1N_2\,] + u_1(t)N_1, \qquad (3.12.9a)$$

$$\dot{N}_2 = N_2[-5 + 0.7N_1 - 0.2N_2] + u_2(t)N_2. \qquad (3.12.9b)$$

The nontrivial equilibrium of the undisturbed system is at (10,10). Let the disturbance functions, $u_1(t)$ and $u_2(t)$, be piecewise continuous and satisfy the a priori estimates:

$$-0.11 \leqslant u_i(t) \leqslant 0.11, \quad i = 1, 2. \qquad (3.12.10)$$

Let U denote this set of functions, $\{u_1(t), u_2(t)\}$.

A very good Liapunov function for the undisturbed system is

$$V(N) = N_1 - 10 - 10\ln(N_1/10) + 2.94[N_2 - 10 - 10\ln(N_2/10)]. \qquad (3.12.11)$$

Let $V_S = 0.05361$ and $V_Z = 0.23143$. The level surfaces, $V(N) = V_S$ and $V(N) = V_Z$, pass through the points (9,10) and (8,10) respectively. The level surface, $V(N) = V_Z$, is closest to the axes planes at the point (8,10). Let us define the sets, $S(0)$ and $Z(T)$, as follows:

$$S(0) = \{N \mid V(N) \leqslant 0.05361\} \qquad (3.12.12)$$

and

$$Z(T) = \{N \mid V(N) \geqslant 0.23143\}. \qquad (3.12.13)$$

We shall establish that (3.12.9) is nonvulnerable relative to the sets, $S(0)$, $Z(T)$ and U, during the time interval, $[0,T] = [0, 2.38]$.

The time derivative of $V(N)$ along solutions of (3.12.9) is

$$\dot{V}(N:u) = -0.1(N_1 - 10)^2 - 0.062(N_1 - 10)(N_2 - 10) - 0.294(N_2 - 10)^2 - u_1(N_1 - 10) - 2.94\, u_2(N_2 - 10). \quad (3.12.14)$$

By means of a minimizing algorithm (Fletcher, 1975), we find that the global maximum of $\dot{V}(N:u)$ for all $u \in U$ and for all $N \in \{N \mid 0.05361 \leqslant V(N) \leqslant 0.23143\}$ is 0.07439. The corresponding values of u and N are as follows:

$$u_1 = 0.11, \qquad u_2 = -0.11, \qquad N_1 = 10.085, \qquad N_2 = 9.641.$$

We find that in this case,

$$\max \dot{V} = 0.07437 < (V_Z - V_S)/T = 0.07471.$$

Thus, in spite of continual disturbances, any trajectory of (3.12.9) which begins inside the set, $\{N \mid V(N) \leqslant 0.05361\}$, will remain inside the set, $\{N \mid V(N) < 0.23143\}$. Hence the system is nonvulnerable relative to the prescribed conditions.

Suppose we want the trajectories of model (3.12.9) to remain indefinitely inside an acceptable region $S(0)$ if they initiate in it. We have to employ Theorem 3.12.2. The set $S(0)$ can be larger than that displayed in (3.12.12). Let the set

$$S(0) = \{N \mid V(N) \leqslant 0.14931\}. \quad (3.12.15)$$

The level surface, $V(N) = 0.14931$, passes through the point (8.37,10). Let G be any positive number that is larger than 0.14931 and let V_Z be a constant that is greater than G. Suppose it is undesirable for the state of the system to be in the set $Z(\infty)$, where

$$Z(\infty) = \{N \mid V(N) \geqslant V_Z\}.$$

We find that the global maximum of $\dot{V}(N:u)$ for all $u \in U$ and $N \in \{N \mid G \geqslant V(N) \geqslant 0.14931\}$ is -0.00035. This occurs for the following set of values:

$$u_1 = -0.11, \qquad u_2 = 0.11, \qquad N_1 = 9.5114, \qquad N_2 = 10.428$$

As $\dot{V}(N:u)$ is negative in the set $\{N \mid V(N) = 0.14931\}$, any solution of (3.10.9) which begins in $S(0)$ remains in it indefinitely. Hence the system is nonvulnerable.

In this example we have employed a Liapunov-like function whose global minimum occurs at the equilibrium of the undisturbed system. This property of the Liapunov-like function is unnecessary. We only require that the global minimum occurs at a point inside the set $S(0)$.

3.13. DISCRETE TIME MODELS: LOCAL STABILITY

There is an increasing interest in the use of difference equations for the modelling of populations (Van der Vaart, 1973; Innis, 1974 a; Beddington et al.,

1976; Hassell and Comins, 1976; Goh, 1976 b; May, 1976 a). A difference equation representation of a dynamic system is also called a discrete time model. A reason for this upsurge of interest in discrete time models is that they can provide realistic representations of ecosystems when differential equations fail to do so. Moreover, the numerical solution of a differential equation requires that the differential equation should be converted into a difference equation before the digital computer can churn out the solution.

Unfortunately the theory of nonlinear difference equations is very limited in comparison with the theory of nonlinear differential equations. Furthermore, even the simplest nonlinear difference equation can exhibit a very complex range of behavior (May, 1976 a). A solution of a discrete time model is just a sequence of points in the state space. This lack of continuity in the solutions of a difference equation makes any analysis of the solutions of a difference equation in the state space a difficult task.

The standard method for the analysis of the stability of a discrete time model examines the eigenvalues of a certain matrix computed at an equilibrium. Unfortunately this method can guarantee stability only relative to small perturbations of the initial state from an equilibrium. It is only a small step towards understanding the stability of a model ecosystem when it is subjected to realistic perturbations. For models of one or two species we often have graphical insight into a model. This turns a local stability analysis into something useful.

If a discrete time model represents a viable ecosystem, it must be stable relative to finite perturbations of the initial state from an equilibrium or it must have the property that solutions which begin in an acceptable region of the state space will remain in the same region if there are no further perturbations. One way to establish that a discrete time model is globally stable is by means of a Liapunov function. Generally the construction of a good Liapunov function requires considerable ingenuity and experimentation. Fortunately for a class of model ecosystem we can construct effective Liapunov functions by means of a biologically meaningful principle. This principle (Goh, 1977 a) states that a viable population must dissipate energy at high densities and it must absorb energy at low densities. More precisely, such energy exchanges are relative to the continual energy input which is necessary to maintain the population at its equilibrium. In a multi-species system a simple way to construct a Liapunov function is to use a positive linear combination of functions which serve as effective Liapunov functions for a single-species population. This is by no means the only way to construct an effective Liapunov function for a multispecies model.

Here we shall study conditions for local stability in population models. At time t let $N_i(t)$ denote the density of the ith species in an interaction among m species. For convenience we shall use N_i in place of $N_i(t)$. In order to distinguish $N_i(t+1)$ from $N_i(t)$ we shall retain the argument of $N_i(t+1)$. Suppose an interaction among m species can be represented by the set of nonlinear difference equations,

$$N_i(t+1) = G_i[N_1, N_2, \ldots, N_m], \tag{3.13.1}$$

where $i = 1, 2, \ldots, m$.

Let $(N_1^\star, N_2^\star, \ldots, N_m^\star)$ be a positive equilibrium of this model. By definition, we have

$$N_1^\star > 0, N_2^\star > 0, \ldots, N_m^\star > 0 \tag{3.13.2}$$

and

$$G_1(N^\star) = N_1^\star, G_2(N^\star) = N_2^\star, \ldots, G_m(N^\star) = N_m^\star. \tag{3.13.3}$$

The equilibrium $N^\star$ is locally stable if every solution of (3.13.1) which begins near $N^\star$ moves into a decreasing neighbourhood of $N^\star$ as t increases. Let

$$X_i = N_i - N_i^\star \tag{3.13.4}$$

for $i = 1, 2, \ldots, m$. Substitute these equations into (3.13.1) and ignore second and higher order terms in X; we get

$$X_i(t+1) = \sum_{j=1}^{m} \frac{\partial G_i}{\partial N_j} X_j \tag{3.13.5}$$

where $i = 1, 2, \ldots, m$ and the partial derivatives are computed at $N^\star$. In matrix notation this system of equation has the form,

$$X(t+1) = AX(t) \tag{3.13.6}$$

where $(A_{ij}) = (\partial G_i/\partial N_j)$.

Theorem 3.13.1. *Model (3.13.1) has a locally stable equilibrium at $N^\star$ if the magnitudes of all the eigenvalues of the matrix A are less than one.*

Proof. Let us guess that a solution of (3.13.6) is of the form,

$$X(t) = \lambda^t Y, \tag{3.13.7}$$

where λ is a constant and Y is a constant vector. This is a solution if

$$\lambda^{t+1} Y = A\lambda^t Y. \tag{3.13.8}$$

This condition implies that

$$(A - \lambda I)Y = 0, \tag{3.13.9}$$

where I is an identity matrix. Hence $\lambda^t Y$ is a solution of (3.13.6) if λ is an eigenvalue of A and Y is the corresponding eigenvector.

Among ecological models it is extremely rare for the matrix A to have repeated eigenvalues. As (3.13.6) is a linear system, any linear combination of a set of solutions is a solution of the system. Hence, if $\lambda_1, \lambda_2, \ldots, \lambda_m$ are distinct eigenvalues and $Y_1, Y_2, \ldots, Y_m$ are corresponding eigenvectors, the general solution of (3.13.6) is

$$X(t) = \sum_{i=1}^{m} c_i(\lambda_i)^t Y_i, \tag{3.13.10}$$

where $c_1, c_2, \ldots, c_m$ are arbitrary constants.

In general λ is a complex number which can be rewritten in the form,

$$\lambda_j = R_j\ (\cos\theta_j + i\sin\theta_j). \tag{3.13.11}$$

where $i^2 = -1$ and R_j is the magnitude of λ_j. It follows that

$$\lambda_j^t = R_j^t[\cos(t\theta_j) + \mathrm{i}\sin(t\theta_j)]. \tag{3.13.12}$$

Clearly, if $R_j < 1$, $R_j^t \to 0$ as $t \to \infty$. Hence if the magnitudes of all the eigenvalues of A are less than one, all the solutions of (3.13.6) tend to 0 as $t \to \infty$. It follows the equilibrium $N^\star$ of (3.13.1) is locally stable.

Very often we have population models in the form,

$$N_i(t+1) = N_i F_i(N_1, N_2, \ldots, N_m) \tag{3.13.13}$$

where $i = 1, 2, \ldots, m$. This reformulation of a population model provides some useful insights into a population model.

At a positive equilibrium $N^\star$, we have

$$F_1(N^\star) = 1, F_2(N^\star) = 1, \ldots, F_m(N^\star) = 1. \tag{3.13.14}$$

Let δ_{ij} be Kronecker deltas. By definition,

$$\delta_{ii} = 1, \quad \delta_{ij} = 0 \quad \text{if } i \neq j. \tag{3.13.15}$$

Theorem 3.13.2. *Model (3.13.13) has a locally stable equilibrium at $N^\star$ if the magnitudes of all the eigenvalues of the matrix,*

$$A = (\delta_{ij} + N_i^\star \partial F_i(N^\star)/\partial N_j), \tag{3.13.16}$$

are less than one.

Proof. Substitute $X_i = N_i - N_i^\star$ into (3.13.13) and ignore second and higher order terms in X; we get

$$X_i(t+1) = X_i + \sum_{j=1}^{m} N_i^\star \frac{\partial F_i}{\partial N_j} X_j \tag{3.13.17}$$

where $i = 1, 2, \ldots, m$. Compare (3.13.17) with (3.13.6); the result follows immediately.

A model of a single species population is

$$N(t+1) = NF(N). \tag{3.13.18}$$

We have a locally stable positive equilibrium at $N^\star$ if $N^\star > 0$. $F(N^\star) = 1$ and

$$|1 + N^\star \partial F/\partial N| < 1 \Rightarrow -2 < N^\star \partial F/\partial N < 0. \tag{3.13.19}$$

Example 3.13.1. Let N be the density of a single species population in the tth generation and $N(t+1)$ denote the density in the $(t+1)$th generation. Suppose the population has no overlapping generations. A popular model (Ricker, 1954) of this type of population is

$$N(t+1) = N\exp[r(1-N/K)], \tag{3.13.20}$$

where r is a measure of the per capita birth rate, and K is the "carrying capacity" of the environment.

This model has a positive equilibrium at $N = K$. For $N = K$, $\partial F/\partial N = (-r/K)$. Hence the positive equilibrium is locally stable if

$$-2 < K(-r/K) < 0 \Rightarrow 0 < r < 2. \tag{3.13.21}$$

It is shown in the next chapter that the model is globally stable if r satisfies this inequality.

Consider a population with overlapping generations and in which the adults breed once in each generation. Let N denote the number of female adults at time t and let S be the fraction of female adults which survive in one generation. A model of such a population is,

$$N(t+1) = SN + N\exp[r(1-N/K)]. \tag{3.13.22}$$

In this case the parameters, r and K, do not have the same biological interpretations as before. This model has an equilibrium at $N^{\star} = K[1 - (1/r)\ln(1-S)]$. This implies that $\exp[r(1-N^{\star}/K)] = (1-S)$. If r is negative, $N^{\star}$ is positive only if

$$r < \ln(1-S). \tag{3.13.23}$$

By means of (3.13.19), the positive equilibrium is locally stable if

$$-2 < (1-S)[-r + \ln(1-S)] < 0 \Rightarrow \ln(1-S) < r$$

$$\text{and } r < 2/(1-S) + \ln(1-S). \tag{3.13.24}$$

Clearly, (3.13.23) and (3.13.24) are incompatible and hence r must be positive. The function $2/(1-S) + \ln(1-S)$ increases monotonically as S increases from zero to one. When $S = 0.9$, we require that r satisfies the weaker inequality, $0 < r < 17.7$. It follows that model (3.13.22) is more stable than model (3.13.20) if S is nonzero. This increase in stability is due to the fact that the adult population acts like a reservoir which damps out fluctuations in the number of maturing females which are being recruited into the adult population.

Example 3.13.2. A flexible model (Hassell, 1975) of a single-species population with nonoverlapping generations is

$$N(t+1) = \lambda N/(1+aN)^{b} \tag{3.13.25}$$

where λ, a and b are positive constants.

It has a positive equilibrium at

$$N = (\lambda^{1/b} - 1)/a.$$

Condition (3.13.19) implies that this is locally stable if

$$0 < b(1 - \lambda^{-1/b}) < 2.$$

Hassell (1975) established that this model has stable limit cycles when this inequality is violated.

For ecosystems with two or three species of a single-species population with a time delay, the Schur—Cohn criterion is a very useful test of local stability for varying parameter values. For systems with four or more species or a single-species population with a long time delay, the simplest way to state the Schur—Cohn criterion is in terms of inners (Jury, 1974).

In using the Schur—Cohn condition we must first compute the eigenvalue equation of the matrix displayed in (3.13.6) or (3.13.16). Let the eigenvalue equation for a two-species system be

$$\lambda^2 + a_1\lambda + a_2 = 0. \qquad (3.13.26)$$

The magnitudes of all the eigenvalues associated with this equation are less than one if and only if

$$|a_2| < 1 \quad \text{and} \quad |a_1| < |1 + a_2|.$$

Let the eigenvalue equation of a population model be

$$\lambda^3 + a_1\lambda^2 + a_2\lambda + a_3 = 0. \qquad (3.13.27)$$

The magnitudes of all the eigenvalues associated with this equation are less than one if and only if

$$1 - a_3^2 > |a_1 a_3 - a_2| \quad \text{and} \quad |1 + a_2| > |a_1 + a_3| \qquad (3.13.28)$$

Note the first inequality implies $|a_3| < 1$.

Example 13.13.3. We shall use the Schur—Cohn criterion to study how two parameters affect the conditions for local stability in a model of two computing species (Hassell and Comins, 1976). Let us consider the model,

$$N_1(t+1) = N_1(t)/[0.3 + 0.2\,N_1(t) + aN_2(t)]^s,$$

$$N_2(t+1) = N_2(t)/[0.4 + 0.1N_1(t) + 0.5N_2(t)]^2$$

where a and s are positive parameters.

The nontrivial equilibrium $(N_1^\star, N_2^\star)$ is a solution of the system of equations:

$$0.3 + 0.2\,N_1 + aN_2 = 1,$$

$$0.4 + 0.1N_1 + 0.5N_2 = 1 \Rightarrow N_1^\star = (3.5 - 6a),\ N_2^\star = 0.5/(1 - a).$$

This equilibrium is positive if $a < 35/60$.

The positive equilibrium is locally stable if and only if the magnitudes of all the eigenvalues of the matrix

$$\begin{pmatrix} 1-0.2sN_1^\star & -sa\,N_1^\star \\ -0.2N_2^\star & 1-N_2^\star \end{pmatrix}$$

are less than one. Its eigenvalue equation is

$$\lambda^2-(1+Z)\lambda+Z+0.25(1-a)N_1^\star N_2^\star=0$$

where $Z=1-0.2sN_1^\star-N_2^\star$.

The Schur—Cohn stability criterion requires that

$$|Z+0.2s(1-a)\,N_1^\star N_2^\star|<1, \tag{3.13.29}$$

$$|1+Z|<|1+Z+0.2s(1-a)N_1^\star N_2^\star|. \tag{3.13.30}$$

Condition (3.13.30) implies that

$$a<1.$$

This condition is equivalent to the condition, (0.2) (0.5) $-$ 0.1$a<0$. It means that the interspecific interactions are weaker than the intraspecific interactions.

Conditions (3.3.29) and (3.13.30) involve nonlinear functions of the parameters, a and s. We shall examine only these inequalities when $a=0.5$. Condition (3.13.29) implies that

$$|0.1s|<1 \Rightarrow |s|<10.$$

Condition (3.13.30) implies that

$$|1-0.2s|<|1-0.1s| \Rightarrow 0<s<20/3.$$

This completes our analysis of this example.

Example 3.13.4. Clark (1976 b) has proposed and studied an interesting and realistic model of a whale population. Clark's model is also applicable to other single-species populations which have a time delay because of the time it takes a female animal or plant to mature before it can begin to reproduce.

At time t let $N(t)$ be the number of adult and productive females in a population. Consider the model of a single-species population,

$$N(t+1)=SN(t)+G[N(t-2)], \tag{3.13.31}$$

where S is the fraction of female adults which survive per unit time and $G[N(t-2)]$ is the number of maturing females which were born at time $t-2$.

In particular, we shall examine the model,

$$N(t+1)=SN(t)+N(t-2)\exp[r(1-N(t-2)/K)]. \tag{3.13.32}$$

The parameters, r and K, in this model do not have the same biological interpretations as those in model (3.13.20). This model has an equilibrium at

$$N^{\star} = K[1 - (1/r)\ln(1 - S)].$$

Note that $N^{\star}$ decreases as r increases because $\ln(1 - S) < 0$.

To analyse the stability properties of discrete time models in a systematic manner, we convert (3.13.32) into a system of dynamical equations. We get

$$N_1(t + 1) = N_2(t),$$

$$N_2(t + 1) = N_3(t),$$

$$N_3(t + 1) = SN_3(t) + N_1(t)\exp[r(1 - N_1(t)/K)]. \qquad (3.13.33)$$

This new model has an equilibrium at the point, $(N_1, N_2, N_3) = (N^{\star}, N^{\star}, N^{\star})$.

For convenience, let

$$a = (1 - S)[1 - r + \ln(1 - S)].$$

If the eigenvalues of all the eigenvalues of the matrix

$$\left(\frac{\partial G_i}{\partial N_j}\right) = \begin{pmatrix} 0 & 1 & 0 \\ 0 & 0 & 1 \\ a & 0 & S \end{pmatrix}$$

are less than one, model (3.13.33) has a locally stable equilibrium at $(N^{\star}, N^{\star}, N^{\star})$. The eigenvalue equation of this matrix is

$$\lambda^3 - S\lambda^2 - a = 0. \qquad (3.13.34)$$

The Schur—Cohn criterion requires that

$$1 - a^2 > |Sa|, \qquad 1 > |S + a|. \qquad (3.13.35)$$

These conditions involve nonlinear functions of r and S. It is difficult to examine them analytically. If $S = 0$, these conditions require that r satisfies the inequality, $2 > r > 0$. This is as expected because model (3.13.32) reduces to model (3.13.20) when $S = 0$. If $S = 0.5$, these conditions require that the parameter r satisfies the inequality, $2.87 > r > 0$. If $S = 0.9$, these conditions require that $5.17 > r > 0$. This suggests that the stability of the model increases as S increases from zero to one. As in model (3.13.22) this increase in stability is due to a larger adult population acting more effectively to damp out the fluctuations in the number of the maturing females. In model (2.13.22) when $S = 0.9$, the parameter r must satisfy the inequality $17.7 > r > 0$. This suggests that the time delay in the model asserts a destabilising influence in this model. It is interesting to note that the parameter K does not affect the stability of the system as long as it is positive.

If all the parameters in a given model are specified numerically the simplest way of examining the stability of an equilibrium is to compute the eigenvalues of the linearized dynamics. There are many numerical algorithms available for the computation of the eigenvalues of a given matrix.

To examine the stability of an equilibrium in a nonlinear model, we have to study the stability of the linear system,

$$X_i(t+1) = \sum_{j=1}^{m} a_{ij} X_j(t), \tag{3.13.36}$$

where $i = 1, 2, \ldots, m$. Let $A = (a_{ij})$. Note that there is no restriction placed on the signs of $X_1, X_2, \ldots, X_m$ because they are displacements from an equilibrium. A suitable Liapunov function for this system is

$$V(X) = X^T(t)PX(t), \tag{3.13.37}$$

where P is a positive definite symmetric matrix. By definition,

$$\Delta V = V[X(t+1)] - V[X(t)]. \tag{3.13.38}$$

For model (3.13.36) we have

$$\Delta V = X^T(t)[A^T PA - P]X(t).$$

Hence, if there exists a positive definite symmetric matrix P such that $A^T PA - P$ is negative definite, model (3.13.36) is stable. It can be shown that this condition is also necessary. Clearly a sufficient condition for model (3.13.36) to be stable is that there exists a positive diagonal matrix D such that $A^T DA - D$ is negative definite. This condition may be useful for multi-species ecosystems in which each species sustains density-dependent mortalities due to intraspecific interactions and in which the interspecific interactions are relatively weak.

3.14. FINITE AND GLOBAL STABILITY IN DISCRETE TIME MODELS

The literature on a finite region of attraction and global stability in discrete time models is very limited. The number of theory papers exceeds those which describe interesting applications of the theory to models of real systems. The papers by Kalman and Bertram (1960 b), Diamond (1976 a), LaSalle (1976), Goh (1977 c) and Fisher and Goh (1977) contain good descriptions of the theory and interesting applications.

From experience it is very difficult to construct effective Liapunov functions for discrete time models. Fortunately for population models there is a fundamental energy principle which provides useful guidelines for the construction of effective Liapunov functions. This biologically meaningful principle states that a viable population must absorb energy at low densities and it must dissipate energy at high densities. In multispecies interactions a positive linear combination of functions, which are good Liapunov functions for single-species models, is a good candidate to act as a Liapunov function. Another candidate is obtained by forming a positive definite quadratic expression of functions which are effective Liapunov functions for single-species models. Finally, we note that in a nonlinear model we can construct a new candidate to act as a Liapunov function by adding two candidate functions. It is plausible that we can use this process to patch up a good Liapunov function.

Let $N_1, N_2, \ldots, N_m$ be population densities at time t in an ecosystem of m species. Consider the model,

$$N_i(t+1) = G_i[N_1, N_2, \ldots, N_m], \tag{3.14.1}$$

where $i = 1, 2, \ldots, m$ and $G_1, G_2, \ldots, G_m$ are continuous functions of the population densities. We shall assume that this model has a positive equilibrium at $N^\star$. It follows that $N_i^\star > 0$ and $G_i(N^\star) = N_i^\star$ for $i = 1, 2, \ldots, m$.

Let Ω be an open region in the positive orthant and let $N^\star \in \Omega$. The function $V(N)$ is a suitable Liapunov function of model (3.14.1) in the region Ω if it has the following properties:

(i) $V(N) > 0$ for all $N \in \Omega$ and $N \neq N^\star$ and $V(N^\star) = 0$;

(ii) In Ω, the level surface $V(N) = K$ for each positive value of K, is a closed surface and $V(N)$ has a unique minimum at $N^\star$;

(iii) The function,

$$\Delta V(N) = V[G(N)] - V(N). \tag{3.14.2}$$

is nonpositive for all $N \in \Omega$.

Condition (ii) compels us to define Ω in terms of a level surface of $V(N)$. We let $\Omega = \{N \mid V(N) < L\}$ where L is a positive number. Note that condition (ii) is satisfied if $V(N) \to \infty$ as $N_i \to 0+$ or $N_i \to +$ for $i = 1, 2, \ldots, m$ and $V(N)$ has a unique minimum in Ω. Condition (iii) means that ΔV is negative semidefinite in Ω.

Theorem 3.14.1. *The region Ω is a region of attraction of model (3.14.1) if there exists a Liapunov function $V(N)$ in Ω and $\Delta V(N)$ is negative definite.*

Proof. The solution of (3.14.1) which begins at the point $N(0) \in \Omega$ consists of a sequence of points, $N(0), N(1), N(2), \ldots$. Let $S(t) = V[N(t)]$. The solution generates a scalar sequence $\{S(t)\}$ which is bounded below by zero. By assumption $\Delta V(N)$ is negative definite in Ω. This implies that the sequence $\{S(t)\}$ is monotonic decreasing as t increases. It follows that $S(t)$ converges to a limit.

Suppose $S(t) \to l$ as $t \to \infty$, where $l > 0$. The assumption that $\Delta V(N)$ is continuous and is negative definite in Ω implies that it has a nonzero maximum for all $N \in \{N \mid l \leqslant V(N) \leqslant V[N(0)]\}$. Denote this maximum by $-\theta$ where θ is a positive number. We have

$$V[N(t)] = \sum_{s=0}^{t-1} \Delta V[N(s)] + V[N(0)] \leqslant -t\theta + V[N(0)]. \tag{3.14.3}$$

This implies that $V[N(t)] \to -\infty$ as $t \to \infty$. This is impossible as $V(N) \geqslant 0$ for all $N \in \Omega$. The solution which begins at $N(0)$ cannot leave the region Ω. Hence $l = 0$. This establishes that if there exists a Liapunov function in the region Ω, all solutions which begin in it tend to $N^\star$.

Usually we have to determine the region Ω numerically. We first construct a function $V(N)$ which has properties (i) and (ii). The region Ω is determined by finding a region as large as possible in which ΔV is negative definite. This can be carried out by maximizing ΔV for all N which satisfy the condition, $V(N) = K$ where K is a positive constant. The value of K is gradually increased until the corresponding maximum value of ΔV becomes positive. From this process we can pick a positive number $K^\star$ such that ΔV is negative definite in the region, $\Omega = \{N \mid V(N) < K^\star\}$.

Example 3.14.1. At time t, let N be the density of a single-species population with nonoverlapping generations. A model of this type of population is

$$N(t+1) = 2N(1-N). \tag{3.14.4}$$

This model is valid only for values of N satisfying the inequality $1 > N \geqslant 0$. For $N > 1$, $N(t+1)$ is negative.

It has a positive equilibrium at $N^\star = 0.5$. A suitable Liapunov function is

$$V(N) = N - 2N^\star + N^{\star 2}/N. \tag{3.14.5}$$

Clearly $V(N) \to \infty$ if $N \to 0+$ or as $N \to \infty$. Let $F(N) = 2(1-N)$. From (3.14.4) and (3.14.5) we get

$$\Delta V(N) = (N/F)(1-F)(F - N^{\star 2}/N). \tag{3.14.6}$$

When $N/N^\star$ is large, the dominant term in $V(N)$ is N. In this case, $V(N)$ is none other than the energy embodied in the standing biomass of the population. If ΔV is negative it means that the population is dissipating energy. When $N/N^\star$ is small the dominant term in $V(N)$ is $N^{\star 2}/N$. Thus, $V(N)$ is inversely proportional to the energy embodied in the standing biomass of the population. Hence if ΔV is negative it means that the population is absorbing energy from elsewhere.

For model (3.14.3), ΔV is negative definite if $N > 0$ and $2(1-N) < 1/(4N^2)$. This implies that $0.809 > N > 0$. We have $V(0.809) = 0.191$. Hence an estimate of the region of attraction is $\Omega = \{N \mid V(N) < 0.191\}$. We can show that $\Omega = \{N \mid 0.272 < N < 0.809\}$. The largest region of attraction in model (3.14.4) is $\{N \mid 1 > N > 0\}$. Clearly the Liapunov function method gives only a rough estimate of the region of attraction.

Example 3.14.2. Let N_1 and N_2 be state variables at time t. Consider the model,

$$N_1(t+1) = \exp[(\ln N_1)^2 - (\ln N_2)^2], \tag{3.14.7a}$$

$$N_2(t+1) = \exp[2\ln N_1 \ln N_2]. \tag{3.14.7b}$$

This model has a positive equilibrium at $N^\star = (1,1)$.

A suitable Liapunov function is

$$V(N) = (\ln N_1)^2 + (\ln N_2)^2. \tag{3.14.8}$$

Note that $V(N) \to \infty$ as $N_i \to 0$ or as $N_i \to \infty$ for $i = 1, 2$, Moreover, $V(N)$ has a unique minimum at $N^\star$.

Along solutions of (3.14.7) we have

$$\Delta V(N) = [(\ln N_1)^2 + (\ln N_2)^2]\,[(\ln N_1)^2 + (\ln N_2)^2 - 1]. \tag{3.14.9}$$

It follows that ΔV is negative definite in the region, $\Omega = \{N \mid V(N) < 1\}$. Hence Ω is an estimate of the region of attraction of $N^\star$ in model (3.14.7). In this specially constructed example Ω is a very good estimate of the largest region of attraction of $N^\star$.

In Sections 3.3 and 3.9, we find that in continuous time models it is useful to relax the condition that $\dot{V}$ is negative definite in Ω, a region of attraction. We can similarly relax the condition that ΔV is negative definite in Ω, a region of attraction in a discrete time model.

Let Ω be an open subregion of the positive orthant. Let $N^\star$ be a positive equilibrium of model (3.14.1). We shall assume that $N^\star \in \Omega$. Let the function $V(N)$ have the following properties: (i) $V(N) > 0$ for all $N \in \Omega$ and $N \neq N^\star$ and $V(0) = 0$; (ii) In Ω, the level surface, $V(N) = K$ for each positive value of K, is a closed surface and $V(N)$ has a unique minimum at $N^\star$. Let the region $\Omega = \{N \mid V(N) < L\}$, where L is a positive number. Under these assumptions the following theorem describes the crucial conditions for the region of attraction of $N^\star$, a positive equilibrium of model (3.14.1).

Theorem 3.14.2. *The region Ω is a region of attraction of model (3.14.1) if in Ω, $\Delta V(N) \leqslant 0$ and $\Delta V(N)$ does not vanish identically along a solution of the model beside the trivial solution, $N = N^\star$.*

A proof of this theorem is given in a paper by LaSalle (1976). We shall apply this theorem to a specially constructed example.

Example 3.14.3. Let N_1 and N_2 be the state variables at time t. Let the region $\Omega = \{N \mid N_1 > 0, N_2 > 0\}$. Consider the model,

$$N_1(t+1) = N_1^{a_{12}}, \tag{3.14.10a}$$

$$N_2(t+1) = N_1^{a_{21}} N_2^{a_{22}} \tag{3.14.10b}$$

where $a_{12} = \sqrt{3}/2$, $a_{21} = 1/\sqrt{2}$ and $a_{22} = 1/\sqrt{8}$. This model has a positive equilibrium at $N^\star = (1,1)$.

A suitable Liapunov function is

$$V(N) = (\ln N_1)^2 + (\ln N_2)^2.$$

Along solutions of (3.14.10) we have

$$\Delta V = -[(1/\sqrt{2})\ln N_1 - (1/\sqrt{8})\ln N_2]^2 \leqslant 0.$$

The function $\Delta V(N)$ does not vanish identically along a solution of (3.14.10) beside the trivial solution, $N = (1,1)$. Hence model (3.14.10) is stable in the region Ω. This means that (3.14.10) is globally stable.

Let $x_1(t) = \ln N_1$ and $x_2(t) = \ln N_2$. We find that model (3.4.10) is equivalent to the model,

$$x_1(t+1) = a_{12}x_2, \tag{3.14.11a}$$

$$x_2(t+1) = a_{21}x_1 + a_{22}x_2, \tag{3.14.11b}$$

in which there is no restriction on the signs of the variables, x_1 and x_2. We can use the Shur–Cohn stability criterion to show that this model is stable.

Let $N_1, N_2, \ldots, N_m$ be population densities at time t. Let Γ denote the positive orthant $\{N \mid N_i > 0, i = 1, 2, \ldots, m\}$. Assume that the model,

$$N_i(t+1) = G_i[N_1, N_2, \ldots, N_m] \tag{3.14.12}$$

where $i = 1, 2, \ldots, m$, has a positive equilibrium at $N^\star$. The function $V(N)$ is a Liapunov function in Γ if it has the following properties:
(i) $V(N) > 0$ for all $N \in \Gamma$ and $N \neq N^\star$ and $V(N^\star) = 0$;
(ii) $V(N) \to \infty$ as $N_i \to 0+$ or as $N_i \to \infty$ for each $i = 1, 2, \ldots, m$ and $V(N)$ has a unique minimum in Γ;
(iii) the function,

$$\Delta V(N) = V[G(N)] - V(N), \tag{3.14.13}$$

is negative for all $N \in \Gamma$ and $N \neq N^\star$.

Theorem 3.14.3. *Model (3.14.12) is globally stable if there exists a Liapunov function $V(N)$ in the positive region.*

The proof of this theorem is similar to that for Theorem 3.14.1. Note that the function $\Delta V(N)$ displayed in (3.14.13) is computed along solutions of model (3.14.12). Condition (iii) means that $\Delta V(N)$ displayed (3.14.13) is computed along solutions of model (3.14.12). Condition (iii) means that $\Delta V(N)$ is negative definite in the feasible region, Γ.

The next theorem gives more general conditions for global stability.

Theorem 3.14.4. *Model (3.14.12) is globally stable if* (a) *there exists a function $V(N)$ which has the properties (i) and (ii) as stated above,* (b) *the condition,*

$$\Delta V(N) = V[G(N)] - V(N) \leqslant 0, \tag{3.14.14}$$

is satisfied for all $N \in \Gamma$ and (c) *$\Delta V(N)$ does not vanish identically along a solution of (3.14.12) beside the trivial solution, $N = N^\star$.*

Example 3.14.4. Consider a single species population with nonoverlapping generations. Let N be the adult population density at time t. We shall establish a simple and robust set of conditions which are sufficient for global stability in the general model,

$$N(t+1) = G(N). \tag{3.14.15}$$

This model is globally stable if (i) there exists a positive equilibrium at $N^\star$, (ii) for $N^\star > N > 0, N^{\star 2}/N > G(N) > N$, (iii) for $N > N^\star, N > G(N) > N^{\star 2}/N$.

Condition (ii) means that if the density is below the equilibrium level in a given generation, the population density is larger in the next generation. In other words, the population does not sustain an Allee effect. Condition (iii) implies that if the density is higher than the equilibrium level in a given generation, the population density is reduced in the next generation because of intraspecific competition.

A suitable Liapunov function is

$$V(N) = [\ln(N/N^\star)]^2. \tag{3.14.16}$$

Along solutions of model (3.14.15) we have

$$\begin{aligned} \Delta V(N) &= [\ln(G/N^\star)]^2 - [\ln(N/N^\star)]^2 \\ &= [\ln(G/N)]\,[\ln(GN/N^{\star 2})]. \end{aligned}$$

For $N^\star > N > 0$, condition (ii) implies that $\ln(G/N) > 0$ and $\ln(GN/N^{\star 2}) < 0$. Hence $\Delta V < 0$ for $N^\star > N > 0$. For $N > N^\star$, condition (iii) implies that $\ln(G/N) < 0$ and $\ln(GN/N^{\star 2}) > 0$. Hence $\Delta V < 0$ for $N > N^\star$. It follows that ΔV is negative definite. We conclude that model (3.14.16) is globally stable. This completes our analysis of the example.

We shall now state a similar set of sufficient conditions for global stability in a multispecies model. Let P be a positive definite symmetric matrix. A Liapunov function for model (3.14.12) is

$$V(N) = \sum_{ij}^{m} [\ln(N_i/N_i^\star)P_{ij}\ln(N_j/N_j^\star)], \tag{3.14.17}$$

where $N^\star$ is a positive equilibrium.

Theorem 3.14.5. *Model (3.14.12) is globally stable if there exists a positive definite matrix P such that the function,*

$$\Delta V = \sum_{ij}^{m} [\ln(G_i/N_i^\star)P_{ij}\ln(G_j/N_j^\star) - \ln(N_i/N_i^\star)P_{ij}\ln(N_i/N_j^\star)],$$

is negative definite in the positive orthant.

There are two problems which have to be solved in order to use this theorem. Firstly, there is no known systematic method of choosing a suitable positive definite and symmetric matrix P, except in the case of linear or related models. Secondly, for $m > 2$, it is very difficult to establish analytically or numerically that a given function is negative definite in the positive orthant. The solution of this problem would profit from advances in methods for computing the global maximum of a given function. This follows because the condition, ΔV is negative definite, is equivalent to the condition that the global maximum of ΔV is unique and is equal to zero. In spite of these problems, given a model of a multispecies system we should always check whether or not Theorem 3.14.5 is applicable.

Note that the function displayed in (3.14.17) is none other than the quadratic function $X^T PX$ under the transformation, $X_i = \ln(N_i/N_i^\star)$ where $i = 1, 2, \ldots, m$. It follows that in the positive orthant $V(N)$ has a global minimum equal to zero at $N^\star$ if P is positive definite. This transformation can produce other Liapunov functions for population models form Liapunov functions $V(X)$ for the model $\dot{X}_i = f_i(X)$, where X_i is not restricted in its sign.

Example 3.14.5. Let N_1 and N_2 be two positive state variables at time t. Consider the model,

$$N_1(t+1) = N_1^{a_{11}} N_2^{a_{12}}, \tag{3.14.18a}$$

$$N_2(t+1) = N_1^{a_{21}} N_2^{a_{22}}, \tag{3.14.18b}$$

where a_{11}, a_{12}, a_{21} and a_{22} are constants. This model has a positive equilibrium at $N^\star = (1,1)$.

Let P be a constant positive definite and symmetric matrix. A suitable Liapunov function for model (3.14.18) is

$$V(N) = P_{11}(\ln N_1)^2 + 2P_{12}(\ln N_1)(\ln N_2) + P_{22}(\ln N_2)^2.$$

Along solutions of model (3.14.18) we have

$$\Delta V = Y^T [A^T PA - P] Y,$$

where $A = (a_{ij})$ and $Y^T = (\ln N_1, \ln N_2)$. Clearly if there exists a positive definite matrix P such that $A^T PA - P$ is negative definite, then $\Delta V(N) < 0$ in the positive orthant and where $N \neq N^\star$.

Example 3.14.6. At time t, let $N(t)$ be the density of the adult population in a single-species population. Suppose it takes 2 years for a member of the population to become an adult and intraspecific competition takes place during the breeding season. A model (Clark, 1976 b) of this type of population is

$$N(t+1) = SN(t) + N(t-1)\exp[r(1 - N(t-1)/K)] \tag{3.14.19}$$

where S is the survival coefficient of the adult population, and r and K are positive parameters.

Let $N_1(t) = N(t-1)$ and $N_2(t) = N(t)$. The above model becomes

$$N_1(t+1) = N_2(t), \tag{3.14.20a}$$

$$N_2(t+1) = SN_2(t) + N_1(t)\exp[r(1 - N_1(t)/K)]. \tag{3.14.20b}$$

Suppose $S = 0.9$, $r = 1.5$ and $K = 0.3952$. Model (3.14.20) has a positive equilibrium at $N^\star = (1,1)$.

By trial and error, we find that a suitable Liapunov function for this model is

$$V(N) = (\ln N_1)^2 - 2(\ln N_1)(\ln N_2) + 3(\ln N_2)^2.$$

We compute the function $\Delta V(N)$ for model (3.14.2) at a large number of points and level surfaces, $\Delta V(N) = d$ where the values of d are varied over a wide range. Some typical members of these level surfaces are displayed in Fig. 3.14.1. From this exercise, we conclude that ΔV is negative definite in the positive region. Hence model (3.14.20) is globally stable in the positive region.

A good candidate to act as a Liapunov function in a single species model is

$$V(N) = (N^s - N^{\star s})/s + N^{\star p+s}/(pN^p) - N^{\star s}/p, \tag{3.14.21}$$

where s and p are positive numbers and $N^\star$ is the positive equilibrium in the model. For example, if $p = s = 1$, we can use this Liapunov function to prove that the model, $N(t+1) = G(N)$, is globally stable if: (i) it has a positive equilibrium at $N^\star$; (ii) for $N^\star > N > 0$, $G(N)$ satisfies the inequality, $N^{\star 2}/N > G(N) > N$; and (iii) for $N > N^\star$, $G(N)$ satisfies the inequality, $N > G(N) > N^{\star 2}/N$.

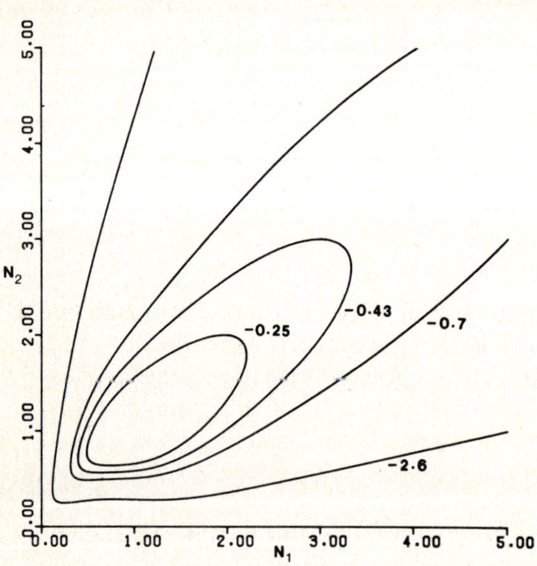

Fig. 3.14.1. Level surfaces of the $\Delta V(N)$ function for model (3.14.20).

Therefore in this case the function (3.14.21) is as effective as the function (3.14.16). From (3.14.21) we deduce that a good candidate to act as a Liapunov function for a model of a multispecies system is

$$V(N) = \sum_{i=1}^{m} c_i\left[\left(N_i^{s_i} - N^{\star s_i}\right)\Big/ s_i + N^{\star p_i + s_i}\Big/\left(p_i N^{p_i}\right) - N^{\star s_i}\big/p_i\right], \qquad (3.14.22)$$

where s_i, p_i and c_i are positive numbers for $i = 1, 2, \ldots, m$.

Another good candidate to act as a Liapunov function for a single-species model is

$$V(N) = (N^s - N^{\star s})/s - N^{\star s}\ln(N/N^\star), \qquad (3.14.23)$$

where s is a positive number. Along solutions of the single-species model, $N(t+1) = NF(N)$, we have

$$\Delta V(N) = N^s(F^s - 1)/s - N^{\star s}\ln F. \qquad (3.14.24)$$

In a viable single-species population when $N/N^\star$ is very large, F must be much smaller than one. Otherwise, $NF(N)$ would be large. This situation is unlikely to persist in a viable single-species population. If $F \ll 1$ when $N/N^\star$ is large, the dominant term in $\Delta V(N)$ is $-N^s/s$. Thus ΔV should be negative when $N/N^\star$ is large. In a viable single-species population which does not sustain an Allee's effect (see Odum, 1971), $F(N)$ should be larger than one when $N/N^\star$ is small. Then the dominant term in $\Delta V(N)$ is $-N^{\star s}\ln F$. In brief, $\Delta V(N)$ should be negative when the density of a population is small or large. This means that the population is absorbing energy at low densities and dissipating energy at high densities. These biological considerations suggest that the function $V(N)$ displayed in (3.14.23) should be a very good candidate to act as a Liapunov function.

Consider the multispecies model,

$$N_i(t+1) = N_i F_i(N_1, N_2, \ldots, N_m), \qquad (3.14.25)$$

where $i = 1, 2, \ldots, m$. A very good candidate to act as a Liapunov function for this model is

$$V(N) = \sum_{i=1}^{m} c_i\left[\left(N_i^{s_i} - N_i^{\star s_i}\right)\Big/ s_i - N_i^{\star s_i}\ln(N_i/N_i^\star)\right] \qquad (3.14.26)$$

where $c_1, c_2, \ldots, c_m; s_1, s_2, \ldots, s_m$ are positive numbers and $N^\star$ is a positive equilibrium of (3.14.25). Along solutions of (3.14.25) we have

$$\Delta V(N) = \sum_{i=1}^{m} c_i\left[N_i^{s_i}\left(F_i^{s_i} - 1\right)\Big/ s_i - N_i^{\star s_i}\ln(N_i/N_i^\star)\right]. \qquad (3.14.27)$$

It follows that model (3.14.25) is globally stable if it has a positive equilibrium at $N^\star$ and the function $\Delta V(N)$ as displayed in (3.13.27) is negative definite in the positive orthant.

Example 3.14.6. Consider a single-species population with nonoverlapping generations. Let N be the adult population density at time t, and r and K be the growth rate and carrying capacity respectively. A very popular model of such a population is

$$N(t+1) = N \exp[r(1 - N/K)]. \tag{3.14.28}$$

This model has a positive equilibrium at $N^\star = K$.

A suitable Liapunov function is

$$V(N) = (N^2 - K^2)/2 - K^2 \ln(N/K). \tag{3.14.29}$$

Along solutions of (3.14.28) we have

$$\Delta V(N) = N^2[\exp\{2r(1 - N/K)\} - 1]/2 - K^2[r(1 - N/K)]. \tag{3.14.30}$$

Model (3.14.28) is globally stable if $\Delta V(N)$ is negative for all $N > 0$ and $N \neq K$. This condition is satisfied if $\Delta V(N)$ has a unique global maximum at $N = K$. It is difficult to establish that $\Delta V(N)$ is negative definite in a purely analytical way. We establish that $\Delta V(N)$ has a unique global maximum at $N = K$. Then we show numerically that outside a small neighbourhood of $N = K$, ΔV is negative.

Condition (3.14.29) implies that

$$\begin{aligned} d(\Delta V)/dN &= N[\exp\{2r(1 - N/K)\} - 1] - N^2(r/K)\exp[2r(1 - N/K)] \\ &\quad + Kr = 0, \quad \text{when } N = K. \\ d^2(\Delta V)/dN^2 &= \exp[2r(1 - N/K)] - 1 - 4(r/K)N\exp[2r(1 - N/K)] \\ &\quad + 2N^2(r/K)^2\exp[2r(1 - N/K)] \\ &= 2r(-2 + r), \quad \text{when } N = K. \end{aligned}$$

Hence $\Delta V(N)$ has a local maximum at $N = K$ if $2 > r > 0$. Note that K plays no role in the stability of the equilibrium.

For the parameter values, $r = 1.9$ and $K = 1.0$, the functions, $V(N)$ and $\Delta V(N)$, are plotted in Fig. 3.14.2. Clearly $\Delta V(N)$, is negative definite. Hence for this value of r, model (3.14.28) is globally stable.

For large values of N/K, the dominant term in (3.14.29) is $N^2/2$. Hence it is a measure of the energy embodied in the standing biomass. If $\Delta V(N)$ is negative at such densities, it means that energy is dissipated by the population. For small values of N/K, the dominant term in (3.14.29) is $K^2\ln(N/K)$. This is again proportional to the energy embodied in the standing biomass. The negative sign in front of this term implies that if $\Delta V(N)$ is negative at such densities, energy is being absorbed by the population from some external sources. In a viable single-species population it is obvious that on balance

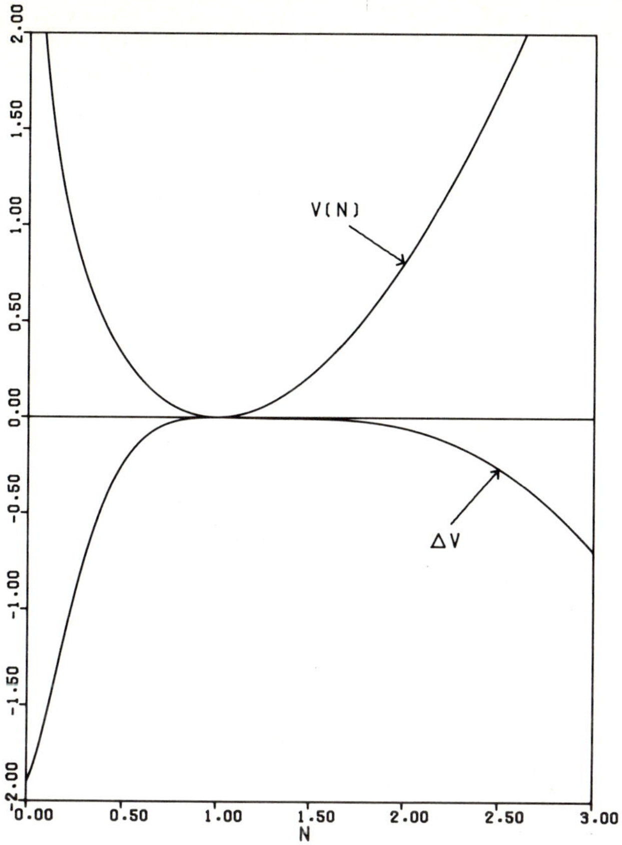

Fig. 3.14.2. Functions $V(N)$ and $\Delta V(N)$ for model $N(t+1) = N\exp[1.9(1-N)]$.

the population must absorb energy when its density is low and it must dissipate energy when its density is high.

This completes the global stability analysis of model (3.14.28). It is widely used to describe fish and insect populations (Ricker, 1954; May, 1976 a). In Example 3.15.1 we shall show how to estimate a region of ultimate confinement in this model when $r > 2$.

3.15. REGION OF ULTIMATE CONFINEMENT AND NONVULNERABILITY

In the last two sections we have examined local, finite and global stability in discrete time models. These types of stability are associated with a single equilibrium in a model. They tell us that if the initial state of a model is displaced from an equilibrium to another point in a local, finite or global neighbourhood of the equilibrium and the system is thereafter left alone, the natural dynamics of the system will drive the state into a decreasing neighbourhood of the equilibrium as time progresses.

Currently there is a widespread interest in a more complex dynamical behavior in discrete time models (Li and Yorke, 1975; Beddington et al., 1975; May, 1976 a; Guckenheimer et al., 1977; Fisher and Goh, 1977; Goh and Agnew, 1978). This complex behavior includes limit cycles of various periods, regions of ultimate confinement, and chaos. The concept of a region of ultimate confinement is useful for it provides sufficient conditions for the persistence of an ecosystem which is subjected to finite perturbations of the initial state.

Let $N_1, N_2, \ldots, N_m$ be population levels at time t in an ecosystem of m species. Suppose a model of the ecosystem is given by the set of difference equations,

$$N_i(t+1) = G_i[N_1, N_2, \ldots, N_m] \tag{3.15.1}$$

where $i = 1, 2, \ldots, m$.

By definition, a region in a state space is an open and simply connected point set. Let Γ be the positive orthant $\{N \mid N_i > 0, i = 1, 2, \ldots, m\}$, and let Y be a proper subregion of Γ. Let Ω be a proper subregion of Γ such that Y is entirely contained in Ω. Moreover the boundary of Y does not intersect the boundary of Ω. The region Y is a region of ultimate confinement if: (a) every solution of model (3.15.1) which begins in Y remains indefinitely in it and (b) every solution of model (3.15.1) which begins in the set, $\Omega - Y$, ultimately enters and remains indefinitely in Y. In this section we shall limit ourselves to the case when $\Omega = \Gamma$.

Let $N^\star = (N_1^\star, N_2^\star, \ldots, N_m^\star)$ be a point in a region Y. Generally $N^\star$ is not an equilibrium of model (3.15.1). Moreover, the model is allowed to have more than one equilibrium in the region Y. We shall employ a Liapunov-like function to determine a region of ultimate confinement. Hence we shall generally obtain only conservative estimates of the smallest region of ultimate confinement.

A function $V(N_1, N_2, \ldots, N_m)$ is a Liapunov-like function for model (3.15.1) in the region Γ if it has the following properties:

(i) $V(N) > 0$ for all $N \in \Gamma$ and $N \neq N^\star$ and $V(N^\star) = 0$;

(ii) $V(N) \to \infty$ as $N_i \to 0+$ or $N_i \to \infty$ for $i = 1, 2, \ldots, m$ and $V(N)$ has a unique minimum in Γ.

By definition, the function

$$\Delta V(N) = V[G(N)] - V(N). \tag{3.15.2}$$

Let X be the point set $\{N \mid \Delta V(N) \geqslant 0\}$. Let ΔV_{max} be the maximum value of $\Delta V(N)$ for all $N \in X$. Let V_0 be the maximum value of $V(N)$ for all $N \in X$. It follows that the region $\{N \mid V(N) \leqslant V_0\}$ is the smallest region which completely contains the set X. Let $V^\star = V_0 + \Delta V_{max}$ and $Y = \{N \mid V(N) < V^\star\}$.

Theorem 3.15.1. *Let $V(N)$ be a Liapunov-like function for model (3.15.1) in the region Γ. The region $Y = \{N \mid V(N) < V^\star\}$, is a region of ultimate confinement of the model.*

If X is not an isolated point, the maximum value of $\Delta V(N)$ must occur at an interior point of X. Generally X is a union of disjointed closed subsets of Γ. Hence it could have a very complex structure. It follows that it is more practical to determine the maximum value of $\Delta V(N)$ for all $N \in X$ by maximizing $\Delta V(N)$ for all $N \in \Gamma$. This relaxation of the admissible set in the optimization problem is justified because $\Delta V(N)$ is negative in $\Gamma - X$. We can use an optimization algorithm (Fletcher, 1975) to determine $\Delta V_{\max}$ numerically.

Similarly the complexity of the set X makes it difficult to find the maximum $V(N)$ for all $N \in X$. A more practical approach is as follows: Choose a positive number c. Maximize $\Delta V(N)$ for all $N \in \{N \mid V(N) \geqslant c\}$. The parameter c is adjusted by trial and error until the maximum of $\Delta V(N)$, subject to $V(N)$, subject to $V(N) \geqslant c$, is equal to zero or just below zero. The corresponding value of c is then called V_0.

A Liapunov-like function is none other than a generalized distance function from a specified point. In this section the specified point is denoted by $N^\star$. We shall establish that Theorem 3.15.1 is valid by examining the behavior of the solution of model (3.15.1) which passes through a typical point in each of the sets which we have defined.

At time t, consider a point $N(t)$ in X. From (3.15.1) and (3.15.2) we have

$$V[N(t+1)] = \Delta V[N(t)] + V[N(t)] < \Delta V_{\max} + V_0 = V^\star. \qquad (3.15.3)$$

We have strict inequality because the maximum of $\Delta V(N)$ occurs in the interior of X, and $V(N) = V_0$ is the smallest $V(N)$ level surface which contains entirely the set X. At time t, consider a point $N(t)$ in the set, $\{N \mid V(N) < V^\star\} - X$. At such a point, $\Delta V(N) < 0$. This implies that $V[N(t+1)] < V^\star$. It follows that a solution of model (3.15.1) which passes through a point $N \subset \{N \mid V(N) < V^\star\}$ at time t cannot get to or go across the hypersurface, $V(N) = V^\star$, at a subsequent time, $\tau \geqslant t + 1$. Finally we note that as $\Delta V(N)$ is negative for all $N \in \{N \mid V(N) \geqslant V^\star\} \cap \Gamma$, all solutions which begin in this set will ultimately enter the region $\{N \mid V(N) < V^\star\}$. Therefore the region, $Y = \{N \mid V(N) < V^\star\}$, is a region of ultimate confinement.

The main difficulty in using Theorem 3.15.1 is the problem of constructing a suitable Liapunov-like function. For a given model we should first try and see if functions which are Liapunov functions for other models provide a suitable Liapunov-like function. In particular, we should examine whether the functions displayed in (3.14.7), (3.14.22) and (3.14.26) include a suitable Liapunov-like function for determining a region of ultimate confinement. We can also exploit the fact that in the determination of a region of ultimate confinement, it is not necessary for a Liapunov-like function to have its global minimum at an equilibrium of the given model.

Let $N^{\star\star}$ be a point in a region of ultimate confinement which is not an equilibrium of model (3.15.1), We can derive Liapunov-like functions from the functions displayed in (3.14.7), (3.14.22) and (3.14.26) if $N^\star$ is replaced by $N^{\star\star}$. We note that any Liapunov function $V(Y)$ for a model of the form,

$\dot{Y}_i = f_i(Y)$ where Y_i is unrestricted in sign, can be turned into a Liapunov-like function for a population model by means of the transformation, $Y_i = \ln(N_i/N_i^{**})$ where $i = 1, 2, \ldots, m$. Finally we note that if $\{c_i, s_i, K_i, H_i\}$ is a set of positive constants, a Liapunov-like function for model (3.15.1) is

$$V(N) = \sum_{i=1}^{m} c_i \left[\left(N_i^{s_i} - K_i^{s_i}\right)/s_i - H_i K_i^{s_i} \ln(N_i/K_i)\right] - W, \tag{3.15.4}$$

where W is a constant. This function has a unique minimum at $N^{**} = (K_i H_i^{1/s_i})$. The constant W is chosen so that $V(N^{**}) = 0$.

Example 3.15.1. Consider a single-species population with nonoverlapping generations. With an appropriate choice of scale, a model of such a population is

$$N(t+1) = N \exp[r(1-N)]. \tag{3.15.5}$$

This model has an equilibrium at $N^* = 1$. From example 3.14.7, we know that this model is globally stable if $2 > r > 0$.

For values of r satisfying $2 < r < 2.692$, model (3.15.5) possesses stable limit cycles of period 2^n where n is an integer (May, 1976 a). For $r > 2.692$, it has chaotic solutions.

From (3.15.4) a suitable Liapunov-like function for this model is

$$V(N) = (N^s - 1)/s - H\ln(N) - W, \tag{3.15.6}$$

where s, H and W are constants. Along solutions of model (3.15.5), we have

$$\Delta V(N) = N^s \{\exp[rs(1-N)] - 1\}/s - Hr(1-N). \tag{3.15.7}$$

We shall illustrate the usefulness of Theorem 3.15.1 by examining two cases, (a) $r = 2.3$ and (b) $r = 3.3$. When $r = 2.3$, the model has a two-point limit cycle which goes from $N_1 = 0.408$ to $N_2 = 1.59$. Model (3.15.5) is of the form, $N(t+1) = G[N(t)]$. The points, N_1, N_2 and $N = 1$, are equilibrium points of the second order system,

$$\begin{aligned} N(t+1) &= G[G[N(t-1)]] \\ &= G[N(t-1)] \exp\{r(1 - G[N(t-1)])\}, \end{aligned} \tag{3.15.8}$$

where $G[N(t-1)] = N(t-1) \exp[r(1 - N(t-1))]$. In this case, we find by trial and error that $s = 3.69$ and $H = 1.1$ provide a good estimate of the smallest region of ultimate confinement. For these values of s and H, the minimum of $V(N)$ occurs at $N^{**} = 1.026$ and this implies $W = -0.0013$. The functions, $V(N)$ and $\Delta V(N)$, are plotted against N in Fig. 3.15.1. The value of V_0 was found to be 0.727 at $N = 1.5915$ (represented by A in the figure) and $\Delta V_{max} = 0.192$ at $N = 0.544$ (B). At the points, C and D, $V(N) = V^* = V_0 + \Delta V_{max}$. CD is a region of ultimate confinement, i.e. $Y = \{N \mid 0.341 < N < 1.656\}$.

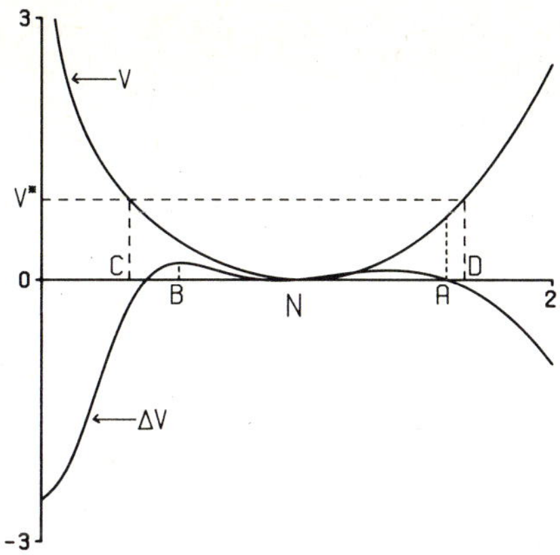

Fig. 3.15.1. CD is a region of ultimate confinement in a single-species model with a two-point limit cycle.

For $r > 2.692$, model (3.15.5) has chaotic solutions. Even then it is possible to establish that the solutions are ultimately confined in a region. Fig. 3.15.1 shows the functions, V and ΔV, plotted against $\ln(N)$ because the values of N are too small to be shown on its normal scale. We employ the parameter values, $s = 1.51$ and $H = 0.52$. The point A defines the value of $\ln(N)$ for which $V = V_0$ and ΔV achieves its maximum value of ΔV_{max} at B. The region of ultimate confinement is given by the interval (C,D) where $V(N) = V^\star = V_0 + \Delta V_{max}$ at the points C and D. This implies that a region of ultimate confinement is $Y = \{N \mid 0.0012 < N < 3.37\}$. This completes our analysis of the example.

The concept of a region of ultimate confinement which we have examined requires that the system is not perturbed after the initial instant. On the

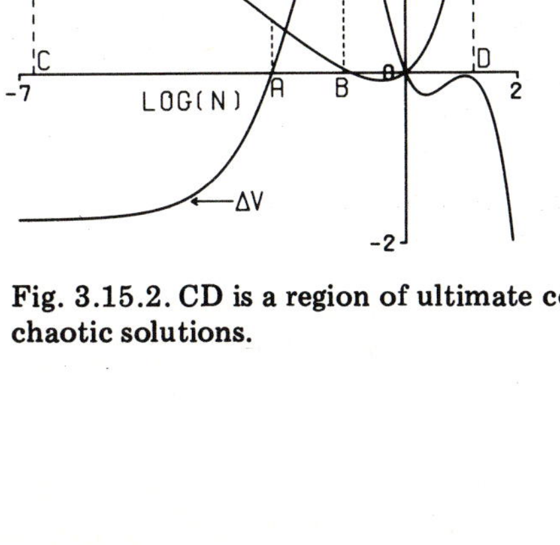

Fig. 3.15.2. CD is a region of ultimate confinement for a single-species model with chaotic solutions.

other hand, ecosystems in the real world are continually disturbed by unpredictable and finite forces. With the continual growth in the human population such distrubances will occur more frequently and on a larger scale. A useful concept for understanding an ecosystem's ability to persist in spite of continual disturbances is the concept of nonvulnerability. So far this concept has been developed only for continuous time models. Here we shall outline some initial steps in the development of the concept of nonvulnerability for discrete time models.

Let $u_1(t), u_2(t), \ldots, u_n(t)$ represent the net effects on the system dynamics of continual disturbances which act on an ecosystem during the time interval $[t, t+1]$. These disturbances (harvesting, pollution, migrating populations, etc.) alter the system dynamics so that model (3.15.1) of a given ecosystem must be replaced by the set of equations,

$$N_i(t+1) = P_i[N_1, N_2, \ldots, N_m; u_1, u_2, \ldots, u_n], \tag{3.15.9}$$

where $i = 1, 2, \ldots, m$. In the real world disturbing forces are finite and hence it is reasonable to assume that these disturbance functions satisfy the constraints,

$$-\zeta_r \leqslant u_r(t) \leqslant \xi_r \tag{3.15.10}$$

where $r = 1, 2, \ldots, n$ and $\{\zeta_r, \xi_r\}$is a set of constants. We denote the set of disturbance functions which satisfy these constraints by U.

The essential elements of a definition of nonvulnerability in a given model are (i) a specified time interval $[0,T]$, (ii) a set $S(0)$ which contains desirable states of the system, (iii) a set $Z(T)$ which contains undesirable states and (iv) a given class of disturbance functions, U. By definition, system (3.15.9) is vulnerable during the time interval $[0,T]$ if there exists an admissible set of functions $\{u_1(t), u_2(t), \ldots, u_n(t)\}$ which could drive the system from an initial state in $S(0)$ to the set $Z(T)$ during the time interval $[0,T]$. Otherwise, the system is said to be nonvulnerable.

This problem can be cast into a discrete time optimal control problem. In a given problem it may be relatively simple to show that the system is vulnerable. All we need to do is to find a set of functions $\{u_1, u_2, \ldots, u_n\}$ which will drive the system from $S(0)$ to $Z(T)$ in the time interval $[0,T]$. However, it is extremely difficult to use optimal control theory to show that it is nonvulnerable in systems with more than two state variables.

Let $V(N)$ denote a suitable Liapunov-like function which can be used to establish global stability or to find a region of ultimate confinement for model (3.15.9) when there are no disturbances, i.e. $u_1(t) = u_2(t) = \ldots = u_n(t) \equiv 0$. It is convenient to define the sets, $S(0)$ and $Z(T)$, in terms of a $V(N)$ function which is a generalized distance function. Let $S(0) = \{N \mid V(N) \leqslant V_s\}$ and $Z(T) = \{N \mid V(N) \leqslant V_z\}$ where V_s and V_z are positive numbers such that $V_z > V_s$. Along solutions of (3.15.9), we have

$$\Delta V(N,u) = V[P(N,u)] - V(N). \tag{3.15.11}$$

We can get a relatively simple method for establishing nonvulnerability by generalizing the method contained in Theorem 3.15.1. Firstly we determine a level surface, $V(N) = V_1$, such that the global maximum of $\Delta V(N,u)$ for all $N \in \{N \mid V(N) \geqslant V_1 \text{ and } N > 0\}$ and $u \in U$ is negative. The value of V_1 is chosen by trial and error. Denote the set $\{N \mid V(N) \leqslant V_1\}$ by X. Let the global maximum of $\Delta V(N,u)$ for all $N \in X$ and $u \in U$ be ΔV_{max}. Clearly ΔV_{max} occurs at an interior point of X. Let $V^\star = V_1 + \Delta V_{max}$.

Theorem 3.15.2. *Model (3.15.9) is nonvulnerable relative to the sets,* $S(0) = \{N \mid V(N) \leqslant V_s\}$ *and* $Z(T) = \{N \mid V(N) \geqslant V_z\}$, *during the time interval* $[0,T]$ *if* $V_z > V^\star$ *and* $V_z > V_s$. *T is any positive number.*

Proof. By arguments similar to those for Theorem 3.15.1, we can show that in spite of the continual disturbances any solution of (3.15.9) which begins in the set $\{N \mid V(N) < V^\star\}$ remains indefinitely in it. Moreover, any solution which initiates in the set $\{N \mid V(N) \geqslant V^\star \text{ and } N > 0\}$ will ultimately enter and remain in the set $\{N \mid V(N) < V^\star\}$. Hence the system cannot be driven from the set $\{N \mid V(N) < V^\star\}$ to the set $Z(T)$.

If $V^\star \geqslant V_s$, Theorem 3.15.2 follows immediately. If $V_s > V^\star$, we find that in spite of the continual disturbances any solution of (3.15.9) which begins in the set $\{N \mid V_s \geqslant V(N) \geqslant V^\star\}$ cannot be driven to the set $Z(T)$ because $\Delta V(N,u)$ is negative for all values of N in this set. As in the previous case the system cannot be driven from $\{N \mid V(N) < V^\star\}$ to the set $Z(T)$. We conclude that the system cannot be driven from set $S(0)$ to the set $Z(T)$. Hence the system is nonvulnerable.

Example 3.15.2. At time t, let N be the density of a population with non-overlapping generations. Let u be the mortality sustained by the population because of continual environmental disturbances during the time interval $[t, t + 1]$. A model of the population is

$$N(t + 1) = (1 - u)N \exp[r(1 - N)]. \tag{3.15.12}$$

We shall assume that $0 \leqslant u(t) \leqslant 0.2$ for all values of t. If $r = 2.3$, a suitable Liapunov-like function from Example 3.15.1 is

$$V(N) = (N^s - 1)/s - H\ln(N) - W, \tag{3.15.13}$$

where $s = 3.69$, $H = 1.1$ and $W = -0.0013$.

For $V_1 = 1.5$, $\Delta V(N,u)$ is negative for all $u \in U = \{u \mid 0 \leqslant u \leqslant 0.2\}$ and $N \in \{N \mid V(N) \geqslant V_1 \text{ and } N > 0\}$. Note that $V(N) = 1.2$ at the points, $N = 0.2$ and $N = 1.81$. The maximum value of $\Delta V(N,u)$ for all $u \in U$ and $N \in \{N \mid V(N) \leqslant V_1\}$ is $\Delta V_{max} = 0.33337$ for the values, $N = 1.395$ and $u = 0.2$. We can let $V^\star = V_1 + \Delta V_{max} = 1.83$. But in order that the result is secure from small numerical errors we set $V^\star = 1.84$. Thus in spite of the continual disturbances a region of ultimate confinement is $\{N \mid V(N) < V^\star\} = \{N \mid 0.146 < N < 1.886\}$.

If V_S and V_Z are positive numbers such that $V_Z \geqslant 1.84$ and $V_Z > V_S$, model (3.15.12) is nonvulnerable relative to the sets U, $S(0) = \{N \mid V(N) \leqslant V_S\}$ and $Z(T) = \{N \mid V(N) \geqslant V_Z\}$, during the time interval $[0,T]$ where T is any positive number.

3.16. A DISCRETE TIME MODEL OF TWO COMPETING SPECIES

In this section we shall examine a discrete time model of two competing species (Fisher and Goh, 1977). The purpose of this analysis is to illustrate the usefulness of some of the methods we have discussed in the previous sections.

Let N_1 and N_2 be population densities at time t. Let $r_1, r_2, \alpha_{11}, \alpha_{12}, \alpha_{22}$, K_1 and K_2 be positive parameters. A model of two competing species (May, 1974) is

$$N_1(t+1) = N_1 \exp\{r_1[K_1 - \alpha_{11}N_1 - \alpha_{12}N_2]/K_1\}, \tag{3.16.1a}$$

$$N_2(t+1) = N_2 \exp\{r_2[K_2 - \alpha_{21}N_1 - \alpha_{22}N_2]/K_2\}, \tag{3.16.1b}$$

The nontrivial equilibrium at $(N_1^\star, N_2^\star)$ satisfies the equations,

$$K_1 - \alpha_{11}N_1 - \alpha_{12}N_2 = 0, \tag{3.16.2a}$$

$$K_2 - \alpha_{21}N_1 - \alpha_{22}N_2 = 0. \tag{3.16.2b}$$

It follows that

$$N_1^\star = (K_1\alpha_{22} - K_2\alpha_{12})/\Delta, \qquad N_2^\star = (K_2\alpha_{11} - K_1\alpha_{21})/\Delta, \tag{3.16.3}$$

where $\Delta = \alpha_{11}\alpha_{22} - \alpha_{12}\alpha_{21}$. By definition $(N_1^\star, N_2^\star)$ is positive if $N_1^\star > 0$ and $N_2^\star > 0$.

There are too many free parameters in this model. Following May (1974) we shall make the simplifying assumption that $K_1 = K_2 = 1$, $r_1 = r$, $r_2 = 2r$, $\alpha_{11} = \alpha_{22} = 1$ and $\alpha_{12} = \alpha_{21} = \alpha$. This choice of parameters allows us to examine in a simple manner some of the effects of interspecific interactions and the growth rates. In this case we have

$$(N_1^\star, N_2^\star) = (1/(1+\alpha), 1/(1+\alpha)). \tag{3.16.4}$$

For convenience, let $N^\star = 1/(1+\alpha)$.

By Theorem 3.13.2, this equilibrium is locally stable if the magnitudes of all the eigenvalues of the matrix

$$A = \begin{pmatrix} 1 - N^\star r & -N^\star \alpha r \\ -2N^\star r\alpha & 1 - 2rN^\star \end{pmatrix} \tag{3.16.5}$$

are less than one. The Shur—Cohn stability criterion requires that

$$\alpha < 1 \tag{3.16.6}$$

and

$$r \leqslant [3 - \sqrt{(1 + 8\alpha^2)}]/[2(1-\alpha)]. \qquad (3.16.7)$$

When r violates condition (3.16.7), there is a regime of stable cycles which gives way to chaotic solutions as the value of r increases.

A suitable Liapunov function for establishing global stability is

$$V(N_1, N_2) = [(N_1^2 - N_1^{\star 2})/2 - N_1^{\star 2}\ln(N_1/N_1^{\star})] + c[(N_2^2 - N_2^{\star 2})/2 - N_2^{\star 2}\ln(N_2/N_2^{\star})] \qquad (3.16.8)$$

where c is a positive constant. For $r = 1.1$ and $\alpha = 0.5$, we can let $c = 0.5$. Along solutions of model (3.16.1), we have

$$\Delta V(N) = (N_1^2/2)[F_1^2(N_1, N_2) - 1] - N_2^{\star 2}\ln[F_1(N_1, N_2)] + c\{(N_2^2/2)[F_2^2(N_1, N_2) - 1] - N_1^{\star 2}\ln[F_2(N_1, N_2)]\} \qquad (3.16.9)$$

if we rewrite (3.16.1) in the form, $N_1(t+1) = N_1 F_1(N_1, N_2)$ and $N_2(t+1) = N_2 F_2(N_1, N_2)$.

It is difficult to show analytically that $\Delta V(N)$ is negative definite in the feasible region. However, we can show numerically that it is so by plotting the level surfaces of $\Delta V(N)$. Fig. 3.16.1 shows a number of level surfaces of the $\Delta V(N)$ function when $r = 1.1$, $\alpha = 0.5$ and $c = 0.5$. Clearly in this case $\Delta V(N)$ is negative definite in the positive region. From this we conclude that model (3.16.1) is globally stable for this particular set of parameters.

For some values of r and α which violate condition (3.16.7), model (3.16.1) has stable limit cycles. In particular, when $\alpha = 0.5$ and $r = 1.5$,

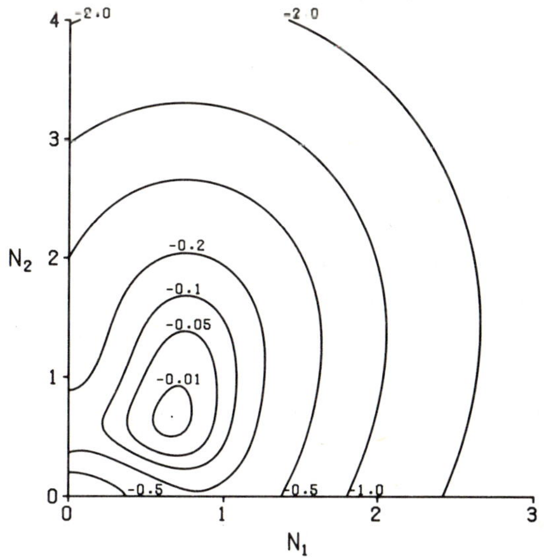

Fig. 3.16.1. Some level surfaces of $\Delta V(N)$ in a two-species competition with a globally stable equilibrium.

it has a two-point limit cycle at the points, (0.45, 0.2) and (0.88, 1.13). In this case we shall show by means of Theorem 3.15.1 that the system has a region of ultimate confinement.

A suitable Liapunov-like function is

$$V(N) = [(N_1^{s_1} - N_1^{s_1})/s_1 - H_1 N_1^{\star s_1} \ln(N_1/N_1^\star)]$$
$$+ c_2 [(N_2^{s_2} - N_2^{\star s_2})/s_2 - H_2 N_2^{\star s_2} \ln(N_2/N_2^\star)] - W \quad (3.16.10)$$

where s_1, s_2, H_1, H_2 and c are positive numbers. Along solutions of (3.16.1), we have

$$\Delta V(N) = (N_1^{s_1}/s_1)(F_1^{s_1} - 1) - H_1 N_1^{\star s_1} \ln F_1$$
$$+ c_2[(N_2^{s_2}/s_2)(F_2^{s_2} - 1) - H_2 N_2^{\star s_2} \ln F_2]. \quad (3.16.11)$$

For $r = 1.5$ and $\alpha = 0.5$, we find by trial and error that a suitable set of parameters is $s_1 = 1.04$, $s_2 = 2.43$, $H_1 = 0.97$, $H_2 = 0.72$ and $c_2 = 0.22$. We find that the set X in which $\Delta V(N)$ is non-negative is made up of two disjoint sets as shown in Fig. 3.16.2. The smallest level surface of the Liapunov-like function (3.16.10) which entirely contains the set X is $V(N) = V_0 = 0.265$. Note that $W = -0.002$.

We require the maximum value of $\Delta V(N)$ for all $N \in X$. As $\Delta V(N)$ is negative at points N outside the set X and which are in the positive orthant, it is more convenient to maximize $\Delta V(N)$ for all N in the feasible region. This gives $\Delta V_{max} = 0.052$ at a point inside the set X. By applying Theorem 3.15.1 to model (3.16.1) with this particular set of parameters we find that it has a region of ultimate confinement, $Y = \{N \mid V(N) < V^\star\}$ where $V^\star = V_0 + \Delta V_{max} = 0.317$. In Fig. 3.16.2 the level surface, $V(N) = V^\star$, appears to touch the

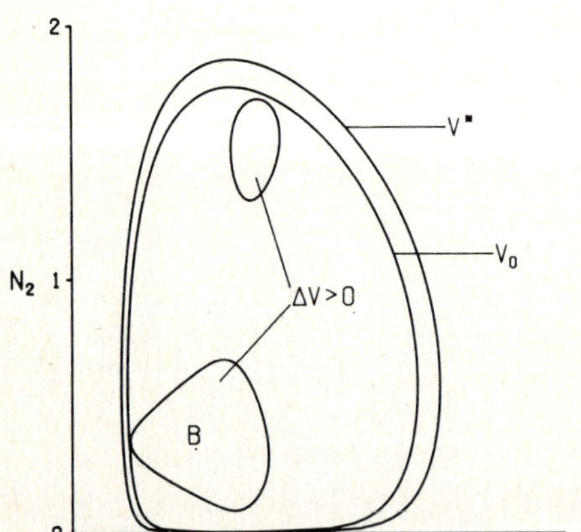

Fig. 3.16.2. $V = V^\star$ is the boundary of a region of ultimate confinement in a model of a two-species competition with a two-point limit cycle.

N_1 axis. In fact the minimum value of N_2 on this level surface is approximately equal to 0.0017.

Another Liapunov-like function which can be used to determine a region of ultimate confinement in model (3.16.1) is

$$V(N) = Z^{T}PZ \tag{3.16.12}$$

where P is a positive definite matrix and

$$Z_i = \ln[N_i F_i(N_1, N_2)/N_i^{\star}], \quad i = 1, 2.$$

For the parameter values, $r = 1.5$ and $\alpha = 0.5$, we find that a suitable P matrix is $P_{11} = 1$, $P_{12} = -0.39$ and $P_{22} = 0.2$. It gives a region of ultimate confinement comparable to that given by the Liapunov-like function (3.16.10).

CHAPTER 3: SELECTED REFERENCES

Local stability

Willems (1970); May (1974).

Finite and global stability

LaSalle and Lefschetz (1961); Hahn (1967); Willems (1970); Goh (1977 a).

Single-species models

Volterra (1938); Pianka (1972); Gilpin and Ayala (1973); Schoener (1972); Swann and Vincent (1977).

Exploited single-species models

Schaefer (1954); Goh (1969/1970, 1977 c); Silliman (1971); Brauer and Sanchez (1975); Clark (1976 a); Agnew (1979).

Two-species Lotka—Volterra models

Lotka (1925); Volterra (1931); Slobodkin (1966); Goh (1976 b, 1979); Hsu (1978 c).

Prey—Predator models

Lotka (1925); Volterra (1931); Gause (1934); Kolmogoroff (1936); Rosenweig and MacArthur (1963); Rosenzweig (1969, 1971); Kilmer (1972); May (1974); Gilpin (1974, 1975 a); Albrecht et al. (1976); Bulmer (1976); Goh (1978 b); Hsu (1978 a, c); Kazarinoff and Van den Driessche (1978); Harrison (1979 b).

Nonlinear two-species models

Gilpin and Ayala (1973); Rescigno and Richardson (1973); May (1974, 1976 b); Schoener (1974); Goh and Agnew (1977).

Nonvulnerability, resilience and stochastic models

Kalman and Bertram (1960 a); LaSalle and Lefschetz (1961); Yoshizawa (1966); Kushner (1967); Lewontin (1969); Astrom (1970); Goh (1971, 1975, 1976 b, 1978 d); Lashiher and Storey (1972); Holling (1973); Botkin and Soble (1974); Innis (1974 b); May (1974); Tuckwell (1974); Wu (1974); Feldman and Roughgarden (1975); Ludwig (1975); Turelli (1977, 1978); Harrison (1979 a); Tuljapurkar and Semura (1979); Vincent and Anderson (1979); Vincent and Skowronski (1979).

Discrete time models: local stability

Kalman and Bertram (1960 b); Willems (1970); Jury (1974); May (1974); Beddington et al. (1976); Clark (1976 b); Goh and Agnew (1978); Hassell (1978).

Finite and global stability in discrete time models

Kalman and Bertram (1960 b); Diamond (1976 a); LaSalle (1976, 1980); Goh (1977 b); Fisher and Goh (1977); Fisher and Kloeden (1978); Fisher et al. (1979); Goh and Agnew (1979).

Region of ultimate confinement in discrete time models

Kalman (1956); Beddington et al. (1975); Li and Yorke (1975); Diamond (1976 b); May (1976 a); May and Oster (1976); Fisher and Goh (1977); Guckenheimer et al. (1977); Marotto (1978).

Chapter 4

FISHERIES

4.1. INTRODUCTION

Fish resources are very important renewable resources. It is estimated that the present total world catch of fish is 70-million metric tons. The exploitation of fish populations provides some classic examples of disastrous exploitation of a renewable resource. A few large fisheries have been destroyed by overfishing (Gulland, 1974).

In this chapter, six management models of a single-species fishery are examined. The first is a dynamic pool model. The logistic equation is a well known example of this type of population. The second model combines a dynamic pool model of a fish population with the dynamics of the efforts which are used by several firms or countries in harvesting the fish resource. The third model is for a population with nonoverlapping generations (Ricker, 1954). This model is often used in the management of a salmon population. The fourth model is the Beverton—Holt (1957) model for harvesting a fishery with many year-classes in order to maximize the total biomass yield. It leads to the prescription of an optimal size limit which depends on the effort that is used in the harvesting of the fish population. The fifth model is a discrete time version of the Beverton—Holt model. It was developed by Goh (1977 b) for a fishery with a limited fishing season. The sixth model is for a fish or whale population which has a well defined delayed recruitment.

In practice the management of a fishery is a decision with multiple objectives (Larkin, 1977; Gulland, 1978). Some of the desirable objectives in the management of fish resources are as follows: (1) the provision of good biomass yield; (2) the conservation of the fish population; (3) the conservation of genetic variability of the fish population; (4) the provision of good economic returns; (5) the provision of steady employment; and (6) the provision of recreation. The formulation of good harvesting policies which take into account these objectives is a complex and difficult task even if the dynamics of a fish population is known accurately and if the objectives are fully quantified. One practical way around the mathematical difficulties in this complex problem is to convert tentatively all the objective functions except for one into constraints and optimize the remaining objective function. If the policy obtained in this manner is unsatisfactory, we should reset the constraints on the objective functions and optimize the main objective function. This process is repeated until a satisfactory policy from the overall point of view is obtained.

The maximum sustainable yield (MSY) policy, which maximizes the biomass yield, should be viewed as a first step in this process of formulating a harvesting policy which is satisfactory from an overall point of view. It provides the outer limit of the biomass yield under ideal conditions.

Many fisheries, especially those which are managed by international commissions, are managed by prescribing a constant quota for the catch which is shared by the participating fishermen. This widespread practice has an undesirable feature. Generally an MSY equilibrium which is created by a constant quota harvesting is unstable. In fact there is a contradiction of terms. This type of harvesting policy is sustainable only in an ideal mathematical sense, but in practice, it is not sustainable. The instability of this type of equilibrium can be overcome by using a variable quota policy which is prescribed and implemented in a feedback manner. But this requires considerable effort in monitoring the population and its dynamics.

If feasible, a more desirable harvesting policy is to prescribe a constant effort harvesting policy which generates the MSY equilibrium. This type of policy has a built-in stability mechanism. The effort that is used in the harvesting of a fish population is a measure of the number of standardized units of boats and the time that they are in service. With a constant effort, a lesser quantity of fish is caught if the population is below the MSY equilibrium, and a greater quantity is caught if it is above the MSY equilibrium. If a fish population moves around in schools, this stability mechanism of a constant effort harvesting policy may not be very effective; in extreme cases, it is necessary to prescribe a stabilizing variable quota policy which is implemented in a feedback manner.

The management of fisheries which takes into account the effects of discount factors has been studied by Clark (1976 a, 1979). An interesting consequence of a high discount rate is that it is optimal for a sole owner of a fishery to mine the fish population and drive it to extinction. When the discount rate is equal to zero it may also be optimal to drive a fish population to extinction. Goh (1969/1970) showed that for maximum biomass yield, it is optimal to drive a fish population to extinction if the planning horizon is short.

4.2. THE LOGISTIC MODEL

Suppose all members of a fish population have identical ecological properties. This implies that age differences among members of the population are not important. Under this assumption the population can be modelled by a nonlinear first order differential equation. In fisheries science this type of model is often called a dynamic pool model. The most popular dynamic pool model is the logistic model. It was applied to fish populations by Schaefer

(1954). Optimal harvesting policies for harvesting a fish population described by a logistic model were developed by Goh (1969/1970), Clark (1973 a) and Cliff and Vincent (1973).

Let $N(t)$ be the population at time t. The logistic model of an unexploited fish population is

$$\dot{N} = (rN/K)(K - N). \tag{4.2.1}$$

The parameters r and K are called the intrinsic rate of increase and the carrying capacity, respectively.

The logistic model of an unexploited population has been criticised by many ecologists as being inadequate. However, the logistic model has the same qualitative behavior as members of a very large class of single-species population models which are described by a nonlinear differential equation. Consider the model

$$\dot{N} = NF(N) \tag{4.2.2}$$

where: (i) $F(N)$ is autonomous, which means it does not contain t explicitly, (ii) $F(N)$ is a continuous function of N, (iii) $F(K) = 0$, and (iv) $F(N) < 0$ when $N > K$ and $F(N) > 0$ when $K > N > 0$. Model (4.2.1) and model (4.2.2) both have a unique globally stable equilibrium in the positive region. This means that if a perturbation shifts the state to a positive point ($N > 0$) and the system is thereafter left alone, the natural dynamics would move the state into a small neighbourhood of $N = K$ in a finite time interval. Hence, the common criticism of the logistic model, that the rate of increase per individual $\dot{N}/N$ is only a linear function of density, is not a serious one.

There are two simple ways to prescribe how a fish population should be harvested. we can prescribe the harvesting rate or the fishing effort. If the technology for harvesting a fish population is very efficient, the impact of harvesting on the population can be specified in terms of the harvesting rate $u(t)$ and

$$\dot{N} = (rN/K)(K - N) - u. \tag{4.2.3}$$

This model is appropriate for whale populations. In some fisheries, it is the fishing effort that can be directly manipulated; the fishing effort is a measure of the number of boats and the number of hours each boat is in service. As the boats are usually different, there are some technical problems in converting the number of boats and the number of hours that are in service into standardized units. However, since the instantaneous fishing mortality rate is directly proportional to the fishing effort, we shall use it as the control variable in place of the fishing effort. It follows that a model of an exploited fishery is

$$\dot{N} = (rN/K)(K - N) - EN \tag{4.2.4}$$

where E is a measure of the fishing effort.

It is not widely realised or emphasized that there is a very important qualitative difference between model (4.2.3) and model (4.2.4) if u and E are constants. This difference is especially important if E and u are chosen so that the corresponding steady state would provide the so-called "maximum sustainable yield". It appears that this serious oversight in fisheries has been caused by the standard approach of examining only the steady state yields and ignoring the question of whether the steady state is stable or not. The "maximum sustainable yield" (MSY) policy for model (4.2.3) is

$$u = rK/4. \tag{4.2.5}$$

This gives an unstable equilibrium at $N^{\star} = K/2$. The maximum sustainable policy for model (4.2.4) is

$$E = r/2. \tag{4.2.6}$$

This gives a globally stable equilibrium at $N^{\star} = K/2$. The region for the global stability is $\{N \mid N > 0\}$. However, if the control variables u and E are used in a *feedback* manner as prescribed by optimal control theory, the qualitative difference between the two models disappears (Goh, 1969/1970).

We shall examine the use of optimal control theory to formulate optimal policies for the exploitation of a population described by the logistic model. For this purpose we shall employ model (4.2.3), and assume that the control variable is bounded. We have

$$0 \leqslant u(t) \leqslant u_{max} \tag{4.2.7}$$

where u_{max} is a constant. A difficulty in the formulation of the problem is to decide what the time horizon in the problem should be. A standard approach in economic theory is to introduce a discount factor and let the time horizon be infinite. Another approach is to fix the time horizon temporarily, develop the optimal control policies and examine what happens when the time horizon is expanded. If $[0, T]$ is a fixed time interval, it is necessary to specify what should happen at time T. An advantage of this approach is that it forces the decision maker to specify what the state of the system should be at time T. The economist's approach avoids this issue. We can tentatively prescribe a target, and alter it if the optimal policies which follow are unsatisfactory. Let

$$N(T) \geqslant a \tag{4.2.8}$$

where a is a positive constant.

At time $t = 0$, let the density of the fish population be known. We have

$$N(0) = N_0. \tag{4.2.9}$$

In practice, this means that the state variable can be measured. What we are really interested in is not just the solution for a particular initial state but solutions for all possible initial states. In other words, we seek optimal feedback solutions. This is because we want policies which are robust to large but infrequent perturbations of the state variable.

The objective is to maximize the total biomass yield during the time interval $[0, T]$. As age differences among members of the population are ignored, the objective can be replaced by the maximization of the total number of fish caught during $[0, T]$. We have

$$J[u] = \int_0^T u\,dt. \qquad (4.2.10)$$

The optimal control problem is to choose an admissible control function $u(t)$ which will maximize the function $J[u]$.

There are two types of solutions; the occurrence of either one depends on whether $u_{max} > rK/4$ or $u_{max} < rK/4$. The mathematical analysis is given in the next section. In case (1), $u_{max} > rK/4$. The optimal feedback policy is displayed in Fig.4.2.1. The admissible state space is divided into two regions by the trajectory ABC. Typically if (t, N) is a point above ABC, the optimal control is $u = u_{max}$ until the trajectory intersects AB where $N = K/2$ or until the trajectory intersects CD at time T. The path AB is traversed by means of the singular optimal control (MSY policy), $u = rK/4$. At B the optimal control switches to $u = u_{max}$ and this is maintained until $t = T$. If $a = 0$, the time taken to move from B to C is $(K/br)\arctan(K/2b)$ where $b = \sqrt{[K(4u_{max} - Kr)/4r]}$. As T is specified, point B can be obtained by integrating (4.2.3) in reverse time.

If (t, N) is a point below ABC, the optimal control is $u = 0$ until the trajectory intersects the path ABC. If the point of intersection is along AB, singular control (MSY policy) $u = rK/4$ is used until B when the optimal control

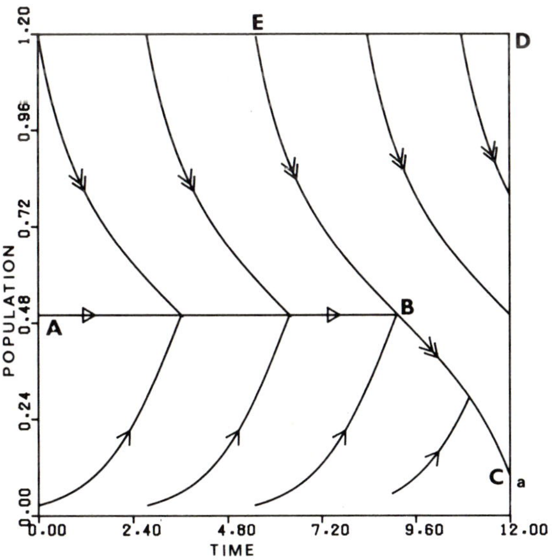

Fig.4.2.1. Optimal feedback control policy: ↠ denotes a $u = u_{max}$ trajectory, ⇾ denotes the singular control trajectory, and → denotes a $u = 0$ trajectory.

variable switches to u_{max} until $t = T$. If the point of intersection is along BC the optimal control switches from $u = 0$ to $u = u_{max}$.

In case (ii) $u_{max} < rK/4$. The singular control (MSY policy) violates the constraint (4.8.7). It is therefore inadmissible. The positive state space is divided into two regions by ABC which is generated by $u = u_{max}$. Above and along ABC, the optimal control is $u = u_{max}$ without any switch. Below ABC the optimal control is $u = 0$. This optimal feedback policy is displayed in Fig.4.2.2. We have $A \equiv (0, N^{*}(0))$ where

$$c = (1/2)\sqrt{(K - 4u_{max}K/r)}, \tag{4.2.11}$$

$$d = \frac{(K/2 + c)}{(K/2 - c)} \exp(2cT/K), \tag{4.2.12}$$

$$N(0) = (K/2) - c(d + 1)/(d - 1). \tag{4.2.13}$$

This defines uniquely the trajectory ABC. It is interesting to observe that there are two steady states. The equilibrium at $K/2 + c$ is stable, relative to all initial states in the region $\{N \mid N > (K/2) - c\}$. It is interesting that in spite of the constant rate of harvest u_{max}, the population is increasing in the region $\{N \mid K/2 - c < N < K/2 + c\}$. The equilibrium at $K/2 - c$ is unstable. Below $K/2 - c$ it is optimal to drive the population to extinction.

Consider again the case when $u_{max} > rK/4$. When $T \to \infty$, the singular optimal control (MSY policy) $u = rK/4$ in theory forms the whole of the optimal policy except for an initial period. In practice the path AB is unstable and therefore cannot be maintained with the singular control. Any perturbation

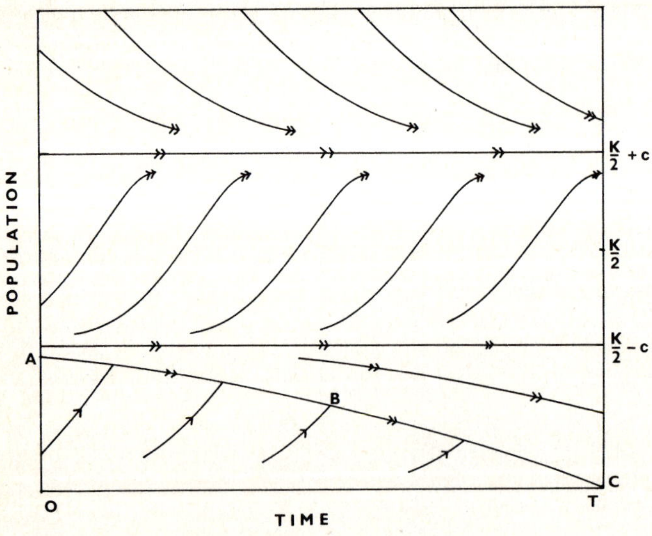

Fig.4.2.2. Optimal feedback control policy when $u_{max} < rK/4$: ↠ denotes a $u = u_{max}$ trajectory, and → denotes a $u = 0$ trajectory.

which causes the population to fall below $K/2$ will lead quickly to the extinction of the fish population unless a rest period or a reduced harvesting program is imposed to allow the population to rise to the $K/2$ level. In under-developed countries this stop—go policy is undesirable for social and economic reasons.

A more practical and robust policy is to limit the value of u_{max} so that $u_{max} < rK/4$. When $(rK/4 - u_{max})$ is large, the optimal control policy for this choice of u_{max} would be robust. However, a price for building this robustness into the optimal control policy is a corresponding loss of total yield. A small modification to the optimal control policy is to set $u = u_{max}$ whenever $N > K/2 - c$ and $u = 0$ whenever $N < K/2 - c$. This is a simpler policy. This leads to a stable policy (see Fig.4.2.3).

If model (4.2.4) is used, the optimal feedback control policy is similar to that discussed above (Cliff and Vincent, 1973). But an important difference is that the "maximum sustainable yield" (MSY) policy $E = r/2$ creates a globally stable equilibrium. It is then a desirable policy because it provides steady employment for the fishermen, and a steady supply of fish; moreover the fish population is stable relative to a wide range of uncertainties.

Clark (1976 a) has emphasized that economic considerations are crucial in the management of a fishery. In particular, if the maximum per capita growth rate of a population is less than the instantaneous discount rate and the cost of harvesting is neglected, then Clark (1976 a) showed that the optimal economic policy is to drive the population to extinction.

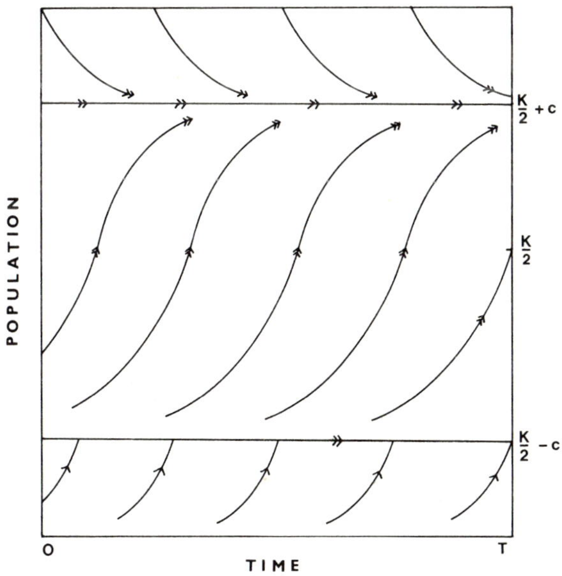

Fig.4.2.3. A stable suboptimal feedback control policy: ↠ denotes a $u = u_{max}$ ($< rK/4$) trajectory, and → denotes a $u = 0$ trajectory.

If economic returns from the harvesting of a fish population are given top priority, then the objective function of (4.2.10) should be replaced by the total discounted net economic revenue

$$J = \int_0^T e^{-\delta t}[v - c/N]\,u\,dt \qquad (4.2.14)$$

where v is the price per unit fish, δ is the instantaneous discount rate and c/N is the average cost in catching one fish. If $r > \delta$ the optimal harvesting policy is qualitatively similar to that in Fig.4.2.1.except for a complex switching regime near $t = T$. The singular control equilibrium is given by

$$N^* = (K/4)\,[g + \sqrt{(g^2 + 8c\delta/vKr)}], \qquad (4.2.15)$$

$$u = (r/K)N^*(K - N^*) \qquad (4.2.16)$$

where $g = [(c/vK) + 1 - \delta/r]$. For large values of T, the behavior of the optimal solutions for values of t close to T is of no practical significance.

4.3. OPTIMAL CONTROL OF THE LOGISTIC MODEL

Let $N(t)$ be the population of time t. The logistic model of the natural dynamics of the population is

$$\dot{N} = (rN/K)(K - N). \qquad (4.3.1)$$

The three major assumptions for this model are (i) all individuals are equivalent, (ii) the parameters r and K are constants, and (iii) there is no time lag in the response of the actual rate of increase per individual to changes in population density. Assumption (i) implies that age and sex differences are ignored. Assumption (ii) implies that the environment is constant. Assumption (iii) is a common assumption in differential equation models of a population.

Initially let

$$N(0) = N_0. \qquad (4.3.2)$$

Model (4.3.1) can be rewritten in the form

$$\frac{dN}{rN} + \frac{dN}{r(K - N)} = dt$$

$$\Rightarrow N = \frac{K}{1 + (K/N_0 - 1)\exp(-rt)}. \qquad (4.3.3)$$

The field of solutions when N_0 is varied is displayed in Fig.4.3.1. Clearly the equilibrium $N = K$ is globally stable. The logistic model has the same qualitative behavior as a large class of nonlinear models.

Let (i) $F(N)$ be autonomous, (ii) $F(N)$ be continuous, (iii) $F(K) = 0$ and (iv) $F(N) < 0$ if $N > K$, and $F(N) > 0$ if $K > N > 0$. Consider the model

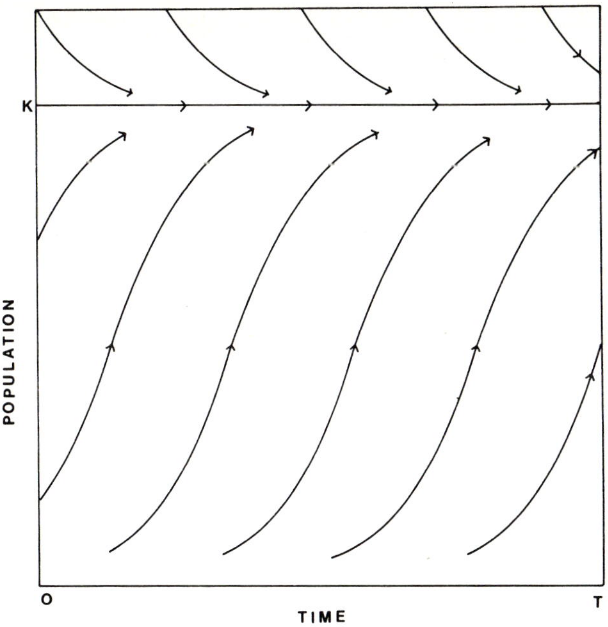

Fig. 4.3.1. Solutions of the logistic model $\dot{N} = (r/K)N(K - N)$.

$$\dot{N} = NF(N). \tag{4.3.4}$$

It has an equilibrium at $N = K$. We shall prove that this equilibrium is globally stable in the feasible region; the proof will imply that it has the same qualitative behavior as the logistic model (4.3.1).

A general method to establish global stability is Liapunov's Method, which requires the construction of a suitable Liapunov function. A Liapunov function has essentially two properties: (i) it has a global minimum at the equilibrium and (ii) the time derivative of the function computed along solutions of the system must be negative. A Liapunov function for model (4.3.4) is

$$V(N) = N - K - K \ln(N/K). \tag{4.3.5}$$

We have

$$\partial V/\partial N = (N - K)/N. \tag{4.3.6}$$

Hence, $\partial V/\partial N > 0$ when $N > K$, and $\partial V/\partial N < 0$ when $K > N > 0$. It follows that V has a unique global minimum at $N = K$ on the admissible set $\{N \mid N > 0\}$. Moreover $V(N) \to \infty$ as $N \to 0+$, and $V(N) \to \infty$ as $N \to \infty$. Hence $N = K$ is globally stable in the positive region $\{N \mid N > 0\}$. Computing $\dot{V}(N)$ along solutions of (4.3.4) we get

$$\dot{V} = (N - K)\, F(N). \tag{4.3.7}$$

As $F(N) < 0$ when $N > K$, and $F(N) > 0$ when $K > N > 0$, $\dot{V}$ is negative for all positive values of N except when $N = K$.

Briefly, the optimal control formulated in the previous section is derived as follows:

System: $\dot{N} = (rN/K)(K - N) - u.$ (4.3.8)

Initially: $N(0) = N_0.$ (4.3.9)

Terminally: $T = \text{constant}, \quad N(T) \geqslant a.$ (4.3.10)

Constraint: $0 \leqslant u \leqslant u_{max}.$ (4.3.11)

Objective: $\max \int_0^T u dt.$ (4.3.12)

Solution: Let $p(t)$ be the costate variable. By definition, the Hamiltonian function

$$H[N, u, p] = p_0 u + p[(r/K) N(K - N) - u] \tag{4.3.13}$$

where $p_0 \geqslant 0$. The costate equation is

$$\dot{p} = -\partial H/\partial N = -(pr/K)(K - 2N). \tag{4.3.14}$$

The terminal condition $N(T) \geqslant a$ is equivalent to the condition $N(T) = a$ or $N(T) > a$. When $N(T) = a$, the transversality condition gives

$$p(T) = \text{constant}. \tag{4.3.15}$$

When $N(T) > a$, the transversality condition gives

$$p(T) = 0. \tag{4.3.16}$$

Eq. (4.3.14) is a first order homogenous equation in $p(t)$. Hence the terminal condition (4.3.16) implies

$$p(t) \equiv 0 \tag{4.3.17}$$

for all $t \in [0, T]$. The terminal condition (4.3.15) provides no immediate information.

The optimality conditions are

$$\begin{aligned} u &= u_{max} && \text{only if } \partial H/\partial u > 0, \\ u &= 0 && \text{only if } \partial H/\partial u < 0, \\ u_{max} &> u > 0 && \text{only if } \partial H/\partial u = 0. \end{aligned} \tag{4.3.18}$$

Assuming that $p_0 = 1$, we have

$$\partial H/\partial u = 1 - p(t). \tag{4.3.19}$$

The control variable u appears linearly in the Hamiltonian function $H(N, u, p)$.

This indicates that the problem may have an optimal singular control. Along a singular extremal,

$$\partial H/\partial u = 1 - p(t) \equiv 0, \tag{4.3.20}$$

$$(\mathrm{d}/\mathrm{d}t)(\partial H/\partial u) = (pr/K)(K - 2N) = 0, \tag{4.3.21}$$

$$(\mathrm{d}^2/\mathrm{d}t^2)(\partial H/\partial u) = -(2pr^2/K^2)N(K - N) + 2pru/K = 0 \tag{4.3.22}$$

$$\Rightarrow \partial/\partial u(\mathrm{d}^2/\mathrm{d}t^2)(\partial H/\partial u) = 2pr/K = 2r/K > 0. \tag{4.3.23}$$

Note that in this problem u maximizes $H(N, u, p)$, as our objective is to maximize the total biomass yield. From eq. (4.3.21) the singular extremal is $N = K/2$. Employing conditions (4.3.20) to (4.3.22), we deduce that the singular control is $u = rK/4$. Condition (4.3.23) shows that the generalized Legendre condition for singular control is satisfied.

At time $\theta \in [0,T]$, let the control variable switch from one level to another. The costate function $p(t)$ must satisfy the condition

$$p(\theta-) = p(\theta+). \tag{4.3.24}$$

Our job now is to make use of the conditions we have written down to synthesize the optimal feedback control policy. If $p(t)$ satisfies (4.3.17), then condition (4.3.18) implies that the optimal control $u = u_{max}$ without any switch in the control variable. Integrating backwards from every point in the target $\{N \mid N(T) > a\}$, we find that the maximum control and no switch policy is only feasible for some initial states. The boundary of the region in the state space which contains such initial states is EBC as shown in Fig.4.2.1. Next, we draw the singular extremal AB in Fig.4.2.1. This is done under the assumption that $u_{max} > rK/4$. Condition (4.3.23) shows that this singular extremal satisfies the generalized Legendre condition for optimality. For initial states above AB, the simplest way to get to the singular extremal is to employ the control $u = u_{max}$ until the trajectory intersects AB. For initial states below ABC, the simplest way to get to ABC is to employ $u = 0$ until the trajectory intersects ABC. This exhausts all the initial states in the feasible region. The complete feedback control policy is displayed in Fig.4.2.1. Its optimality can be established by using Green's Theorem (Mancill, 1950; Miele, 1962).

Consider the case when the singular control violates the constraint (4.3.11). We have $u_{max} < rK/4$. Employing condition (4.3.16), we integrate (4.3.8) and (4.3.14) backwards from points in the target $\{N \mid N(T) > a\}$. This generates optimal trajectories along which $u = u_{max}$ without any switch in the control variable. Such trajectories will cover a large portion of the state space. This is shown in Fig.4.2.2. The lower boundary of this region is the extremal ABC. The integrand of the objective function is $L = u$. Clearly the functional is maximized if it is possible to use the maximum control for all $t \in [0, T]$. However, this simple argument is not valid for more complicated functionals. For states below ABC, it is advantageous to get to the trajectory ABC quickly. This can be achieved by setting $u = 0$ until the trajectory intersects ABC.

4.4. STABILITY OF BIONOMIC EQUILIBRIUM

Let us suppose a large number of fishermen or firms is exploiting a fish population and the cost per unit of effort is the same for each fisherman or firm. If the fishery is not regulated, a plausible model of the fishery is

$$\dot{N} = NF(N) - EN, \tag{4.4.1a}$$

$$\dot{E} = kE(pN - c) \tag{4.4.1b}$$

where N is the fish population, $NF(N)$ is the natural growth rate, E is a measure of the fishing effort, p is the price of one fish, c is the cost per unit of effort, and c and k are positive constants. Eq. (4.4.1b) implies that the rate of increase of the effort is proportional to the rate of net economic revenue.

This model has equilibria at $(N^\star, E^\star) = (c/p, F(c/p))$, $(N', E') = (0, 0)$, and $(N'', E'') = (K, 0)$ where K satisfies $F(K) = 0$.

Suppose $pK - c < 0$, $F(N) < 0$ for $N > K$, and $F(N) > 0$ for $N < K$. In this case, every solution which begins in the positive quadrant will converge to $(K, 0)$. Therefore if $pN - c > 0$ and $E > 0$ at $t = 0$, only a transient fishing industry is possible because it is not worthwhile to exploit the fish continuously. This result can be established rigorously by using the Liapunov function

$$V(N, E) = d[N - K - K \ln (N/K)] + d|E|$$

where $d = 1$ and $d = d/(pk)$.

If $F(N) < 0$ for small values of N, then every solution which begins in a small neighbourhood of (0, 0) will converge to (0, 0). This can be established by using the Liapunov function

$$V(N, E) = |N| + |E|/(pk).$$

In this case, the exploitation of the fish population for low initial population level will drive it to extinction.

On the other hand, consider the case where the equilibrium is positive. If $F(N) - F(c/p) < 0$ for $N > c/p$, and $F(N) - F(c/p) > 0$ for $N < c/p$, then the equilibrium $(c/p, F(c/p))$ is globally stable. This means that every solution which begins in the positive quadrant converges to the equilibrium. Here the dynamics of the fishery is robust. This result can be established by using the Liapunov function

$$V(N, E) = d_1[N - N^\star - N^\star \ln (N/N^\star)] + d_2[E - E^\star - E^\star \ln (E/E^\star)]$$

where $d_1 = 1$, $d_2 = 1/(pk)$, $N^\star = c/p$ and $E^\star = F(c/p)$.

If the population experiences difficulties in breeding at low densities (Allee effect), it is possible that $F(N) - F(c/p) > 0$ for some $N > c/p$, and $F(N) - F(c/p) < 0$ *for all* values of $N < c/p$. In this particular case, the equilibrium $(c/p, F(c/p))$ is unstable. Mathematically the population cannot be driven to extinction during a finite time interval. But for small values of N, the model for the fish population dynamics breaks down; and in practice, the fish population can be driven to extinction by overfishing in a finite time.

Next consider the case in which the fish population is exploited by several firms or countries, each of which has a different price and cost structure. Suppose the fishery dynamics can be described by the equations,

$$\dot{N} = NF(N) - (E_1 + E_2 + \ldots + E_m)N, \tag{4.4.2a}$$

$$\dot{E}_i = k_i E_i (p_i N - c_i), \quad i = 1, 2, \ldots, m, \tag{4.4.2b}$$

where k_i, p_i and c_i are positive constants.

By definition the first harvester is the most efficient harvester if $c_1/p_1 < c_i/p_i$ for $i = 2, \ldots, m$. Here every solution of (4.4.2) which begins in the set $S = \{(N, E) \mid N > 0, E_1 > 0 \text{ and } E_i \geqslant 0, i = 2, \ldots, m\}$ will converge to the equilibrium $(c_1/p_1, F(c_1/p_1), 0, \ldots, 0)$, if $F(N) - F(c_1/p_1) < 0$ *for* $N > c_1/p_1$, and $F(N) - F(c_1/p_1) > 0$ for $N < c_1/p_1$. *In this case the most efficient harvester will eliminate the other harvesters.* This result can be established by using the Liapunov function

$$V(N, E) = [N - N^\star - N^\star \ln (N/N^\star)] + d_1[E_1 - E_1^\star - E_1^\star \ln (E_1/E^\star)] + \sum_{i=2}^{m} d_i |E_i|$$

where $d_i = 1/(p_j k_j)$ for $j = 1, 2, \ldots, m$.

Along solutions of the model in the positive orthant

$$\dot{V}(N, E) = (N - N^\star)[F(N) - F(c_1/p_1)] + \sum_{i=2}^{m} (c_1/p_1 - c_i/p_i).$$

It follows that $\dot{V}(N, E) \leqslant 0$. Moreover $\dot{V}(N, E)$ does not vanish identically along a nontrivial solution. Hence every solution of (4.4.2) which begins in the positive orthant converges to $(c_1/p_1, F(c_1/p_1), 0, \ldots, 0)$. Similar analysis can be carried out for the other subsets of S. In other words the equilibrium $(c_1/p_1, F(c_1/p_1), 0, \ldots, 0)$ is globally sector stable.

4.5. STOCK RECRUITMENT MODEL

In this section we shall consider some concepts for the management of a population with nonoverlapping generations in order to maximize the long-term biomass yield. These concepts are useful for the management of a salmon population even though the latter usually has several overlapping generations. To first approximation the dynamics of each year-class in a salmon population behaves like that of a population with nonoverlapping generations.

Let $N(t)$ be the number of fish in the tth generation. A model of this type of population is

$$N(t+1) = N(t)F[N(t)] \tag{4.5.1}$$

A well known example as proposed by Moran (1950) and Ricker (1954) is

$$N(t+1) = N(t)\exp[r(1-N(t)/K)]. \tag{4.5.2}$$

Other examples are: (i) $F(N) = c/(1 + aN)$, where a, b, c are positive constants, (Hassell, 1975); and (ii) $F(N) = a \exp -bN^2$ (Goh and Agnew, 1978).

This type of model is used in the management of a salmon population. Typically it takes 4 years for a salmon to return to a river to spawn and it dies after spawning. Thus a salmon population can be managed as a collection of four separate subpopulations each of which behaves like a population with nonoverlapping generations. In practice there is a small diffusion between the adjoining subpopulations.

By definition the group of fish which is vulnerable to fishing at time t is known as the recruitment. Let $u(t)$ be the number of fish that is harvested during the time interval $[t, t + \Delta t]$ where Δt is a small number. Let $H(t)$ be the post-harvest population (the escapement) at time $t + \Delta t$, and $N(t)$ be the pre-harvest population at time t. Suppose spawning takes place after the harvesting period. Using (4.5.1) and the definitions of $H(t)$ and $u(t)$, we get

$$H(t+1) = H(t)F[H(t)] - u(t+1). \tag{4.5.3}$$

If $(H^\star, u^\star)$ is a steady state of (4.5.3), then

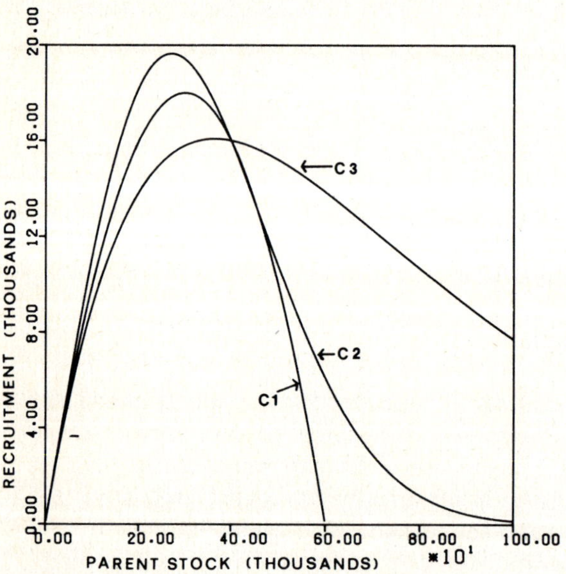

Fig.4.5.1. Three stock recruitment functions: $C1$ represents $R = 0.12S(1 - S/600)$; $C2$ represents $R = 0.12S \exp[-(\ln 3)(S/400)^2]$; and $C3$ represents $R = 0.12S \exp[-(\ln 3)(S/400)]$.

$$H^\star = H^\star F(H^\star) - u^\star. \quad (4.5.4)$$

The corresponding yield (number of fish harvested) per unit time is

$$u^\star = H^\star F(H^\star) - H^\star. \quad (4.5.5)$$

The function $u^\star(H^\star)$ is a maximum only if $du^\star/dH^\star = 0$. Therefore a necessary condition for a maximum steady state yield is

$$d[H^\star F(H^\star)]/dH^\star = 1. \quad (4.5.6)$$

Using discrete time optimal control theory, it can be shown that the pair $(u^\star, H^\star)$ which satisfies (4.5.5) and (4.5.6) also satisfies the necessary conditions for a long-term optimal policy which maximizes the yield. The pair $(u^\star, H^\star)$ which satisfies (4.5.6) is called the MSY policy.

If the handling time is negligible and the effects of the fish schooling are insignificant, then a constant effort harvesting can be approximated by a constant fraction harvesting. If $E(t)$ is the fraction of a cohort which is harvested impulsively at recruitment, then (4.5.3) gives

$$H(t+1) = [1 - E(t+1)]H(t)F[H(t)]. \quad (4.5.7)$$

We shall show that the MSY equilibrium is locally stable if it is created by a constant fraction harvesting policy and that it is unstable if it is generated by a constant quota harvesting policy (Goh, 1977 c).

Let $E^{\star\star}$ be the constant fraction harvesting which generates the MSY equilibrium. We have $E^{\star\star} = u^{\star\star}/H^{\star\star}$ where $u^{\star\star}$ satisfies (4.5.5) and (4.5.6). A necessary and sufficient condition for $H^{\star\star}$ to be a locally stable equilibrium of (4.5.7) is

$$|(d/dH)(1 - E^{\star\star})HF(H)| < 1. \quad (4.5.8)$$

Using (4.5.6) this condition is equivalent to

$$|(1 - E^{\star\star})| < 1. \quad (4.5.9)$$

Clearly this condition is satisfied because $1 > E^{\star\star} > 0$. We conclude that a MSY equilibrium which is generated by a constant fraction harvesting is locally stable.

Let $u^{\star\star}$ generate an MSY equilibrium at $H^{\star\star}$. Under the constant quota $u^{\star\star}$ harvesting policy the population dynamics is described by

$$H(t+1) = H(t)F[H(t)] - u^{\star\star}. \quad (4.5.10)$$

If $x(t) = H(t) - H^{\star\star}$, the linearized dynamics of (4.5.10) is

$$x(t+1) = x(t) - ax^2(t) \quad (4.5.11)$$

where $a = -(d^2/dH^2)(HF(H))$ at $H^{\star\star}$. For most population models, a is positive. If $1/a > x(0) > 0$, then $x(t) \to 0$ as $t \to \infty$. If $x(0) < 0$, then $x(t) \to -\infty$ as $t \to \infty$. It follows that $H^{\star\star}$ is an unstable equilibrium of (4.5.10). Thus a constant quota harvesting policy which creates a MSY equilibrium is unstable.

In practice the recruitment to the fish population varies from year to year. A way to manage the population in the presence of this type of uncertainty is to control the escapement. Firstly the function $HF(H)$ is obtained by fitting it to stock (parent population) and recruitment data. From this the MSY equilibrium H^{**} is computed. The value of H^{**} is also called the optimal escapement for maximum biomass yield. By definition the recruitment at time $(t + 1)$ is $R(t + 1) = H(t)F[H(t)]$. A harvesting policy for this type of population is

$$u(t) = \begin{cases} R(t) - H^{**} & \text{if } R(t) > H^{**}, \\ 0 & \text{if } R(t) < H^{**}. \end{cases} \tag{4.5.12}$$

The dynamics of the harvested population is

$$H(t + 1) = \begin{cases} H^{**} & \text{if } R(t + 1) > H^{**}, \\ H(t)F[H(t)] & \text{if } R(t + 1) < H^{**}. \end{cases} \tag{4.5.13}$$

If $H(t + 1) > \mathrm{H}(t)$ for all $H(t) < H^{**}$ and $H(t + 1) > 0$ for all $H(t) > H^{**}$, then by Theorem 4.6.3, model (4.5.13) is globally stable. Clearly this is a very robust harvesting policy. Another desirable feature of this harvesting policy is that the population is rapidly driven to H^{**}.

4.6. STABILITY OF A STOCK RECRUITMENT MODEL

In this section, several sets of sufficient conditions for stability of an equilibrium of a stock recruitment model for a population with nonoverlapping generations are established (Fisher et al., 1979; Goh, 1980).

All existing stock recruitment models for unexploited populations have continuous stock-recruitment relations; this is not necessarily so with a harvested population. For an example, let us assume that it is difficult to estimate the current recruitment to a fish population; and assume further that the recruitment in the previous generation is known from catch data. Suppose u^* denotes a constant quota policy which generates a desirable equilibrium at H^* for model (4.5.3). Then a reasonable harvesting policy for model (4.5.3) is

$$u(t + 1) = \begin{cases} u^* & \text{if } H(t) \geqslant H^*, \\ 0 & \text{if } H(t) < H^*. \end{cases} \tag{4.6.1}$$

With this delayed feedback harvesting policy, the right hand side of (4.5.3) is a discontinuous function of $H(t)$. Mathematically this type of equilibrium is said to be unstable, because $H(t) = H^* - \epsilon$, where ϵ is a small positive number, gives rise to $H(t + 1)$ which differs from H^* by a relatively large number. Nevertheless H^* can be an attractor. This means that every solution of (4.5.3) which begins near H^* will converge to H^* as $t \to \infty$. For harvested population, this may be a desirable dynamical property of the population.

By definition the stock recruitment model

$$N(t+1) = G[N(t)] \tag{4.6.2}$$

has an equilibrium at $N^\star$ if $G(N^\star) = N^\star$. The equilibrium $N^\star$ is globally stable if every solution of (4.6.2) which begins in the set $\{N \mid N > 0\}$ remains in it and converges to $N^\star$ as $t \to \infty$. Moreover every solution which begins close to $N^\star$ remains close to $N^\star$, If the model has this behavior only in a finite neighbourhood of $N^\star$, the equilibrium is said to have a finite region of stability.

Theorem 4.6.1. *Let $Q(N)$: $[N^\star, \infty) \to (0, N^\star]$ be a continuous and strictly monotonic decreasing function such that $Q(N^\star) = N^\star$ and $Q(N) \to 0+$ as $N \to \infty$. Let $P(N)$: $(0, N^\star] \to [N^\star, \infty)$ be the inverse function of $Q(N)$. The equilibrium $N^\star$ of (4.6.2) is globally stable if* (i) $G(N^\star) = N^\star$, $N^\star > 0$; (ii) $P(N) > G(N) > N$ *for all* $N \in (0, N^\star)$; *and* (iii) $N > G(N) > Q(N)$ *for all* $N > N^\star$ *(see Fig.4.6.1).*

Proof. This theorem is proved by showing that the function

$$V(N) = \begin{cases} P(N) - N & \text{for } N^\star > N > 0, \\ N - Q(N) & \text{for } N \geqslant N^\star \end{cases} \tag{4.6.3}$$

is a Liapunov function of (4.6.2).

The graph of $P(N)$ is the image of that of $Q(N)$ in the "mirror" described

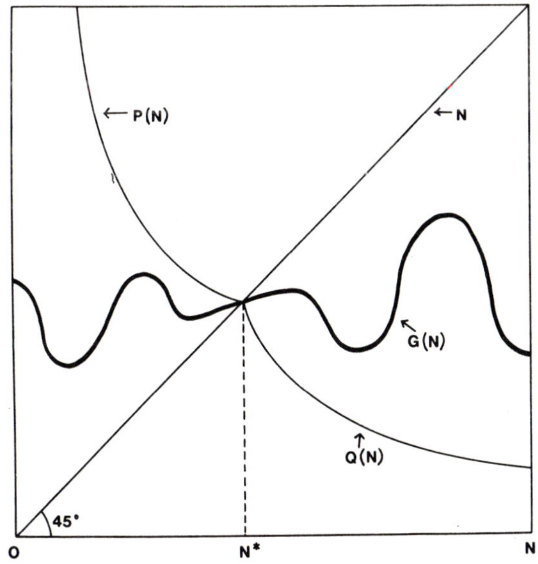

Fig.4.6.1. $P(N) > G(N) > N$ for all $N \in (0, N^\star)$, and $N > G(N) > Q(N)$ for all $N \geqslant N^\star$. The equilibrium $N^\star$ of $N(t+1) = G(N)$ is globally stable.

by the line $y(N) = N$. Thus $P(N)$ inherits from $Q(N)$ the following properties: (i) it is continuous; (ii) it is strictly monotonic decreasing; (iii) $P(N^\star) = N^\star$ and (iv) $P(N) \to \infty$ as $N \to 0+$. It follows that $V(N)$ of (4.6.3) has the properties: (i) it is a continuous function; (ii) it is strictly monotonic decreasing for $N^\star > N > 0$ and it is strictly monotonic inrreasing for $N \geqslant N^\star$; and (iii) $V(N) \to \infty$ as $N \to \infty$ or as $N \to 0+$.

By definition, the function

$$\Delta V(N) = V[G(N)] - V(N). \tag{4.6.4}$$

We show that $\Delta V(N)$ is negative definite by examining five cases: (1) $N^\star > N > 0$ and $P(N) > G(N) > N^\star$; (2) $N^\star > N > 0$ and $N^\star > G(N) > N$; (3) $N > N^\star$ and $N > G(N) > N^\star$; (4) $N > N^\star$ and $N^\star > G(N) > Q(N)$; and (5) $N \neq N^\star$ and $G(N) = N^\star$.

In case (1), we have

$$\Delta V(N) = [G(N) - Q(G(N))] - [P(N) - N]$$

$$< [G(N) - Q(P(N))] - [P(N) - N]$$

because (i) $P(N) > G(N)$ and (ii) $Q(N)$ is a strictly monotonic decreasing function. Furthermore $Q(P(N)) = N$ because $P(N)$ is the inverse function of $Q(N)$. It follows that

$$\Delta V(N) < G(N) - P(N) < 0. \tag{4.6.5}$$

In case (2), we have

$$\Delta V(N) = [P(G(N)) - G(N)] - [P(N) - N]$$

$$= [P(G(N)) - P(N)] - [G(N) - N] < 0, \tag{4.6.6}$$

because (i) $G(N) > N$, and (ii) $P(N)$ is a strictly monotonic decreasing function.

Similarly it can be shown that $\Delta V(N) < 0$ in cases (3) and (4). In case (5), $\Delta V(N) = 0 - V(N) < 0$. It follows that $\Delta V(N)$ is negative definite for $N > 0$. Therefore $V(N)$ is a Liapunov function. We conclude that $N^\star$ is globally stable.

Corollary 4.6.1. *Let r be a positive constant and* RAS $= \{N \mid V(N) < r\}$ *where $V(N)$ is given in (4.6.3). The set* RAS *is a finite region of attraction of $N^\star$ if* (i) $G(N^\star) = N^\star$, (ii) $P(N) > G(N) > N$ *for all* $N \in \text{RAS} \cap \{N \mid N^\star > N > 0\}$, *and* (iii) $N > G(N) > Q(N)$ *for all* $N \in \text{RAS} \cap \{N \mid N > N^\star\}$.

Corollary 4.6.2. *Let* $G'(N^\star+) = \lim \mathrm{d}G/\mathrm{d}N$ *as* $N \to N^\star+$; *and similarly, let* $G'(N^\star-) = \lim \mathrm{d}G/\mathrm{d}N$ *as* $N \to N^\star-$. *Let c be a positive constant. The equilibrium $N^\star$ of (4.6.2) is locally stable if* (i) $G(N^\star) = N^\star$; (ii) $1 > G'(N^\star+) > -c$; *and* (iii) $1 > G'(N^\star-) > -1/c$.

Proof. Let $Q(N) = N^* - c(N - N^*)$ and $P(N) = N^* - (N - N^*)/c$. The function $P(N)$ is the inverse function of $Q(N)$. If $G(N)$ satisfies conditions (ii) and (iii), then there exists a small positive number ϵ such that (a) $P(N) > G(N) > N$ for all $N \in (N^* - \epsilon, N^*)$; and (b) $N > G(N) > Q(N)$ for all $N \in (N^*, N^* + \epsilon)$. Hence N^* is locally stable.

Corollary 4.6.3. *Let $Q(N)$ and $P(N)$ be the functions as defined in Theorem 4.6.1. The equilibrium N^* is unstable if* (i) $G(N^*) = N^*$; (ii) $G(N) > P(N)$ *or* $G(N) < N$ *for all* $N \in (0, N^*)$; *and* (iii) $G(N) > N$ *or* $G(N) < Q(N)$ *for all* $N \in (N^*, \infty)$.

Theorem 4.6.2. *Let $b > N^*$ be a constant. Let $Q(N)$: $[N^*, b] \to [0, N^*]$ be a continuous and strictly monotonic decreasing function such that $Q(N^*) = N^*$ and $Q(b) = 0$. Let $P(N)$: $[0, N^*] \to [N^*, b]$ be the inverse function of $Q(N)$. The equilibrium N^* of (4.6.2) is globally stable if* (i) $G(N^*) = N^*$; (ii) $P(N) > G(N) > N$ *for all* $N \in (0, N)$; and (iii) $N > G(N) > Q(N)$ *for all* $N \in (N^*, b)$ *and* $N > G(N) > 0$ *for all* $N \in (b, \infty)$ *(see Fig. 4.6.2).*

Proof. Consider the function

$$V(N) - \begin{cases} P(N) - N & \text{for } N \in (0, N^*), \\ N - Q(N) & \text{for } N \in [N^*, b), \\ N & \text{for } N \in [b, \infty). \end{cases} \tag{4.6.7}$$

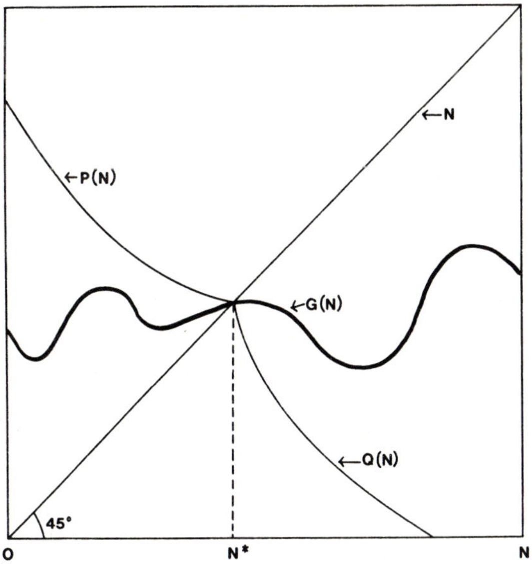

Fig.4.6.2. $P(N) > G(N) > N$ for all $N \in (0, N^*)$, and $N > G(N) > Q(N)$ for all $N \geqslant N^*$. The equilibrium N^* of $N(t + 1) = G(N)$ is globally stable.

Using arguments similar to those in the proof of Theorem 4.6.1, we deduce that $\{N \mid V(N) < b\}$ is a region of attraction of $N^\star$. We have $\{N \mid V(N) < b\} = \{N \mid 0 < N < b\}$.

If the conditions of the theorem are satisfied, then $\Delta V(N) < 0$ for all $N \geqslant b$. This implies that every solution of (4.6.2) which begins at a finite point in the interval (b, ∞) remains bounded and positive for all $t \geqslant 0$ and ultimately enters the set $\{N \mid 0 < N < b\}$. We conclude that every solution of (4.6.2), which begins in the set $\{N \mid N > 0\}$, converges to $N^\star$ as $t \to \infty$.

The function $V(N)$ of (4.6.7) does not tend to infinity as $N \to 0+$. Therefore it is not "radially unbounded". Thus it is not a Liapunov function according to the classical definition of a Liapunov function (see Kalman and Bertram, 1960 b). However, according to a recent definition by LaSalle (1976), a Liapunov function has to satisfy only two conditions. These are (i) $V(N)$ is continuous on a given set S and (ii) $\Delta V(N) \leqslant 0$ for all $N \in S$. To distinguish this new definition from the classical one, we shall call a Liapunov function which satisfies the new definition a *weak* Liapunov function. Thus $V(N)$ of (4.6.7) is a weak Liapunov function.

If a weak Liapunov function is used, it is essential to give another argument which ensures that every solution of a model remains bounded and stays in a given set for all values of $t \geqslant 0$.

Theorem 4.6.3. *The equilibrium $N^\star$ of (4.6.2) is a global attractor if* (i) $G(N^\star) = N^\star$; (ii) $N^\star > G(N) > N$ *for all* $N \in (0, N^\star)$; *and* (iii) $N > G(N) > 0$ *for all* $N \in (N^\star, \infty)$ *(see Fig.4.6.3).*

Proof. We shall use the vector function

$$w(N) = \begin{cases} \begin{pmatrix} N^\star - N \\ N^\star \end{pmatrix} & \text{if } N^\star > N > 0, \\ \begin{pmatrix} N - N^\star \\ N \end{pmatrix} & \text{if } N \geqslant N^\star. \end{cases} \tag{4.6.8}$$

Following LaSalle (1976), we construct a scalar Liapunov function $V(N)$ from a vector function $w(N)$ by letting

$$V(N) = \max\{w_1(N), w_2(N)\}. \tag{4.6.9}$$

Eqs. (4.6.8) and (4.6.9) give

$$V(N) = \begin{cases} N^\star & \text{if } N^\star > N > 0, \\ N & \text{if } N \geqslant N^\star. \end{cases} \tag{4.6.10}$$

The function $\Delta V(N)$ is shown to be negative semidefinite by considering the auxiliary vector difference function

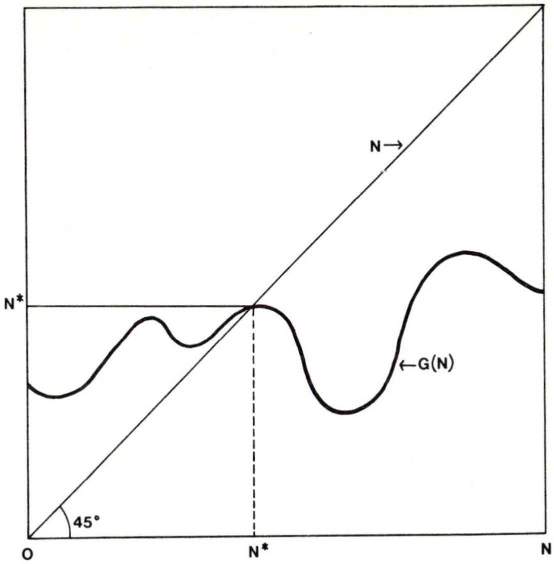

Fig.4.6.3. $N^\star > G(N) > N$ for all $N \in (0, N^\star)$, and $N > G(N) > 0$ for all $N \geqslant N^\star$. The equilibrium $N^\star$ of $N(t+1) = G(N)$ is globally stable.

$$\mathring{w}(N) = w(G(N)) - V(N)\,u \tag{4.6.11}$$

where u is a vector with components $u_1 = 1$ and $u_2 = 1$. We have

$$\begin{aligned}\max_i \{\mathring{w}_i(N)\} &= \max_i \{w_i(G(N)) - V(N)\} \\ &= \max_i \{w_i(G(N))\} - V(N) \\ &= \Delta V(N).\end{aligned} \tag{4.6.12}$$

It follows that $\Delta V(N) \leqslant 0$ if $\mathring{w}(N) \leqslant \mathbf{0}$, where $\mathbf{0}$ denotes a zero vector.

For $N^\star > N > 0$ and $N^\star > G(N) > N$, we have

$$\mathring{w}(N) = \begin{bmatrix} N^\star - G(N) \\ N^\star \end{bmatrix} - \begin{bmatrix} N^\star \\ N^\star \end{bmatrix} = \begin{bmatrix} -G(N) \\ 0 \end{bmatrix} \leqslant \mathbf{0}. \tag{4.6.13}$$

It follows that every solution of (4.6.2) which begins in the open interval $(0, N^\star)$ remains in it for all $t \geqslant 0$. In addition the condition $\Delta w_1(N) = -G(N) < 0$ for all $N \in (0, N^\star)$ implies that every solution which begins in the interval $(0, N^\star)$ converges monotonically to $N^\star$ as $t \to \infty$.

For $N > N^\star$, it can be shown that $\mathring{w}(N) < \mathbf{0}$ by considering separately the cases: (i) $N > G(N) > N^\star$, (ii) $N^\star > G(N) > 0$, and (iii) $G(N) = N^\star$. Hence $\Delta V(N) < 0$ for $N > N^\star$. This implies that every solution of (4.6.2) which begins at a finite point in the open interval $(N^\star, \infty)$ is bounded for all $t \geqslant 0$.

Conditions (ii) and (iii) imply that every solution which begins at a positive point remains positive for all $t \geqslant 0$. We conclude that every solution of (4.6.2) converges to $N^\star$ as $t \to \infty$.

If $G(N)$ is continuous at $N^\star$ and it satisfies the conditions of Theorem 4.6.3, then $N^\star$ is globally stable. However, if $G(N)$ is discontinuous at $N^\star$, the equilibrium is locally unstable even though it is a global attractor. In other words, every solution which begins at a positive point converges to $N^\star$ as $t \to \infty$, but a solution which begins near $N^\star$ may not remain close to $N^\star$ for all values of $t \geqslant 0$.

Theorem 4.6.4. *Let $b > N^\star$ be a constant. Let $Q(N)$: $[b, \infty) \to (0, N^\star]$ be a strictly monotonic decreasing and continuous function such that $Q(b) = N^\star$ and $Q(N) \to 0+$ as $N \to \infty$. Let $P(N)$: $(0, N^\star] \to [b, \infty)$ be the inverse function of $Q(N)$. The equilibrium $N^\star$ is a global attractor if* (i) $G(N^\star) = N^\star$; (ii) $P(N) > G(N) > N$ *for all* $N \in (0, N^\star)$; (iii) $N > G(N) > N^\star$ *for all* $N \in (N^\star, b]$; *and* (iv) $N > G(N) > Q(N)$ *for all* $N \in (b, \infty)$ *(see Fig.4.6.4).*

This theorem can be proved by using arguments which are similar to those in the proofs of Theorems 4.6.1 and 4.6.3, and by using the vector function

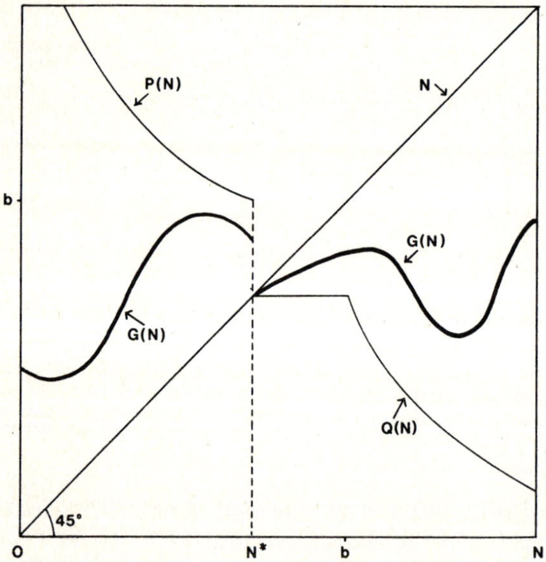

Fig.4.6.4. $P(N) > G(N) > N$ for all $N \in (0, N^\star)$, and $N > G(N) > Q(N)$ for all $N \geqslant N^\star$. The equilibrium $N^\star$ of $N(t + 1) = G(N)$ is a global attractor.

$$w(N) = \begin{cases} \begin{pmatrix} P(N) - N \\ b - N \\ N^\star - N \end{pmatrix} & \text{for } N^\star > N > 0, \\ \begin{pmatrix} b - N^\star \\ b - N^\star \\ N - N^\star \end{pmatrix} & \text{for } b \geqslant N \geqslant N^\star, \\ \begin{pmatrix} N - Q(N) \\ b - N^\star \\ N - N^\star \end{pmatrix} & \text{for } N > b. \end{cases}$$

Corollary 4.6.4. *Let $b < N^\star$ be a positive constant. Let $Q(N)$: $[N^\star, \infty) \to (0, b]$ be a strictly monotonic decreasing and continuous function such that $Q(N^\star) = b$ and $Q(N) \to 0+$ as $N \to \infty$. Let $P(N)$: $(0, b] \to [N^\star, \infty)$ be the inverse function of $Q(N)$. The equilibrium $N^\star$ is a global attractor if* (i) $G(N^\star) = N^\star$; (ii) $P(N) > G(N) > N$ *for all* $N \in (0, b]$; (iii) $N^\star > G(N) > N$ *for all* $N \in (b, N^\star)$; *and* (iv) $N > G(N) > Q(N)$ *for all* $N \in (N^\star, \infty)$.

The proof of this corollary is similar to that for Theorem 4.6.4.

4.7. GLOBAL STABILITY OF THE RICKER MODEL

One of the best known stock recruitment models is

$$N(t + 1) = N(t) \exp [r(1 - N(t)/K)] \tag{4.7.1}$$

where r and K are positive constants. This equation was used by Moran (1950) to demonstrate oscillatory behavior in population models and by Ricker (1954) in the management of a salmon population. We shall demonstrate the usefulness of the results in Sections 4.5 and 4.6 by applying them to model (4.7.1).

May (1976 a) showed that (4.7.1) has (i) a stable equilibrium at $N^\star = K$ if $2 > r > 0$; (ii) stable cycles if $2.692 > r > 2$; and (iii) chaotic solutions if $r > 2.692$.

From (4.5.4) we deduce that under constant fraction harvesting we have

$$H(t + 1) = (1 - E) H(t) \exp [r(1 - H(t)/K] \tag{4.7.2}$$

where $H(t)$ denotes the post-harvest population in the tth period and E is a positive fraction. Model (4.7.2) has a steady state at $H^\star = 1 - 1/[r(1 - E)]$ which is positive if $1 - 1/r > E \geqslant 0$. Comparing (4.7.1) with (4.7.2), we deduce that increasing the value of E stabilizes the population, provided that $E < 1 - 1/r$.

Condition (4.5.10) implies that the MSY constant effort harvesting policy always creates a locally stable equilibrium. For (4.7.2) the MSY equilibrium is unique and globally stable. Another interesting observation is that the MSY equilibrium is stable even if the unexploited population model has chaotic behavior. For example, consider model (4.7.1). It has chaotic behavior if $r = 4$ and $K = 1$; its MSY equilibrium under constant effort harvesting is at $H^\star = 0.2381$ and the corresponding value of E is 0.9525; this MSY equilibrium is globally stable.

This observation on the stability of the MSY equilibrium provides a way to distinguish between fluctuations in a population with nonoverlapping generations which are caused by chaotic behavior (i.e. they are caused by natural dynamics), and those which are caused by random disturbances. If an unexploited population has chaotic behavior, a sufficiently high constant effort harvesting could damp out the fluctuations. However, if the fluctuations are caused by random disturbances, then no constant effort harvesting could damp out the fluctuations completely.

If economic factors are ignored, the long-term optimal harvesting policy for maximum yield is to maintain the population at the MSY level. As pointed out by Clark (1976 a, p. 248) the optimal harvesting policy which incorporates economic factors could be very complex.

We shall show that model (4.7.1) is globally stable if $2 > r > 0$. This implies that (4.7.2) is globally stable if $H^\star = 1 - 1/[r(1-E)] > 0$ and $2 > r + \ln(1-E) > 0$.

Using Theorem 4.6.1, we construct first a strictly monotonic decreasing function $Q(N)$ such that $G(N) = N \exp[r(1-N)] > Q(N)$ for all $N > K$. As $2 > r > 0$, we choose

$$Q(N) = N \exp 2(1 - N/K) \tag{4.7.3}$$

for $N \geqslant K$. Clearly $G(N) > Q(N)$ for $N > K$. Unfortunately $P(N)$, the inverse function of $Q(N)$, cannot be expressed as an explicit function. It is obtained numerically by making use of the property that $P(N)$ is the image of $Q(N)$ in the "mirror" $y = N$.

For $K > N > 0$ and $2 > r > 0$, we have

$$N \exp[2(1-N/K)] > N \exp[r(1-N/K)]. \tag{4.7.4}$$

Thus a sufficient condition for $P(N) > G(N)$ for all $N \in (0, K)$ and $r \in (0, 2)$ is that $P(N) > N \exp[2(1-N/K)]$ for all $N \in (0, K)$. It can be shown graphically that $P(N) > N \exp[2(1-N/K)]$ for all $N \in (0, K)$. Using Theorem 4.6.1 we conclude that (4.7.1) is globally stable if $2 > r > 0$.

Using Theorem 4.6.2 we can show analytically that (4.7.1) is globally stable if $2 > r > 0$. Let $Q(N) = 2K - N$ for $2K \geqslant N \geqslant K$. The inverse function of $Q(N)$ is $P(N) = 2K - N$ for $K \geqslant N \geqslant 0$.

For $N \geqslant K$ let $H(N) = G(N) - Q(N)$. We get

$$H(N) = N \exp[r(1-N/K)] - (2K - N) \tag{4.7.5}$$

$$\Rightarrow H'(N) = (1 - rN/K) \exp [r(1 - N/K)] + 1 \quad (4.7.6)$$

$$\Rightarrow H''(N) = (r/K)(rN/K - 2) \exp [r(1 - N/K)] > 0 \quad (4.7.7)$$

if $2 > r > 0$ and $N > K$. The conditions, $H(K) = 0$, $H'(K+) > 0$, and $H''(N) > 0$ for all $N > K$, imply that $H(N) > 0$ for $N > K$. Hence $G(N) > Q(N)$.

For $K \geqslant N > 0$, let $H(N) = P(N) - G(N)$. We get

$$H(N) = 2K - N - N \exp [r(1 - N/K)] \quad (4.7.8)$$

$$\Rightarrow H'(N) = -1 - (1 - rN/K) \exp [r(1 - N/K)] \quad (4.7.9)$$

$$\Rightarrow H''(N) = (r/K)(2 - rN/K) \exp [r(1 - N/K)] > 0 \quad (4.7.10)$$

if $2 > r > 0$ and $K \geqslant N > 0$. The conditions, $H(K) = 0$, $H'(K-) < 0$ and $H''(N) > 0$ *for all* $N \in (0, K]$ imply that $H(N) > 0$ for all $N \in (0, K)$. Hence $P(N) > G(N)$.

For $N \geqslant 2K$, $G(N) = N \exp [r(1 - N/K)]$ is clearly positive. Using Theorem 4.6.2, we conclude that model (4.6.2) is globally stable.

4.8. THE BEVERTON—HOLT MODEL: OPTIMAL SIZE LIMIT

By definition a year-class in a fish population is the cohort of fish born in the same year. Usually a fish population contains several year-classes; in other words, it has an age structure. The Beverton—Holt model (1957) was developed for formulating the optimal harvesting policy which maximizes the biomass yield from a fish population with an age structure under the assumption that a constant effort is used in harvesting the fish population. The optimal policy prescribes that an optimal size limit should be imposed. This requires that only fish that are larger than the optimal size limit should be harvested.

By definition, the recruitment to a fish population in a given year is the year-class which becomes vulnerable to fishing for the first time during that year. In the Beverton—Holt model the relationship between stock and recruitment is completely ignored. This is a reasonable approach for the management of a fish species which has a high fecundity. For many commercially important fish species, a mature female can lay hundreds of thousands or even millions of eggs per season. In this type of fish population the stock and recruitment is highly variable (see Cushing and Harris (1973)) and cannot be used to predict the recruitment which is spawned by a given stock.

The optimal harvesting policy which is obtained from a Beverton—Holt model of a fish population is independent of the recruitment of the different year-classes in the population. It follows that *it is applicable even if the recruitment fluctuates greatly from year to year.* This ability of the Beverton—Holt model to cope with uncertainties in the recruitment is not widely appreciated. It is often incorrectly stated that the optimal harvesting policy for a Beverton—Holt model of a fishery is only valid for a steady state (constant) recruitment.

Assume tentatively that we can harvest each year-class independently of the other year-classes in a fish population. This assumption will be removed at the end of the analysis. We shall now consider the problem of harvesting a single-year class so as to maximize the biomass yield.

Let the available fishing effort which is applied to the year-class be a constant. The fishing effort is a measure of the number of boats and the time that they are in service. We shall assume that the effects of the fish schooling on the mass contact law for the fishing mortality rate are negligible.

At time t, let $N(t)$ be the number of fish in the year-class. Let t also denote the age of the year-class. During an initial phase in the life of a fish in the year-class, its survival is determined mainly by environmental factors. It is difficult to predict these environmental factors accurately enough to build a predictive model of the dynamics of the year-class during the initial phase. Fortunately in most fisheries the natural mortality rate of the year-class becomes relatively constant from a certain age. This roughly coincides with the age of recruitment which occurs when the year-class becomes vulnerable to fishing. This age is denoted by t_r. Let T be the age at which most of the fish in the year-class, in the absence of fishing, would have died from natural causes; or the age when the year-class leaves the fishing area permanently.

In the absence of fishing, the dynamics of the year-class is

$$\dot{N} = -MN, \quad t_r \leqslant t \leqslant T, \tag{4.8.1}$$

where M is the natural mortality rate. At time t_r, let

$$N(t_r) = R. \tag{4.8.2}$$

The parameter R is called the recruitment of the year-class. In many fisheries, the value of R for different year-classes varies tremendously; a ten-fold or a hundred-fold variation in the value of R from year to year is not unknown.

Let $u(t)$ be a feasible fishing mortality rate. We have

$$\dot{N} = -[M + u(t)]N. \tag{4.8.3}$$

Let the available fishing effort which can be applied to the year-class be limited. This gives the constraint

$$0 \leqslant u(t) \leqslant F \tag{4.8.4}$$

where F is a constant. As we shall see later on, the parameter F is the fishing mortality rate that is *actually* imposed on the year-class when the size of an average fish in the year-class is larger than or equal to the optimal size limit.

In most fisheries there is no constraint on the number of fish that must be allowed to escape from the fishing at age T. One exception to this is a salmon fishery. Thus for most fisheries we have the condition

$$N(T) > 0 \tag{4.8.5}$$

but otherwise $N(T)$ is unconstrained.

Let $W(t)$ be the average weight of a fish in the year-class at time t. The total biomass yield from the year-class is

$$J = \int_{t_r}^{T} u(t)N(t)W(t)\,\mathrm{d}t. \tag{4.8.6}$$

The optimal control problem is to choose an admissible control function $u(t)$ which maximizes the total biomass yield from the year-class.

The optimal harvesting policy is

$$u(t) = \begin{cases} 0 & \text{if } t_r \leqslant t < t^\star, \\ F & \text{if } t^\star \leqslant t < T, \end{cases} \tag{4.8.7}$$

where $t^\star$ is the age at which an average fish in the year-class attains the optimal size limit.

Under optimal harvesting

$$W(t)N(t) > \int_{t}^{T} FW(s)\,N(s)\,\mathrm{d}s \tag{4.8.8}$$

for all $t > t^\star$. This implies that

$$W(t) > \int_{t}^{T} FW(s)\,\exp\,[-(M + F)(s - t)]\,\mathrm{d}s \tag{4.8.9}$$

for all $t > t^\star$. Condition (4.8.8) states that for $t > t^\star$, the total biomass of the year-class at time t is greater than the expected biomass yield during the period $[t, T]$, if the maximum fishing effort is applied to the year-class during this period.

The optimal control program given in (4.8.7) and (4.8.9) is independent of the state variable $N(t)$ and the recruitment R. This property makes the optimal policy independent of the fluctuations in the recruitment from year to year.

Condition (4.8.9) provides a method to compute $t^\star$. Let

$$p(t) = \int_{t}^{T} u(s)\,W(s)\,\exp\,[-(M + u(s))(s - t)]\,\mathrm{d}s \tag{4.8.10}$$

where $u(s)$ is given in (4.8.7). Plot on the same graph paper the functions $W(t)$ and $p(t)$. The function $p(t)$ is plotted backwards from the point $(T, p(T)) = (T, 0)$ with $u(T) = F$. The value of t at the point where the graph of $p(t)$ intersects the graph of $W(t)$ is equal to $t^\star$ (see Fig.4.8.1).

The size of an average fish in the year-class at the age $t^\star$ is the optimal size limit for the fish population. The optimal harvesting policy for the fish population which contains many year-classes is to prescribe an optimal size limit. This means that only fish which are larger than the optimal size limit should be harvested. This can be implemented by prescribing a mesh size, if nets are used to harvest the fish.

To illustrate the usefulness of this theory, we shall apply it to the North Sea plaice fishery. An average mature female lays about 100 000 eggs in a spawning season. The fish is of marketable size at the age of 4 years. The life spans of an average male fish and an average female are 13 years and 22 years respectively (Beverton and Holt, 1957). We shall let $T = 15$. The optimal biomass yield is not sensitive to the choice of T since the value of the fishing mortality rate is relatively large.

In this example we shall employ the data: $t_r = 3.7$, $T = 15$, $F = 0.73$ and $M = 0.1$. The growth function is

$$W(t) = 2867[1 - \exp -0.095(t + 0.82)]^3. \qquad (4.8.11)$$

The criterion in (4.8.9) does not require a smooth analytical growth function. A piece-wise linear growth would suffice. The value of $t^\star$ which is computed by using (4.8.9) is insensitive to small errors in $W(t)$. In Fig.4.8.1 the growth function $W(t)$ and the costate function $p(t)$ of (4.8.10) are plotted backwards from the terminal time T. The point of intersection of $W(t)$ and $p(t)$ defines

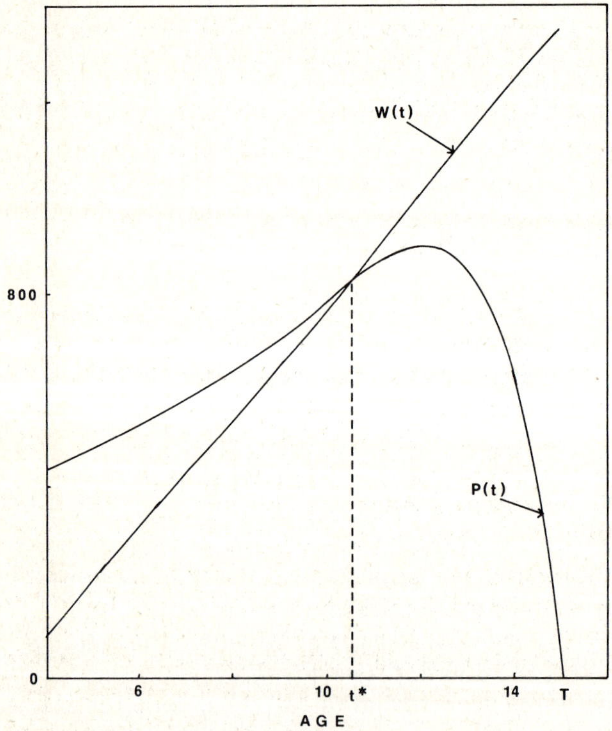

Fig.4.8.1. Intersection of the weight function $W(t)$ and the costate variable $p(t)$ determines the age $t^\star$ when the length of an average fish is equal to the optimum size limit. The functions are computed in reverse time.

$t^\star$ which in turn defines the optimal size limit. From this exercise we get $t^\star = 10.4$. Using the length function

$$L(t) = 68.5[1 - \exp -0.095(t + 0.82)] \qquad (4.8.12)$$

we deduce that the optimal size limit is 45 cm.

4.9. OPTIMAL CONTROL OF THE BEVERTON—HOLT MODEL

This section contains details of the mathematical analysis of the Beverton—Holt model. The purpose in presenting these details is to show how optimal control theory can be applied to an interesting biological problem. The symbols are defined in the previous section.

In brief, the optimal control problem is as follows:

System: $\dot{N} = -(M + u)N, \quad t_r \leqslant t \leqslant T.$ (4.9.1)

Initially: $N(t_r) = R.$ (4.9.2)

Terminally: $N(T) > 0.$ (4.9.3)

Control constraint: $0 \leqslant u \leqslant F.$ (4.9.4)

Objective: $\max \int_{t_r}^{T} u(t)\, W(t)\, N(t)\, dt.$ (4.9.5)

By definition the Hamiltonian function is

$$H(t, N, u, p) = uW(t)N - p(M + u)N. \qquad (4.9.6)$$

The function $p(t)$ is called the costate variable and $N(t)$ is called the state variable. The costate equation is

$$\dot{p} = -\partial H/\partial N = p(M + u) - uW(t). \qquad (4.9.7)$$

In applying the transversality condition, $N(T)$ is considered as unconstrained because it belongs to the interior of its constraint set. It follows that

$$p(T) = 0. \qquad (4.9.8)$$

The optimality conditions require the optimal control function to satisfy the conditions

$$u = \begin{cases} 0 & \text{only if } \partial H/\partial u < 0, \\ F & \text{only if } \partial H/\partial u > 0. \end{cases} \qquad (4.9.9)$$

This means that the optimal control variable maximizes H with respect to u. There is no singular control because the conditions

$$\partial H/\partial u = (W(t) - p)N \equiv 0, \qquad (4.9.10)$$

$$(d/dt)(\partial H/\partial u) = (\dot{W} - MW)N \equiv 0 \qquad (4.9.11)$$

are not satisfied.

For most fish populations, $W(t)$ is a monotonic increasing function. Furthermore, for biological reasons, we expect only a single switch in the control variable. This is because the optimal harvesting program provides a balance between a loss in biomass yield due to natural mortalities and a gain in biomass yield from the growth of the fish which belong to the year-class. In any case, for a given set of data, we can explore numerically and show that there is only a single switch in the optimal control function.

To avoid loss of yield through natural mortality of old fish, we should use the maximum effort at T, i.e. $u(T) = F$. The optimal control program should be

$$u(t) = \begin{cases} 0 & \text{if } t_r \leqslant t < t^\star, \\ F & \text{if } t^\star < t \leqslant T. \end{cases} \tag{4.9.12}$$

Eq. (4.9.7) gives

$$[p(s) \exp -(M + u)s]_t^T = \int_t^T [-u(s)\, W(s) \exp -(M + u(s))s]\, ds. \tag{4.9.13}$$

Conditions (4.9.8) and (4.9.13) give

$$p(t) = \int_t^T u(s)\, W(s) \exp\, [-(M + u(s))(s - t)]\, ds. \tag{4.9.14}$$

Conditions (4.4.9) and (4.9.10) imply

$$u(t) = \begin{cases} 0 & \text{only if } W(t)\, N(t) < p(t)\, N(t), \\ F & \text{only if } W(t)\, N(t) > p(t)\, N(t). \end{cases} \tag{4.9.15}$$

Conditions (4.9.12) and (4.9.15) imply that for $t > t^\star$,

$$W(t)\, N(t) > N(t) \int_t^T FW(s) \exp\, [-(M + F)(s - t)]\, ds$$

$$= \int_t^T FW(s)\, N(s)\, ds. \tag{4.9.16}$$

It follows that for $t > t^\star$, we have

$$W(t) > \int_t^T FW(s) \exp\, [-(M + F)(s - t)]\, ds. \tag{4.9.17}$$

This completes the proof of conditions (4.8.8) and (4.8.9).

We shall now establish that the specification of the optimal size limit provides the optimal policy for biomass yield if (i) the fishery contains many year-classes; and (ii) the applied fishing effort is a constant.

Suppose the fish population contains many year-classes. If nets are used it is not possible to exploit each year-class selectively. Thus the problem has a very intricate constraint. Assume for a moment that it is possible to exploit each year-class separately. This means that the nonselectivity constraint is relaxed and hence the set of admissible options is enlarged. For a collection of year-classes, each of which can be harvested separately, the optimal policy is to exploit each year-class optimally. The crucial point in this analysis is that this relaxed optimal control program can be implemented when the fish population contains many year-classes which cannot be harvested selectively. It can be implemented by specifying the optimum size limit which is the average length of a fish at age $t^{\star}$ of (4.9.12). This requires that only fish which are larger than the optimal size limit should be harvested. This policy gives the maximum total biomass yield for a fixed fishing effort.

If economic factors are considered, and a high discount rate is applicable, the optimal harvesting policy can be very complex (Clark, 1976 a).

4.10. FISHERY WITH A LIMITED FISHING SEASON

In the harvesting of a fish population with many year-classes, a standard objective is to maximize the biomass yield. If the applied fishing effort is fixed, this objective can be achieved by prescribing an optimal size limit. In this section we shall develop a method for computing the optimal size limit when the fishing season is limited to the same period in each year. A fishing season may be limited by weather conditions, migration patterns of the fish population, or a policy of conservation.

Another method for computing the optimal size limit for a fishery with a limited fishing season uses the continuous time Beverton—Holt model with an averaged fishing mortality rate. For example, let 12% of a year-class be harvested during a fishing season which is limited to one month in a year. Using the continuous time Beverton—Holt model, it is assumed that this fishing mortality is equivalent to a continuous fishing mortality rate of 1% per month. For practical purposes, it appears that these two methods for computing the optimal size limit for a fishery with a limited fishing season give the same optimal size limit.

By definition a year-class is the cohort of fish born in a given year. First we shall develop the optimal harvesting program for a single year-class. At time t let the age of a year-class be t, and let the number of fish in the year-class be $N(t)$. Let the limited fishing season take place during the period $[t, t + \Delta t]$ in the unit time interval $[t, t + 1]$. Let $s = 1/\Delta t$. Usually the dynamics of a year-class is very variable during the early stages of its life. Fortunately the dynamics of the year-class from the age of recruitment can usually be described by a constant linear differential equation (Beverton and Holt, 1957). Let T be the life span of an average fish in the year-class.

In the absence of fishing, the dynamics of the year-class can be described by

$$N(t + \Delta t) = (1 - m)\, N(t)$$

where $m = 1 - \exp(-M\Delta t)$ and M is the instantaneous natural mortality rate. It follows that

$$N(t + 1) = (1 - m)^{s-1} N(t + \Delta t). \tag{4.10.1}$$

If $u(t)$ is the fishing mortality coefficient during the fishing season $[t, t + \Delta t]$, then

$$N(t + \Delta t) = [1 - m - u(t)]N(t). \tag{4.10.2}$$

Combining eqs. (4.10.1) and (4.10.2), we get

$$N(t + 1) = f[u(t)]N(t) \tag{4.10.3}$$

where $f[u(t)] = (1 - m)^{s-1}[1 - m - u(t)]$ and $t = t_r, t_r + 1, \ldots, T - 1$.

If R denotes the recruitment to the year-class at time t_r, then

$$N(t_r) = R. \tag{4.10.4}$$

At the end of its life span, $N(T)$ must satisfy the constraint

$$N(T) > 0. \tag{4.10.5}$$

Usually the fishing mortality is limited because the number of boats and the number of hours they are in service are limited. This leads to

$$0 \leqslant u(t) \leqslant b \tag{4.10.6}$$

where b is a constant. If F denotes the instantaneous fishing mortality rate which is in fact applied to harvest the year-class, then

$$N(t + \Delta t) = N(t) \exp[-(M + F)\Delta t]. \tag{4.10.7}$$

Conditions (4.10.2), (4.10.6) and (4.10.7) give

$$b = \exp(-M\Delta t) - \exp[-(M + F)\Delta t]. \tag{4.10.8}$$

Let $W(t + \theta\Delta t)$ be the weight of an average fish in the year-class at the age of $t + \theta\Delta t$ where θ is roughly equal to 0.5. We shall assume that the objective of the management is the maximization of the total biomass yield from the year-class. The problem is to choose the sequence of fishing levels $u(t_r)$, $u(t_r + 1), \ldots, u(T - 1)$ which maximizes the objective function. The total biomass yield is

$$J = \sum_{t_r}^{T-1} [u(t)\, N(t)\, W(t + \theta\Delta t)]. \tag{4.10.9}$$

We shall minimize $-J$ because in our standard discrete time optimal control problem, a function is minimized. A maximization problem can always be converted into a minimization problem by using the identity

$\max J = -\min(-J).$

The Hamiltonian function is

$$H[N, u, p] = p(t+1)\,f[u(t)]N(t) - u(t)\,N(t)\,W(t+\theta\Delta t). \tag{4.10.10}$$

The costate equation is

$$p(t) = \partial H/\partial N(t) = p(t+1)\,f[u(t)] - u(t)\,W(t+\theta\Delta t). \tag{4.10.11}$$

Multiplying this equation by $N(t)$ and using (4.10.3), we get

$$p(t)N(t) - p(t+1)N(t+1) = -u(t)\,W(t+\theta\Delta t)N(t). \tag{4.10.12}$$

If t of (4.10.12) is replaced by $t+1, t+2, \ldots, T-1$, we generate a system of equations, which, when added together, give

$$p(t)N(t) - p(T)N(T) = -\sum_{i=t}^{T-1} u(s)\,W(i+\theta\Delta t)\,N(i). \tag{4.10.13}$$

At the terminal time, $N(T) > 0$. Thus $N(T)$ belongs to the interior of its constraint set. For the purpose of applying the transversality condition, $N(T)$ is unconstrained. Hence

$$p(T) = 0. \tag{4.10.14}$$

Eqs. (4.10.13) and (4.10.14) give

$$p(t)N(t) = \sum_{i=t}^{T-1} u(i)\,W(i+\theta\Delta t)\,N(i). \tag{4.10.15}$$

The optimal control program seeks to balance the gain in yield from the growth of the fish in the year-class with the loss in yield from natural mortalities. This suggests that there should be a single switch in the optimal control program. It follows that there exists $t^\star$ such that the optimal control program is

$$u(t) = 0, \quad t = t_r, t+1, \ldots, t^\star - 1,$$
$$u(t) = b, \quad t = t^\star, t^\star + 1, \ldots, T-1. \tag{4.10.16}$$

We seek a criterion which will enable us to compute $t^\star$.

Using (4.10.10) we get

$$\partial H/\partial u(t) = -(1-m)^{s-1}p(t+1)N(t) - W(t+\theta\Delta t)N(t). \tag{4.10.17}$$

The optimality conditions require that $u(t) = 0$ only if $\partial H/\partial u(t) > 0$, and $u(t) = b$ only if $\partial H/\partial u(t) < 0$. Eq. (4.10.17) implies that $u(t) = b$ only if

$$W(t+\theta\Delta t) > -(1-m)^{s-1}p(t+1). \tag{4.10.18}$$

For $t \geqslant t^\star$, eq. (4.10.15) gives

$$p(t+1)\,f(b)N(t) = -\sum_{i=t+1}^{T-1} bW(i+\theta\Delta t)[f(b)]^{i-t}N(t).$$

It follows that

$$p(t+1) = -\sum_{i=t+1}^{T-1} bW(i+\theta\Delta t)[f(b)]^{i-t-1}. \tag{4.10.19}$$

Eqs. (4.10.18) and (4.10.19) imply that $t \geqslant t^\star$

$$W(t+\theta\Delta t) > (1-m)^{s-1}\sum_{i=t+1}^{T-1} bW(i+\theta\Delta t)[f(b)]^{i-t-1}. \tag{4.10.20}$$

This inequality criterion provides a method for computing $t^\star$.

Let the function

$$G(t+\theta\Delta t) = -(1-m)^{s-1}p(t+1). \tag{4.10.21}$$

Note that the argument of $G(t+\theta\Delta t)$ is $t+\theta\Delta t$. To compute $t^\star$, the functions $W(t+\theta\Delta t)$ and $G(t+\theta\Delta t)$ are plotted together (see Fig.4.10.1). The function $G(t+\theta\Delta t)$ is plotted backwards from the terminal point $(T-1+\theta\Delta t, G(T-1+\theta\Delta t)) = (T-1+\theta\Delta t, 0)$. The value of t at the point where $G(t+\theta\Delta t)$ is just below $W(t+\theta\Delta t)$ is $t^\star+\theta\Delta t$. We can use this property to compute $t^\star$. A more refined method for computing $t^\star$ is discussed elsewhere (Goh, 1977 b).

The optimal control program given in (4.10.16) provides the optimal harvesting program for a single year-class. It can also be used to formulate the optimal harvesting policy for a fish population which contains many year-classes.

Suppose the same effort is used to harvest the fish population during the limited fish season in each year. The fact that it is impossible to utilise nets to harvest each year-class selectively constitutes an intricate constraint in the problem of harvesting optimally a fish population with many year-classes. For the sake of argument, let us assume that it is possible to harvest each year-class selectively. This means the nonselectivity constraint is relaxed and hence the admissible set of options is enlarged. For a collection of year-classes which can be exploited separately, the optimal policy is to exploit each year-class separately. The crucial point in this argument is that this optimal harvesting policy can be implemented even if it is not possible to harvest each year-class separately. It can be implemented by specifying the optimal size limit. This requires that only fish which are larger than the optimal size limit should be harvested. This optimal policy maximizes the biomass yield for a fixed fishing effort.

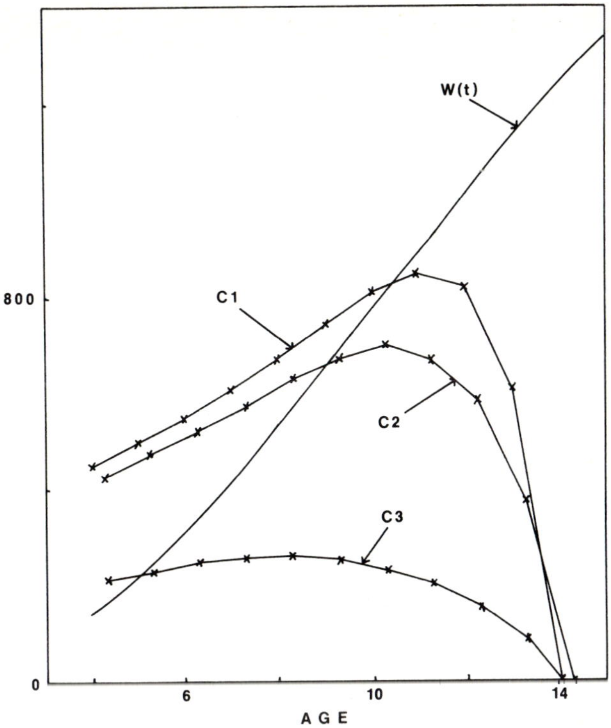

Fig.4.10.1. Intersections of the weight function $W(t)$ and the curves, $C1$, $C2$ and $C3$, give refined estimates of the ages for optimum size limits when: (i) the fishing season is unlimited; (ii) the fishing season is limited to 6 months in a year; and (iii) the fishing season is limited to one tenth of a year in each year.

For an example, we shall analyse a hypothetical fishery, using some of the data for the North Sea plaice. This hypothetical fishery is analysed instead of a real fishery because we want to study the effects of limiting the fishing season in a fishery.

Let $W(t)$ and $L(t)$ denote the growth and length functions respectively. We have

$$W(t) = 2867[1 - \exp -0.095(t + 0.82)]^3, \tag{4.10.22}$$

$$L(t) = 68.5[1 - \exp -0.095(t + 0.82)]. \tag{4.10.23}$$

The instantaneous natural mortality rate is $M = 0.10$ and the instantaneous fishing mortality rate is 0.73.

Suppose the fishing season is limited to one tenth of a year in each year. In this case, $\Delta t = 0.1$, $\theta = 0.5$ and $s = 10$. With $m = 0.10$ and $F = 0.73$, eqs. (4.10.1) and (4.10.8) give $m = 0.00995$ and $b = 0.0697$. For a typical year-class, let the limited fishing season begin at the ages of 3.4, 4.4, . . ., 14.4. We adjust the life span of an average fish from 15 years to 15.4 years; the effects of this adjustment on the total biomass yield would be negligible. This small adjustment is made so that the dynamics of the year-class can be described by (4.10.3) with $t_r = 3.4$ and $T = 15.4$.

The functions $W(t)$ and $G(t + \theta\Delta t)$ of (4.10.21) for this hypothetical fishery are plotted in Fig.4.10.1. The function $G(t + \theta\Delta t)$ is just below $W(t + \theta\Delta t)$ at $t = 5.4 + (0.5)(0.1)$. It follows that $t^\star = 5.4$. Using (4.10.23) we deduce that the optimal size limit is 30.6 cm. In Section 4.8 it is shown that the optimal size limit is 45 cm and the corresponding value of $t^\star$ is 10.4, if fishing at the given rate of $F = 0.73$ is carried out continuously during the year. Thus the optimal size limit should be reduced if the fishing season is shortened. It is interesting to note that the optimal size limit is also 30.6 cm if fishing is carried out continuously during the year with one tenth of the given fishing effort, namely, $F = 0.073$.

4.11. FISHERY WITH DELAYED RECRUITMENT

For many species the population contains many year-classes, and recruitment to the breeding population takes place several years after birth. For this type of population the Beverton—Holt model is an incomplete dynamical model because it does not incorporate a relationship between the parent stock and its progeny. It provides an adequate model for a population with a high fecundity. But for a whale or seal population which has a low fecundity, the relationship between stock and recruitment plays a crucial role in the management of the species. This type of population could be described by a nonlinear delay difference equation. This type of model is used by the International Whaling Commission in the management of whale populations. This type of model and its generalizations will become important in the management of harvested populations.

To simplify the analysis all the mature reproductive females are lumped into a single group. At time t, let $N(t)$ be the number of reproductive females. It is assumed that the population has a constant sex ratio. It is also assumed that in a harvested population, there is usually a negligible number of old and nonreproductive females. At time t, let $Y_m(t)$ where $m = 1, 2, \ldots, k$, denote the number of females which are m years old. A model of this type of population is

$$Y_1(t + 1) = F_1[N(t)],$$

$$Y_2(t + 1) = F_2[Y_1(t)],$$

$$\cdots\cdots\cdots\cdots\cdots\cdots$$

$$N(t + 1) = S[N(t)] + F_{k+1}[Y_k(t)], \tag{4.11.1}$$

where $F_1, F_2, \ldots, F_{k+1}, S$ are linear or nonlinear functions, and $t = 0, 1, 2, \ldots$.

Eliminating $Y_1, Y_2, \ldots, Y_k$ we get

$$N(t+1) = S[N(t)] + F[N(t-k)] \tag{4.11.2}$$

where $F(\cdot) = F_{k+1}F_k \cdots F_1(\cdot)$. This delay difference equation was used by Clark (1976 b) for formulating optimal policies in the harvesting of whale populations.

Model (4.11.2) is equivalent to

$$N_1(t+1) = N_2(t),$$
$$N_2(t+1) = N_3(t),$$
$$\ldots\ldots\ldots\ldots\ldots$$
$$N_{k+1}(t+1) = S[N_{k+1}(t)] + F[N_1(t)]. \tag{4.11.3}$$

This system has an equilibrium at $N^\star = (n^\star, n^\star, \ldots, n^\star)$ where $n^\star = S(n^\star) + F(n^\star)$.

The linearized dynamics of the last equation in (4.11.3) is

$$N_{k+1}(t+1) - n^\star = s(N_{k+1} - n^\star) + a(N_1(t) - n^\star) \tag{4.11.4}$$

where $s = \partial S/\partial N_{k+1}$ and $a = \partial F/\partial N_1$ at $N = N^\star$.

Using Rouche's theorem, Clark (1976 b) showed that a sufficient condition for $N^\star$ to be locally stable is

$$|a| < 1 - s \tag{4.11.5}$$

where $1 > s > 0$. This sufficient condition can also be established by using the Liapunov function

$$V(N) = |N_1 - n^\star| + |N_2 - n^\star| + \ldots + d|N_{k+1} - n^\star| \tag{4.11.6}$$

where $d = \min\,[1/|a|, 1/(1-s)]$ and where $|a|$ denotes the absolute value of a. If (4.11.5) is satisfied, the equilibrium $N^\star$ is locally stable irrespective of the length of the time delay.

For a real population to persist, it must be stable relative to finite perturbations of its initial state. Thus it is desirable to show that the equilibrium $N^\star$ of (4.11.3) is stable relative to finite perturbations.

If $k = 1$, model (4.11.3) is a two-variable system. In this case it is possible to establish that $\Delta V(N)$ is negative semidefinite for a wide range of the functions, $S(N_2)$, $F(N_1)$ and $V(N)$ where $V(N)$ is a Liapunov function. For $k > 1$, model (4.11.3) is a system with more than two variables. In this case, it is usually very difficult to verify that $\Delta V(N)$ is negative semidefinite. Fortunately for an important subclass of models, it is possible to reduce $\Delta V(N)$ to a function of two variables.

We shall show that the maximum steady state yield (MSSY) equilibrium of (4.11.2) when the population is subjected to a constant fraction harvesting policy, satisfies condition (4.11.5), and therefore, is locally stable. The Liapunov function $V(N)$ of (4.11.6) and similar functions may be used to establish a finite region of attraction for the MSSY equilibrium.

Recall that $N(t)$ is the number of reproductive females at time t. Let $u(t)$ be the number of reproductive females which is harvested during the time interval $[t, t + \Delta t]$ where Δt is a small positive number. Suppose reproduction takes place during the time interval $[t + \Delta t, t + 1]$. A model of the harvested population is

$$N(t + 1) = S[N(t) - u(t)] + F[N(t - k) - u(t - k)]. \qquad (4.11.7)$$

It is convenient and desirable to use the post-harvest population (the escapement) as the state variable. Let $H(t)$ denote the post-harvest population of reproductive females at time $t + \Delta t$. We have

$$H(t + 1) = N(t + 1) - u(t + 1). \qquad (4.11.8)$$

Eqs. (4.11.7) and (4.11.8) give

$$H(t + 1) = S[H(t)] + F[H(t - k)] - u(t + 1). \qquad (4.11.9)$$

If the handling time of a catch is negligible and the schooling effect is also negligible, then a constant effort harvesting can be approximated by a constant fraction harvesting. In this case

$$u(t) = EN(t) \qquad (4.11.10)$$

where E is a constant. Eqs. (4.11.9) and (4.11.10) give

$$H(t + 1) = (1 - E)[S(H(t)) + F(H(t - k))]. \qquad (4.11.11)$$

By definition $H^\star$ is a steady state if

$$H^\star = (1 - E)[S(H^\star) + F(H^\star)]. \qquad (4.11.12)$$

The corresponding yield per unit time is

$$Y^\star = E[S(H^\star) + F(H^\star)]. \qquad (4.11.13)$$

Combining (4.11.12) with (4.11.13) we get

$$Y^\star = [S(H^\star) + F(H^\star) - H^\star]. \qquad (4.11.14)$$

A necessary condition for $H^{\star\star}$ to be a maximum steady state yield (MSSY) equilibrium is

$$S'(H^{\star\star}) + F'(H^{\star\star}) - 1 = 0 \qquad (4.11.15)$$

where the prime denotes differentiation.

The equilibrium $H^{\star\star}$ satisfies (4.11.5) if

$$|(1 - E^{\star\star})\, F'(H^{\star\star})| < 1 - (1 - E^{\star\star})|S'(H^{\star\star})|. \qquad (4.11.16)$$

For most stock recruitment models, $S'(H^{\star\star})$ satisfies the inequality

$$0 < S'(H^{\star\star}) < 1. \qquad (4.11.17)$$

Conditions (4.11.15) to (4.11.17) give

$$(1-E^{**})(1-S') < 1-(1-E^{**})S'$$
$$\Rightarrow 1-E^{**} < 1. \tag{4.11.18}$$

This is satisfied because $1 > E^{**} > 0$. It follows that the MSSY equilibrium which is created by a constant fraction harvesting is locally stable.

We shall describe a method for constructing a Liapunov function for computing a finite region of stability of an equilibrium which satisfies (4.11.5). Let b, c and p be positive constants. The constants b and c are chosen by trial and error so that a certain function is negative semidefinite in a region as large as possible. For convenience let

$$U(N_i) = (N_i^2 - n^{\star 2})/2 - n^{\star 2} \ln (N_i/n^\star), \tag{4.11.19}$$

$$R(N_1, N_{k+1}) = S(N_{k+1}) + F(N_1), \tag{4.11.20}$$

$$Z(N_{k+1}) = N_{k+1}^2(N_{k+1} - n^\star)^2, \tag{4.11.21}$$

$$V(N) = \sum_{i=1}^{k} U(N_i) + bU(N_{k+1}) + cZ(N_{k+1}), \tag{4.11.22}$$

$$W(N_1, N_{k+1}) = -U(N_1) - (b-1)\,U(N_{k+1}) + bU[R(N_1, N_{k+1})]$$
$$+ c[Z(R(N_1, N_{k+1})) - Z(N_{k+1})], \tag{4.11.23}$$

$$M(N_1, N_{k+1}) = U(N_1) + bU(N_{k+1}) + cZ(N_{k+1}), \tag{4.11.24}$$

$$\Omega = \{N \mid V(N) < p; N_i > 0, i = 1, 2, \ldots, k+1\}, \tag{4.11.25}$$

$$\Gamma = \{(N_1, N_{k+1}) \mid M(N_1, N_{k+1}) < p; N_1 > 0, N_{k+1} > 0\}. \tag{4.11.26}$$

Theorem 4.11.1. *Let $W(N_1, N_{k+1})$ be defined by (4.11.19) to (4.11.23). If there exist positive constants b, c and p such that $W(N_1, N_{k+1})$ is negative definite in the region Γ, then Ω is a region of attraction of the equilibrium $N^\star$ of model (4.11.3).*

Proof. Along solutions of (4.11.3),

$$\Delta V(N) = V[N(t+1)] - V[N(t)]$$
$$= \sum_{i=2}^{k+1} U(N_i) + bU[R(N_1, N_{k+1})] + cZ[R(N_1, N_{k+1})]$$
$$- \sum_{i=1}^{k} U(N_i) - bU(N_{k+1}) - cZ(N_{k+1})$$
$$= W(N_1, N_{k+1}). \tag{4.11.27}$$

By assumption $W(N_1, N_{k+1})$ is negative definite in Γ. For $N_2 = N_3 = \ldots = N_k = n^\star$ we have $V(N) = M(N_1, N_{k+1})$. This equality, eq. (4.11.27) and the assumption that $W(N_1, N_{k+1})$ is negative definite in Γ, imply that $\Delta V(N)$ is negative semidefinite in Ω.

Using an extension of the direct method of Liapunov (Theorem 3.14.2) the proof is completed by showing that $\Delta V(N)$ does not vanish identically along a nontrivial solution of (4.11.3). The assumption that $W(N_1, N_{k+1})$ is negative definite in Γ implies that in the region Ω, the function $\Delta V(N)$ is equal to zero only at a point of the form

$$N = (n^\star, N_2, N_3, \ldots, N_k, n^\star). \qquad (4.11.28)$$

If (4.11.3) has a solution of this form for $t = 0, 1, 2, \ldots k-1$, we have

$$N(0) = (n^\star, N_2(0), N_3(0), \ldots, N_k(0), n^\star),$$

$$N(1) = (n^\star, N_2(1), N_3(1), \ldots, N_{k-1}(1), n^\star, n^\star),$$

$$N(2) = (n^\star, N_2(2), N_3(2), \ldots, N_{k-2}(2), n^\star, n^\star, n^\star),$$

$$\ldots\ldots\ldots\ldots\ldots\ldots\ldots\ldots\ldots\ldots\ldots\ldots$$

$$N(k-1) = (n^\star, n^\star, \ldots, n^\star, n^\star) = N^\star. \qquad (4.11.29)$$

We conclude that besides the equilibrium solution or a solution which reaches $N^\star$ in the manner shown in (4.11.29), the function $\Delta V(N)$ does not vanish identically along a solution of (4.11.3). It follows that Ω is a region of attraction of the equilibrium $N^\star$.

Corollary 4.11.1. *If $W(N_1, N_{k+1})$ is negative definite in the positive quadrant of the (N_1, N_{k+1})-space, then $\Delta V(N)$ is negative semidefinite in the positive orthant $\Omega = \{N \mid N_i > 0, i = 1, 2, \ldots, k+1\}$. In this case model (4.11.3) is globally stable.*

For the sake of simplicity we had used a special expression for defining $U(N_i)$ in (4.11.19). We could let $U(N_i)$ be some other Liapunov function of a scalar first order difference equation. Similarly $Z(N_{k+1})$ of (4.11.21) could be replaced by some other positive definite function of $(N_{k+1} - n^\star)^2$.

We have shown how a generalization of the direct method of Liapunov can be used to establish finite or global stability in a class of nonlinear Leslie models for a population with an age structure. This class of models is important in the management of some fish and whale populations.

A harvested population should be stable relative to finite perturbations of its initial state. Otherwise errors in estimating the population could lead to the destruction of the population. Theorem 4.11.1 may also be used to compute a region of attraction of an equilibrium of a subclass of models for a population which is subjected to quota harvesting.

A proper stabilizing feedback policy for quota harvesting of a population described by (4.11.2) should be a feedback control function of $N(t)$, $N(t-1)$, . . ., $N(t-k-1)$. At present the quota harvesting policy of the International Whaling Commission for a population close to its maximum sustainable yield equilibrium is a function of only the current stock of reproductive females, $N(t)$. This may not provide the stabilizing effect that is expected from the prescribed quota harvesting policy.

In the next section, we shall apply the method developed in this section to models of the Antarctic fin whale population.

4.12. THE ANTARCTIC FIN WHALE POPULATION

We shall show that the unexploited population of the Antarctic fin whale, *Balaenoptera physalus*, is globally stable. We shall compute a region of attraction for the equilibrium in a model of the population when it is subjected to a constant quota harvesting.

Following Ricker (1954) the function $F(N)$ in (4.11.2) is called a stock-recruitment model. The Ricker (1954) and the Beverton—Holt (1957) stock-recruitment models are fitted by inspection to the data provided by Allen (1973) for the southern hemisphere stock of the fin whale population. The quadratic stock-recruitment model $F(N) = (r/K)N(K-N)$ which has been used by the International Whaling Commission is unsuitable for a global stability analysis of the population dynamics. The quadratic stock-recruitment model gives negative recruitment when the parent stock is very large. Biologically this is nonsensical.

Using the Ricker stock-recruitment model we get

$$N(t+1) = sN(t) + AN(t-8) \exp[-BN(t-8)] \tag{4.12.1}$$

where $s = 0.96$, $A = 0.12$ and $1/B = 3 \times 10^5$. This model has an equilibrium at $N = n^\star = (1/B) \ln [A/(1-s)] = (3 \ln 3) \times 10^5$. The stability of this equilibrium is independent of the value of B.

For convenience let $x = N/n^\star$. Using the transformation in (4.11.3) we get

$$x_1(t+1) = x_2(t),$$

$$x_2(t+1) = x_3(t),$$

. . . .

$$x_9(t+1) = sx_9(t) + Ax_1(t) \exp[-Cx_1(t)], \tag{4.12.2}$$

where $C = \ln [A/(1-s)]$. This model has an equilibrium at $x^\star = (1, 1, . . ., 1)$.

A Liapunov function for this normalized model is the function $V(N)$ in (4.11.22). By trial and error we choose $b = 15$ and $c = 1$. For model (4.12.2) the function $W(x_1, x_9)$ in (4.11.23) is negative definite in the positive quadrant of the (x_1, x_9)-space. This is established by plotting some of the level sur-

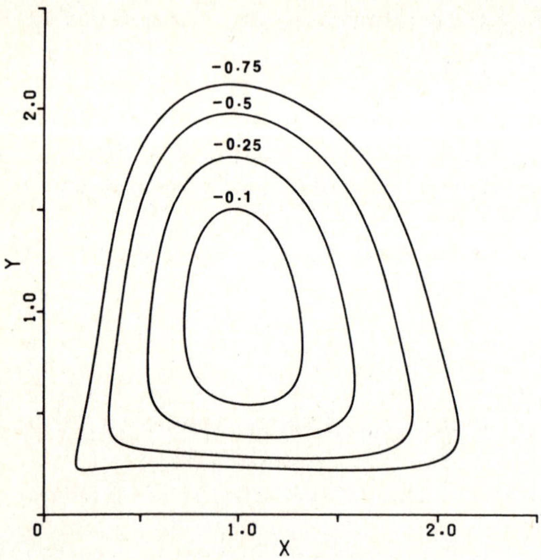

Fig.4.12.1. Level sets of the function $W(x_1, x_9)$ for model (4.12.2).

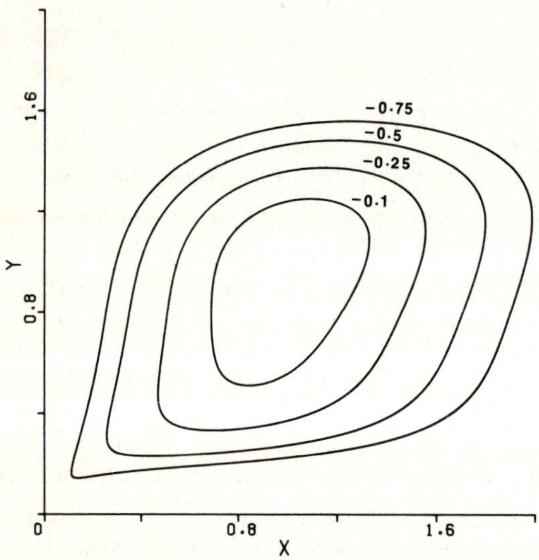

Fig.4.12.2. Level sets of the function $W(x_1, x_9)$ for a system of first order equations which is equivalent to model (4.12.3).

faces of $W(x_1, x_9)$ (see Fig.4.12.1). Let $X = x_1$ and $Y = x_9$. Figure 4.12.1 displays some of the level sets of $W(X, Y)$. Many of the computed level surfaces of $W(X, Y)$ are not displayed. The function $W(X, Y)$ is a continuously differentiable function in the positive quadrant. This justifies the conclusion that $W(X, Y)$ is negative definite in the positive quadrant. It follows that model (4.12.2) is globally stable.

Using the Beverton—Holt stock-recruitment model, eq. (4.11.2) gives

$$N(t+1) = sN(t) + rN(t-8)/[1 + AN(t-8)] \qquad (4.12.3)$$

where $s = 0.96$, $r = 0.12$ and $A = 6 \times 10^{-6}$. The analysis for (4.12.2) is applied to this model with $b = 15$ and $c = 10$ in the Liapunov function (see Fig.4.12.2). These constants are chosen by trial and error. We get the same conclusion that the population is globally stable.

Under the constant quota harvesting of u females per year, model (4.12.1) becomes

$$N(t+1) = sN(t) + AN(t-8)\exp[-BN(t-8)] - u. \qquad (4.12.4)$$

If $u = 3000$ the equilibrium of (4.12.4) is at $N = n^\star = 251\,210$. For convenience let $x = N/n^\star$ and $v = u/n^\star$. Under this change of variables, we get

$$x_1(t+1) = x_2(t),$$

$$x_2(t+1) = x_3(t),$$

$$\ldots\ldots\ldots\ldots\ldots$$

$$x_9(t+1) = sx_9(t) + Ax_1(t)\exp[-Cx_1(t)] - v \qquad (4.12.5)$$

where $C = 0.838$.

The Liapunov function for (4.12.2) is used in the analysis of model (4.12.5). A region in which $W(x_1, x_9)$ is positive is determined. Some of the level surfaces of the function $M(x_1, x_9)$ of (4.11.24) for model (4.12.5) are computed. From this we find that $W(x_1, x_9)$ is negative definite in the region $\{(x_1, x_9) \mid M(x_1, x_9) < 17.98\}$. This implies that $\Delta V(x)$ is negative semidefinite in the region $\{x \mid V(x) < 17.98\}$ where $V(x)$ is given in (4.11.22). It follows that a region of attraction of the equilibrium $x^\star = (1, 1, \ldots, 1)$ is $\{x \mid V(x) < 17.98\}$. This region contains the region $\{x \mid \Sigma(x_i - 1)^2 < (0.71)^2\}$.

4.13. HARVESTING A MULTISPECIES COMMUNITY

The theory of optimal harvesting of a multispecies community is not as well developed as that for a single species population. One reason is that it is very difficult to build a realistic model of a multispecies community which leads to an analytically tractable optimal control problem. Another reason is that it is difficult to apply optimal control theory to a problem with more than two state variables.

A pragmatic approach to this difficult management problem is to build two classes of models. One class of models consists of simplified models which capture some of the key dynamical processes of a harvested multispecies community. These models are subjected to stability and optimal control analysis. The insights obtained in this exercise are then used in the study of the second type of model. The latter is a detailed computer simulation model of the harvested multispecies community.

We shall examine some aspects in the mathematical analysis of a simplified analytical model of a harvested multispecies community. We shall describe several ideas which may be useful for formulating good harvesting policies for a multispecies community in the real world.

Firstly let us consider the problem of harvesting m interacting species so as to maximize the total discounted net revenues. This problem has been studied by Clark (1976 a) add Silvert and Smith (1977). Let N_i be the population of the ith species, and u_i be the rate of harvest of the ith species. Suppose the dynamics of the harvested multispecies community is

$$\dot{N}_i = G_i(N) - u_i, \quad i = 1, 2, \ldots, m. \tag{4.13.1}$$

Let $[0, T]$ be the planning horizon, and at time T, let

$$N_i(T) \geqslant e_i, \quad i = 1, 2, \ldots, m \tag{4.13.2}$$

where e_i are positive constants. The constants $e_1, e_2, \ldots, e_m$ are chosen so that the community is viable after time T if harvesting ceases thereafter.

Let the harvest rates be bounded from above so that

$$b_i \geqslant u_i(t) \geqslant 0, \quad i = 1, 2, \ldots, m, \tag{4.13.3}$$

where $b_1, b_2, \ldots, b_m$ are constants. These constraints are imposed tentatively in order to provide a healthy and robust fishing industry.

The problem is to choose the functions $u_i(t)$ which will drive the system from the initial state

$$N_i(0) = N_{i0}, i \;\; i = 1, 2, \ldots, m, \tag{4.13.4}$$

so as to maximize the present value integral

$$J = \int_0^T e^{-\delta t}\left[\sum_{i=1}^{m} (v_i - c_i/N_i)u_i\right] dt \tag{4.13.5}$$

where v_i is the price for fish of the ith species and $c_i u_i/N_i$ is the cost of harvesting the ith species at the rate of u_i and δ is the discount rate.

This is an optimal control problem. For $m > 2$ it is expensive to compute the optimal feedback control policy. A feedback control harvesting policy for a model of a real system is desirable because it could provide a harvesting policy for the real system which is insensitive to small errors in the measurement of the state vector and in the system dynamics.

The analysis of a single-species model suggests that an optimal harvesting program for a multispecies community should consist of three stages. In the first stage the system is driven rapidly to an optimal steady state. The system is then maintained at the steady state until the value of t is close to T. In the third stage the population levels are driven to lower levels which depend on the cost of harvesting and the terminal constraints. However, this structure in an optimal harvesting program for a multispecies community has not been established rigorously for any problem; and in general, an optimal harvesting program will not have this structure.

If the optimal feedback harvesting policy has such a structure, the first task is to compute the optimal steady state. This steady state is a singular extremal because the control variables appear linearly in the system equations (4.13.1) and the objective function (4.13.5). It can be computed more easily by transforming first the integral (4.13.5). Let $y_i(N)$ satisfy

$$\dot{y}_i = (v_i - c_i/N_i)\dot{N}_i, \quad i = 1, 2, \ldots, m, \tag{4.13.6}$$

such that $y_i(0) = 0$ for $i = 1, 2, \ldots, m$. Eqs. (4.13.1) and (4.13.5) give

$$J = \int_0^T e^{-\delta t} \sum_{i=1}^{m} [(v_i - c_i/N_i)(G_i - \dot{N}_i)]\,dt$$

$$= \sum_{i=1}^{m} e^{-\delta T} y_i(T) + \int_0^T e^{-\delta t} \sum_{i=1}^{m} (v_i - c_i/N_i)(G_i - \delta y_i)\,dt. \tag{4.13.7}$$

For convenience, let

$$R(N, c, \delta) = \sum_{i=1}^{m} (v_i - c_i/N_i)(G_1(N) - \delta y_i(N)). \tag{4.13.8}$$

We shall use optimal control theory to compute the optimal steady state. By definition the Hamiltonian function

$$H = e^{-\delta t} R(N, c, \delta) + \sum_{i=1}^{m} p_i(G_i(N) - u_i). \tag{4.13.9}$$

Suppose there exists an optimal subarc which satisfies the conditions $0 < u_i(t) < b_i$ for $i = 1, 2, \ldots, m$. Let $(N^\star(t), u^\star(t))$ denote such an optimal solution. It must satisfy the conditions

$$\dot{p}_i = -\partial H/\partial N_i = -e^{-\delta t}\partial R/\partial N_i - \sum_{s=1}^{m} p_s \partial G_s/\partial N_i, \tag{4.13.10}$$

$$\partial H/\partial u_i \equiv 0, \quad i = 1, 2, \ldots, m. \tag{4.13.11}$$

Eqs. (4.13.9) and (4.13.11) give

$$\partial H/\partial u_i = -p_i \equiv 0, \quad i = 1, 2, \ldots, m. \tag{4.13.12}$$

The control vector u appears linearly in the Hamiltonian function. Therefore if there is an optimal interior control vector, it must be a singular control. Let $D \equiv d/dt$. By Theorem 2.5.2 necessary conditions for the singular control to be optimal are

$$D(\partial H/\partial u_i) = -\dot{p}_i \equiv 0, \tag{4.13.13}$$

$$(\partial/\partial u_j)D(\partial H/\partial u_i) = e^{-\delta t}\partial^2 R/\partial u_j \partial N_i + \sum_{s=1}^{m} p_s \partial^2 G_s/\partial u_j \partial N_i \equiv 0, \tag{4.13.14}$$

$$D^2(\partial H/\partial u_i) = \sum_{r=1}^{m} e^{-\delta t}\left(\partial^2 R/\partial N_i \partial N_r + \sum_{s=1}^{m} p_s \partial^2 G_s/\partial N_i \partial N_r\right)(G_r - u_r)$$

$$- \delta e^{-\delta t}\partial R/\partial N_i \equiv 0 \tag{4.13.15}$$

for $i, j = 1, 2, \ldots, m$. Conditions (4.13.10), (4.13.12) and (4.13.13) imply that

$$e^{-\delta t}\partial R/\partial N_i \equiv 0, \quad i = 1, 2, \ldots, m. \tag{4.13.16}$$

If the function R is autonomous, then a solution of (4.13.16) gives a steady state solution to the optimal control problem. Let (4.13.16) have a unique positive solution at $N^\star$ and the corresponding constant control vector be $u^\star$.

In this optimal control problem the objective function J in (4.13.5) is to be maximized. Thus the generalized Legendre condition (Theorem 2.5.2) requires that along the singular solution, the matrix whose typical element is

$$(\partial/\partial u_j)D^2(\partial H/\partial u_i) = -e^{-\delta t}\partial^2 R/\partial N_i \partial N_j \tag{4.13.17}$$

where $i, j = 1, 2, \ldots, m$, must be positive semidefinite. Hence the matrix $(\partial^2 R/\partial N_i \partial N_j)$ must be negative semidefinite at $N^\star$. It follows that the function $R(N, c, \delta)$ must be a local maximum at $N^\star$.

Example 4.13.1. Consider the competition model

$$\dot{N}_1 = N_1(3 - 2N_1 - N_2) - u_1,$$

$$\dot{N}_2 = N_2(2 - N_1 - N_2) - u_2.$$

For $u = (0, 0)$ this model has a globally stable equilibrium at (1, 1).

Let $v_i = 1$, $c_i = 0$ for $i = 1, 2$ and $\delta = 0$. The function

$$R = N_1(3 - 2N_1 - N_2) + N_2(2 - N_1 - N_2)$$

has a global maximum at (1/2, 11/2). The corresponding optimal control vector is $u^\star = (3/4, 1/2)$.

Example 4.13.2. Consider the harvested prey-predator model

$$\dot{N}_1 = N_1(1.45 - 0.1N_1 - N_2) - u_1 = G_1 - u_1, \tag{4.13.18}$$

$$\dot{N}_2 = N_2(-0.8 + 0.2N_1 - 0.1N_2) - u_2 = G_2 - u_2, \tag{4.13.19}$$

where N_1 = population of prey and N_2 = population of predator. For $u = (0, 0)$ this model has a globally stable equilibrium at (4.5, 1).

Let $v_1 = 0.095$, $v_2 = 1$, $c_1 = 0$, $c_2 = 0$ and $\delta = 0$. The function

$$R = 0.095N_1(1.45 - 0.1N_1 - N_2) + N_2(-0.8 + 0.2N_1 - 0.1N_2) \qquad (4.13.20)$$

has an interior stationary solution at (7.813, 0.1019). The corresponding control vector is $u = (4.428, 0.07667)$ and $R = 0.4974$. But the matrix

$$\left(\frac{\partial^2 R}{\partial N_i \partial N_j}\right) = \begin{pmatrix} -0.019 & 0.105 \\ 0.105 & -0.2 \end{pmatrix}$$

is indefinite. Hence the singular solution at (7.813, 0.1019) is nonoptimal. This solution is a saddle point of the function R.

The parameter values $v_1 = 0.095$, $v_2 = 1.0$ imply that the predator is more valuable than the prey. Thus we can expect that an optimal solution may be given by $u_1 = 0$ and u_2 is a singular control function.

By definition, the Hamiltonian function is

$$H = R(N_1, N_2) + p_1(G_1 - u_1) + p_2(G_2 - u_2) \qquad (4.13.21)$$

where G_1, G_2 and R are given in (4.13.18) to (4.13.20). Along a partially singular extremal we have

$$\partial H/\partial u_2 = -p_2 = 0, \qquad (4.13.22)$$

$$(\mathrm{d}/\mathrm{d}t)(\partial H/\partial u_2) = 0, \qquad (4.13.23)$$

$$(\mathrm{d}^2/\mathrm{d}t^2)(\partial H/\partial u_2) = 0. \qquad (4.13.24)$$

Eqs. (4.13.22) to (4.13.24) and $G_1 = 0$ are satisfied by the steady state solution, $N^\star = (9.5, 0.5)$ and $u^\star = (0, 0.525)$. The corresponding value of R is equal to 0.525. Clearly this steady state gives a better return than the totally singular solution at $(N_1, N_2) = (7.813, 0.1019)$.

We shall now consider some aspects of the problem of optimally driving a harvested community from an arbitrary admissible state to a steady state singular solution at $N^\star$. For simplicity we shall consider the harvesting of a two-species system under the assumption that the planning horizon $[0, T]$ is very large. In this case the control programs for t close to T can be ignored.

In brief the optimal harvesting problem is as follows:

System: $\dot{N}_i = G_i(N) - u_i, \quad i = 1, 2.$ (4.13.25)

Initially: $N_i(0) = N_{i0}.$ (4.13.26)

Terminally: $N_i(T) \geqslant e_i, \quad T = \text{constant}.$ (4.13.27)

Constraints: $0 \leqslant u_i \leqslant b_i.$ (4.13.28)

Objective: $\max \left[-\sum_{i=1}^{2} \mathrm{e}^{-\delta T} y_i[N(T)] + \int_0^T \mathrm{e}^{-\delta t} R(N, c, \delta)\,\mathrm{d}t \right].$ (4.13.29)

Here y and R are given in (4.13.6) and (4.13.8) respectively.

Suppose $R(N, c, \delta)$ has a global maximum at the positive state at $N^\star$; and the corresponding totally singular control $u^\star = G(u^\star)$ is admissible. Accordingly the singular solution $(N^\star, u^\star)$ satisfies (4.13.16), (4.13.25), (4.13.28) and the generalized Legendre condition.

The costate equations are

$$\dot{p}_i = -e^{-\delta t}\frac{\partial R}{\partial N_i} - p_1\frac{\partial G_1}{\partial N_i} - p_2\frac{\partial G_2}{\partial N_i}, \quad i = 1, 2. \tag{4.13.30}$$

An optimal trajectory must satisfy these equations. Furthermore, at a switching junction, H, p_1 and p_2 must be continuous. These conditions are needed for computing the field of solutions as the initial state is varied.

Typically the system is first driven by bang-bang control (i.e. $u_i = 0$ or $u_i = b_i$) to a partially singular control arc. The system should then be driven along the partially singular control arc until the state reaches the totally singular solution at $N^\star$.

The field of solutions in the positive quadrant of the (N_1, N_2)-space is generated by integrating, in reverse time, the state eqs. (4.13.25), and using the necessary conditions for optimality. Generally, we first compute eight isolated trajectories which are joined to the totally singular solution $N = N^\star$. These trajectories are generated by the control vectors (0, 0), (0, b_2), (b_1, 0), (b_1, b_2), (0, singular), (singular, 0), (b_1, singular) and (singular, u_2). We must also use conditions $p_1 = 0$ and $p_2 = 0$ along the totally singular solution $N = N^\star$ in order to join optimally the eight trajectories to the totally singular solution.

As an example, let us consider how the u = (singular, 0) arc is computed. It is obtained by integrating, in reverse time from the totally singular solution $N = N^\star$, the state eqs. (4.13.25) with $u_2 = 0$ and u_1 satisfying

$$\partial H/\partial u_1 = -p_1 = 0, \tag{4.13.31}$$

$$(d/dt)(\partial H/\partial u_1) = e^{-\delta t}\partial R/\partial N_1 + p_2\partial G_2/\partial N_1 = 0, \tag{4.13.32}$$

$$(d^2/dt^2)(\partial H/\partial u_1) = 0. \tag{4.13.33}$$

The generalized Legendre condition is satisfied if

$$(\partial/\partial u_1)(d^2/dt^2)(\partial H/\partial u_1) = -e^{-\delta t}[\partial^2 R/\partial N_1^2 - (\partial^2 G_2/\partial N_1^2)(\partial R/\partial N_1)/(\partial G_2/\partial N_1)] \geqslant 0 \tag{4.13.34}$$

along the partially singular control solution.

Eqs. (4.13.32) and (4.13.33) can be used to compute u_1 as a function of t and N if condition (4.13.34) is satisfied in the inequality manner. Recall that $u_2 = 0$. With these functions, u_1 and u_2, the state eqs. (4.13.25) are numerically integrated in reverse time from the totally singular solution $N^\star$. The costate variables, p_1 and p_2, along the partially singular solutions are computed by means of (4.13.31) and (4.13.32) instead of integrating the costate equations (4.13.30). This is because the numerical solutions of the costate equations (4.13.30) will not satisfy conditions (4.13.31) in an exact manner.

From each point of the u = (singular, 0) arc, we can construct two bang-bang arcs by integrating, in reverse time, the state and costate equations, (4.13.25) and (4.13.30). We must also use the conditions that p_1 and p_2 are continuous at a switching point. At the beginning of the integration, let $u = (0, 0)$ in one case and $u = (b_1, 0)$ in the other case. Subsequently the control vector is determined by the signs of $\partial H/\partial u_1$ and $\partial H/\partial u_2$.

As an example, consider how the $u = (0, 0)$ arc which joins the totally singular solution at $N^\star$ is computed. The state and costate equations, (4.13.25) and (4.13.30), are integrated in reverse time, using the "initial" conditions, $N = N^\star$, $p = (0, 0)$ and $u = (0, 0)$. Subsequently the values of u_1 and u_2 are determined by the signs of $\partial H/\partial u_1$ and $\partial H/\partial u_2$. In this manner we can compute the four bang-bang arcs which join directly with the totally singular solution $N = N^\star$.

The above method may be used to compute a unique field of solutions in the positive quadrant, provided that the generalized Legendre condition is satisfied along the entire arc of each of the partially singular control solutions and the switching of the control variables occurs only on the singular solutions. From (4.13.34) it is clear that it is possible for the generalized Legendre condition to be satisfied only on a subarc of a partially singular arc even if $R(N, c, \delta)$ has a global maximum at $N^\star$. In this case there could be a complex field of solutions.

However, the method for constructing a complete field of solutions in the positive quadrant of the (N_1, N_2)-space is not applicable if the function $R(N, c, \delta)$ does not have a local maximum at a totally singular solution at $N^\star$. Example 4.13.2 demonstrates that this possibility exists. This case is being investigated.

Consider again the case where $R(N, c, \delta)$ has a global maximum at $N^\star$. Here, one way to drive the system to the totally singular solution at $N^\star$ is to choose admissible control variables, u_1 and u_2, so as to maximize the function

$$\dot{R} = (\partial R/\partial N_1)(G_1 - u_1) + (\partial R/\partial N_2)(G_2 - u_2). \tag{4.13.35}$$

In other words, u_1 and u_2 are chosen so that the projection of the direction of a trajectory in the steepest ascent direction is maximized. However, the function $-R(N, c, \delta)$ does not tend to infinity as $N_i \to 0+$ for $i = 1, 2$. This means that the above way of prescribing u_1 and u_2 may generate solutions which intersect a coordinate axis. Therefore, close to the coordinate axes, the control variables must be prescribed according to some other rules.

From this discussion, it is clear that the optimal policy to drive a system to a totally singular solution is generally very complex. This complexity means that it may be difficult to implement the optimal feedback control. Furthermore the optimal policy may be sensitive to the model of the natural dynamics.

Let $t^{\star}(N(0))$ be the minimum time it takes to drive the system from the state $N(0)$ to the totally singular solution at $N^{\star}$. Let δ be the discount interest rate. If $\delta t^{\star}(N(0)) \gg 1$, the major part of the value of the functional in (4.13.29) is accumulated during the transition from $N(0)$ to $N^{\star}$. It is then desirable to use the optimal feedback control or another feedback policy which approximates the optimal policy.

If $\delta = 0$, the totally singular solution $N^{\star}$ is none other than a modified version of the maximum sustainable yield (MSY) equilibrium. The modifications take into account the relative values of different species and the costs in harvesting them. Here, it is possible to use any stable feedback control policy to drive the system to $N^{\star}$ without making much difference to the value of the objective function in the long run. The robustness of this harvesting policy provides a very cogent argument for using the modified MSY equilibrium in harvesting a multispecies community instead of one which incorporates a discount interest rate.

CHAPTER 4: SELECTED REFERENCES

Fisheries management

Gordon (1954); Ricker (1954, 1975); Schaefer (1954); Beverton and Holt (1957); Goh (1969/1970, 1973 a, 1977 b); Smith (1969); Clark (1973 a, b, 1976 b, 1978, 1979); Gulland (1974, 1977, 1978); Cushing (1975); Palm (1975, 1977); Anderson (1977); Larkin (1977); Silvert and Smith (1977); Clark and Mangel (1978); Clark and Munro (1978); Silvert (1978).

The logistic model

Schaefer (1954); Goh (1969/1970); Pella and Tomlinson (1969); Silliman (1971); Clark (1973 a, 1976 a); Cliff and Vincent (1973); Beddington and May (1977); Schnute (1977).

Stability of bionomic equilibrium

Smith (1969); Clark (1976 a, 1978).

Stock recruitment model

Moran (1950); Ricker (1954); Cushing and Harris (1973); Hassell (1975); Hassell et al. (1976 b); Goh (1977 c); Goh and Agnew (1978); Fisher et al. (1979).

Optimum size limit

Allen (1953); Beverton and Holt (1957); Goh (1973 a, 1977 b).

Models with age structure

Leslie (1945, 1948); Lefkovitch (1965); Booth (1972); Beddington and Taylor (1973); Auslander et al. (1974); Rorres and Fair (1975); Clark (1976 b); Usher (1976); Goh and Agnew (1979); Reed (1979).

Models of whale population

Annual Reports of the International Whaling Commission; Allen (1973); Schevill (1974); Clark (1976 b); Allen and Kirkwood (1977); Gambell (1977); Goh and Agnew (1978, 1980); Beddington (1978).

Harvesting a multispecies community

Larkin (1963, 1966); Rosenzweig (1971); Gilpin (1972); Clark (1976 a); Brauer et al. (1976); Getz (1977); Silvert and Smith (1977).

Chapter 5

COMPLEX ECOSYSTEMS

5.1. INTRODUCTION

Elton (1958) put forward the hypothesis that complex ecosystems are more stable than simple ecosystems. For some years this hypothesis was accepted as a central principle in ecology. May (1974), however, warned against a blind acceptance of this principle, and showed that stability decreases with complexity among mathematical models of ecosystems.

May's criticism has been supported and extended by Roberts (1974), Gilpin (1975 b), Goh and Jennings (1977). They have shown that the requirement that a nonlinear model should have a positive equilibrium imposes a very severe constraint on the parameters of the system when the number of species is large. Moreover, Goh and Jennings (1977) showed that the result of Gardner and Ashby (1970) and May (1974) on the stability of randomly assembled linear systems applies to the subset of Lotka-Volterra models with a positive equilibrium. Thus it is doubly difficult for a Lotka–Volterra model to represent a viable and stable complex ecosystem which is spatially homogenous and well mixed.

In the real world, ecosystems are continually perturbed by large and unpredictable disturbances. Nonvulnerability was developed by Goh (1971, 1975, 1976 a) to describe an ecosystem's ability to withstand large, unpredictable and continual disturbances. Using a result of Lee and Markus (1967), Goh (1975) showed that the set of linear ecosystem models which is completely vulnerable under a single perturbing force is open and dense in the space of such systems. It follows that species richness per se does not confer on a complex linear ecosystem model the property of the "not-putting-all-eggs-in-one-basket" kind.

From these studies on the relationships between stability and complexity, we conclude that real ecosystems must have some very special and mathematically atypical sorts of complexity which permit and enhance stability. It follows that a central problem in theoretical ecology is to elucidate the very special features in a complex ecosystem which promote the persistence of the system.

Stability as a population model is often established by means of an eigenvalue analysis. This establishes only local stability. Ideally, we should test that an ecosystem model is stable relative to finite perturbations. But it is generally very difficult to do so in a population model with more than two species. In Sections 5.3 to 5.5, we shall derive some conditions for global stability in several classes of complex nonlinear ecosystem models. These results are obtained by means of the direct method of Liapunov.

This direct method was discovered in 1892 by A.M. Liapunov (see Liapunov, 1966), but it was ignored by mathematicians for a long time. In the West it became a popular tool for the analysis of nonlinear systems in the nineteen-fifties. This method requires the construction of certain functions called Liapunov functions. For a physical system, the Liapunov method generalizes the idea that a system which continuously dissipates energy until it attains equilibrium is stable.

In an ecosystem the population of each species must be nonnegative. This requires that in ecological theory a two-sided energy principle for constructing a Liapunov function is more appropriate (Goh, 1977 a). A viable single-species population must have net energy dissipation when its population is large; and it must have net energy absorption when its population is low. If a single-species population has this behavior it is said to be self-regulating. A complex ecosystem is robust if it is stable relative to large perturbations of its initial state, and if the qualitative behavior of its dynamics is not sensitive to small variations in the parameters of the system. In Section 5.5, it is shown that an ecosystem model is robust if it is a collection of populations each of which is self-regulating at extreme densities, and is such that from the total system's point of view, the interactions between the populations are weaker than the self-regulating interactions of the populations.

By definition, an ecosystem model is connectively stable if it remains stable under certain classes of finite changes in the parameters which may alter the type of interactions between species. This concept of stability was pioneered by Siljak (1978). In Section 5.6 it is applied to a class of non-linear ecosystem models.

Nonvulnerability describes a system's ability to withstand large and unpredictable disturbances. In Section 5.7 we establish that a class of nonlinear models is nonvulnerable under a prescribed class of disturbances. The bounded but unknown functions which act on a nonlinear ecosystem model can also be interpreted as errors in the system's equations which can be tolerated without a qualitative change in the behavior of the dynamics of the system.

The requirement that a nonlinear ecosystem model has a positive equilibrium imposes a very severe constraint on the parameters of the system. But a given nondegenerate Lotka—Volterra model for m species has 2^m equilibria; usually a significant fraction of these are nonnegative. This suggests that the stability of nonnegative equilibrium is important in the analysis of a complex ecosystem model.

The concept of asymptotic stability cannot always be used directly to establish the stability of a nonnegative equilibrium. For example, the equilibrium $N = 0$ of $\dot{N} = -N^2$ is not asymptotically stable. But every solution of the equation which has a positive initial value remains positive for all finite values of t and converges to $N = 0$ as $t \to \infty$. This motivated the development of the concept of sector stability. By definition, a nonnegative equilibrium is sector stable if every solution of the system which begins in an admissible nonnegative neighbourhood remains in an admissible neighbourhood for all finite values of t, and converges to the nonnegative equilibrium as $t \to \infty$. Some simple conditions for a nonnegative equilibrium to be globally sector stable are obtained. This concept is useful for understanding the dynamics of individual patches in a spatially heterogenous community.

In the real world, the growth rate of a population usually does not respond immediately to changes in its own population or that of an interacting population. This fact introduces time delays into population dynamics. In Section 5.9 some simple conditions for global stability are obtained for a class of population models with continuous time delays.

Spatial heterogeneity plays an important role in the persistence of a complex ecosystem. Suppose a community in a region is an inter-connected network of patches. From the preceding discussions we can expect that each patch in isolation has a sector stable equilibrium which contains a few species. Suppose different combinations of species persist in different patches when the latter are in isolation. If the diffusion coefficients between patches are small, then we expect that the network of patches will allow a relatively large number of species to persist in the region. In Section 5.10 we shall examine some computer simulations which suggest that this type of network is globally stable, if all the patches in isolation are globally sector stable, and the diffusion coefficients between the patches are small. Thus spatial heterogeneity can make a complex ecosystem robust.

5.2. MATHEMATICAL RELATIONSHIPS BETWEEN STABILITY AND COMPLEXITY

Elton (1958) put forward the hypothesis that in Nature a complex ecosystem tends to be more stable than a simple ecosystem. This hypothesis was supported by the observation that populations in a complex tropical ecosystem are more stable than those in an arctic or a desert region. For a number of years this hypothesis was accepted as a central principle in ecology.

However, it was attacked by May (1974), who showed that increasing complexity reduces stability in population models. From studies of specific classes of population models of increasing complexity, May (1974) showed that a simple community model tends to have a comparatively large domain of parameter space wherein it is stable; while on the other hand, a complex community model tends to be fragile because it is stable for parameters within a comparatively small and restricted region of its more highly dimensional parameter space. Following Gardner and Ashby (1970), May (1974) showed that the stability of randomly connected linear systems decreases as the number of species increases, and as the strength and the number of connections among the species increase.

Let the linearized dynamics of an ecosystem model in a neighbourhood of an equilibrium be

$$\dot{x}_i = \sum_{j=1}^{m} a_{ij}x_j, \quad i = 1, 2, \ldots, m. \tag{5.2.1}$$

Let I be an identity matrix, and let $B = (b_{ij})$ and $A = (a_{ij}) = B - I$. The community connectance C is defined as the fraction of non-zero elements in the matrix B. The non-zero elements of B are equally likely to be positive or negative; and have magnitudes which are chosen from a random number distribution with mean equal to zero and a mean square value equal to s^2. Thus s^2 is a measure of the strength of the non-zero interactions.

It was shown by May (1974) that the matrix A will almost certainly be stable if

$$s(mC)^{1/2} < 1; \tag{5.2.2}$$

and it would almost certainly be unstable if (5.2.2) is violated. This analytical result of May confirmed the simulation studies of Gardner and Ashby (1970). It follows that stability tends to decrease as complexity increases in an ecosystem model. An interesting question which follows is: what are the special features in a real complex ecosystem which will allow and enhance stability?

Goh (1974) showed that a linear food web in which each trophic level dissipates energy in the absence of energy inputs will be stable. DeAngelis (1975) discovered a class of randomly assembled food web models in which stability increases with complexity. In these models energy flows in one direction, from the lowest trophic level to the highest trophic level. Energy is absorbed from the environment at the lowest trophic level and is dissipated at all the other trophic levels.

Following Cross (1978) a matrix A is said to be Liapunov—Volterra stable if there exists a positive diagonal matrix $D = \text{diag}(d_1, d_2, \ldots, d_m)$ such that $DA + A^TD$ is negative definite. Let us denote this class of matrices by $L - V$. Let

$$V(x) = (d_1x_1^2 + d_2x_2^2 + \cdots + d_m x_m^2)/2. \tag{5.2.3}$$

Computing $\dot{V}$ along solutions of (5.2.1) we get

$$\dot{V}(x) = (1/2)x^T(DA + A^TD)x. \tag{5.2.4}$$

Hence, if A belongs to $L - V$, model (5.2.1) is stable. Goh (1977 a) showed that a necessary condition for A to belong to $L - V$ is that *all* the principal minors of $-A$ must be positive. Hence the probability for a randomly constructed matrix A to belong to $L - V$ decreases as the number of species increases, and, in a more subtle way, as the magnitudes of the elements of A increase. If an ecosystem consists of a collection of self-regulating species and the sum total of the self-regulating forces dominates the destabilizing effects of the interspecific interactions, then we can expect that it would have at an equilibrium, a matrix A which belongs to $L - V$.

Roberts (1974) made the important point that in order for a non-linear population model to represent a viable spatially homogenous ecosystem it must have a positive (feasible) equilibrium. Roberts (1974), Gilpin (1975 b), and Goh and Jennings (1977) studied randomly assembled Lotka—Volterra models. They concluded that the requirement that the nontrivial equilibrium of a Lotka—Volterra model must be positive imposes a very severe constraint on the parameters of a model if the number of species is large. Goh and Jennings (1977) showed that the result of Gardner and Ashby (1970) and May (1974) for randomly assembled linear systems is applicable to the subset of Lotka—Volterra models each of which has a positive equilibrium. From this discussion it follows that it is doubly difficult for a complex model to represent a viable and stable ecosystem in a spatially homogenous environment.

Consider the Lotka—Volterra model

$$\dot{N}_i = N_i[b_i + \sum_{j=1}^{m} a_{ij}N_j], \quad i = 1, 2, \ldots, m \tag{5.2.5}$$

where $b = (b_i)$ and $A = (a_{ij})$ are constant matrices. The nontrivial equilibrium $N^\star$ of (5.2.5) satisfies the equations

$$b_i + \sum_{j=1}^{m} a_{ij}N_j^\star = 0, \quad i = 1, 2, \ldots, m. \tag{5.2.6}$$

Thus the equilibrium $N^\star$ is the intersection of m hyperplanes. The act of choosing the set of parameters $\{b_i, a_{ij}\}$ in a random manner is equivalent to choosing m hyperplanes in a random manner. If the hyperplanes are not chosen in a biased manner we can expect that the probability that $N^\star$ is positive (i.e. $N_i^\star > 0, i = 1, 2, \ldots, m$) is approximately equal to 2^{-m}.

On the other hand a nondegenerate Lotka—Volterra model of m species has 2^m equilibria. This is because an equilibrium of (5.2.5) can be obtained by setting $N_i = 0$ or $b_i + \Sigma a_{ij}N_j = 0$ for $i = 1, 2, \ldots, m$. For each species we have two choices. Hence, for m species there are 2^m possibilities. Gilpin and Case (1976) showed that a randomly assembled Lotka—Volterra model of competition can be expected to have a significant fraction of locally sector stable nonnegative equilibria. Furthermore there is a high probability that a randomly assembled Lotka—Volterra model has a nonnegative equilibrium with a large domain of attraction if $m \leqslant 20$. For $m \geqslant 8$ the mean number of species present in a locally sector stable nonnegative equilibrium is 2.5.

These properties of Lotka—Volterra models can be expected to occur also in more nonlinear differential equation models of interacting populations. They have important implications for a community which is made up of an interconnected network of patches each of which in isolation is described by a nonlinear differential equation model of the form

$$\dot{N}_i = N_i F_i(N_1, N_2, \ldots, N_m), \quad i = 1, 2, \ldots, m \tag{5.2.7}$$

where $F_1(N), F_2(N), \ldots, F_m(N)$ are continuous functions of N. From the preceding discussions we conclude that it is highly probable that each patch in isolation would contain only a few species. If the environment is spatially heterogenous then different combinations of species will persist in different patches. In this case, Examples 5.10.3 to 5.10.5 clearly demonstrate that the total network of patches would allow a large number of species to persist if the diffusion coefficients between the patches are small.

Hitherto the concept of stability that we have used in this section is related to impulsive and infrequent perturbations of the initial state of the system. But real ecosystems are continually perturbed by unpredictable distrubances. To describe a system's ability to persist in spite of incessant disturbances, Goh (1971, 1975, 1976 a, 1978 d) developed the concept of nonvulnerability. Goh (1975) showed that the set of linear ecosystem models which is completely vulnerable under a single perturbing force is open and dense in the space of all such systems. It follows that species richness per se does not confer on a complex linear model the property of the "not-putting-all-eggs-in-one-basket" kind. Thus there must be special features in the structure of a real ecosystem which reduce the vulnerability of a system. We note that a weakly connected network of patches, some of which are continually perturbed by a disturbance, is less vulnerable than the same community in a homogenous environment. Recently Vincent and Anderson (1979) showed that the addition of a trophic level to a food chain may make the system less vulnerable.

Suppose a model of an ecosystem which is continually perturbed by the forces $u_1, u_2, \ldots, u_p$ is

$$\dot{N}_i = G_i(N_1, N_2, \ldots, N_m, u_1, u_2, \ldots, u_p), \quad i = 1, 2, \ldots, m. \tag{5.2.8}$$

Let $u_1(t), u_2(t), \ldots, u_p(t)$ be piecewise continuous functions of t and $a_k \leqslant u_k \leqslant b_k$ for $k = 1, 2, \ldots, p$ where a_k and b_k are a priori bounds. Denote the admissible set of $\{u_k(t)\}$ by U.

Let $S(0)$ be a desirable region of the $(N_1, N_2, \ldots, N_m)$-space. Let $Z(T)$ be a set of undesirable states. By definition (5.2.8) is vulnerable relative to the sets U, $S(0)$ and $Z(T)$ during the time interval $[0, T]$ if there exists an admissible $u \in U$ which will drive the system from a state $N(0) \in S(0)$ to $Z(T)$ during the time interval $[0, T]$. Otherwise it is said to be nonvulnerable.

Example 5.2.1. Let a logistic population model be continually perturbed by $u(t)$. Suppose we have

$$\dot{N} = N - N^2 - u \tag{5.2.9}$$

where $0 \leqslant u(t) \leqslant 1/8$. The population will persist during the time interval $[0, T]$ where T is any positive number of initially

$$N(0) > \sqrt{2}(\sqrt{2} - 1)/4.$$

Hence, if $S(0) = \{N \mid N > \sqrt{2}(\sqrt{2} - 1)/4\}$ and $Z(T) = \{N \mid N < \sqrt{2}(\sqrt{2} - 1)/4\}$, then (5.2.9) is nonvulnerable.

A subclass of nonlinear population models which are nonvulnerable is studied in Section 5.7. But generally, in the case of more than two species, it is very difficult to test whether or not a population model is vulnerable. To get some insight into how the vulnerability of a system may change as the complexity of the system increases, we shall consider the linearized dynamics of the system.

Let $N^\star \in S(0)$ and $G_i(N^\star, 0) = 0$ for $i = 1, 2, \ldots, m$. Consider the special case when there is a single perturbing force $u(t)$. Let $x_i = \epsilon(N_i - N_i^\star)$ and $v = \epsilon u$ where ϵ is a small positive constant. The linearized dynamics of (5.2.8) is

$$\dot{x}_i = \sum_{j=1}^{m} a_{ij} x_j + b_i v, \quad i = 1, 2, \ldots, m \tag{5.2.10}$$

where $a_{ij} = \partial G_i / \partial N_j$ and $b_i = \partial G_i / \partial u$ at $(N^\star, 0)$. Let us assume that no a priori bounds are imposed on $v(t)$.

The linear system is said to be completely vulnerable if, for any given initial state $x(0)$ and any given final state $x(T)$, there exists a piecewise continuous function $v(t)$ which will drive the system from $x(0)$ to $x(T)$ during the time interval $[0, T]$. This definition parallels that for complete controllability of a linear system (see Barnett, 1975). It follows that a sufficient condition for (5.2.10) to be completely vulnerable is

$$\operatorname{rank}[b \vdots Ab \vdots A^2 b \vdots \ldots \vdots A^{m-1} b] = m. \tag{5.2.11}$$

For convenience, the compound matrix in (5.2.11) is called the vulnerability matrix.

Lee and Markus (1967, p.100) have shown that the set of all pairs (A, b) which satisfies (5.2.11) is open and dense in the space of all pairs (A, b) with the metric

$$\|(A, b)\| = \sum_{ij=1}^{m} |a_{ij}| + \sum_{i=1}^{m} |b_i| \tag{5.2.12}$$

where $|c|$ denotes the absolute value of c. This means that if a pair (A, b)

satisfies (5.2.11), then all variations of (A, b) caused by small changes in the elements of A and b also satisfy (5.2.11). Furthermore if (A, b) does not satisfy (5.2.11), then by small changes in the elements of A and b, we can get a pair which satisfies (5.2.11). Thus linear systems which are not completely vulnerable to a single perturbing force are extremely rare in the world of all possible linear systems each of which is subject to a single perturbing force. It is so rare that if one constructs randomly connected food webs with a single perturbing force, the probability that a randomly connected food web with a single perturbing force is completely vulnerable is always equal to one, irrespective of the number of species. Hence species richness per se does not confer on a system the property of the "not-putting-all-eggs-in-one-basket" kind.

It follows that a high percentage of the elements of A and b for a full model of an ecosystem must be fixed zeros. To keep mathematical models tractable to analysis, population theorists have focused their attention on richly connected simple submodels of an ecosystem. This approach has made us overlook the important role that null interactions between species in different submodels perform in the persistence of the total ecosystem.

We expect that spatial heterogeneity confers on an ecosystem the "not-putting-all-eggs-in-one-basket" property relative to certain disturbances. We shall consider how our type of analysis supports this expectation. We shall also demonstrate that the vulnerability of a system relative to a perturbing force depends on the manner in which the force acts selectively on the dynamics of the system.

Suppose we have an insect population on islands $P1$ and $P2$; during the period $[0, T]$ migrants move only from $P2$ to $P1$. Let an epidemic begin on island $P1$ and wipe out the insect population during the time interval $[0, t_1]$ where $t_1 < T$. During the period $[t_1, T]$ let the insect recolonize island $P1$. Thus during the time interval $[0, T]$ the insect populations on islands $P1$ and $P2$ are not vulnerable to the epidemic.

On the other hand, suppose the epidemic begins on $P2$ and is transmitted to $P1$ by the migrants. This wipes out the insect populations on both islands. Thus the manner in which a perturbing force acts directly on the components of an ecosystem is important. Therefore a novel disturbance to an ecosystem can be a very serious threat to the system because the pattern of interactions in the community would not have evolved to prevent the transmission of the adverse effects throughout the system.

Let us examine caricatures of these examples. Consider the system

$$\begin{pmatrix}\dot{x}_1\\ \dot{x}_2\end{pmatrix} = \begin{pmatrix}a_{11} & a_{12}\\ 0 & a_{22}\end{pmatrix} \begin{pmatrix}x_1\\ x_2\end{pmatrix} + \begin{pmatrix}1\\ 0\end{pmatrix} v(t). \tag{5.2.13}$$

The vulnerability matrix for this system is

$$\left(\begin{array}{c:c}1 & a_{11}\\ 0 & 0\end{array}\right). \tag{5.2.14}$$

Clearly its rank is less than 2 and thus condition (5.2.11) is not satisfied. This means that $v(t)$ does not drive x_2 directly and indirectly. In this case x_2 is secure from the adverse effects of the perturbing force. This corresponds to the case in our example when the epidemic begins on island $P1$.

Consider the system

$$\begin{pmatrix}\dot{x}_1\\ \dot{x}_2\end{pmatrix} = \begin{pmatrix}a_{11} & a_{12}\\ 0 & a_{22}\end{pmatrix}\begin{pmatrix}x_1\\ x_2\end{pmatrix} + \begin{pmatrix}0\\ 1\end{pmatrix} v(t). \tag{5.2.15}$$

The vulnerability matrix is

$$\begin{pmatrix}0 & \vdots & a_{12}\\ 1 & \vdots & a_{22}\end{pmatrix}. \tag{5.2.16}$$

Its rank is equal to 2 if $a_{12} \neq 0$. Hence condition (5.2.11) is satisfied. In this case $v(t)$ acts directly on x_2 and indirectly on x_1. This corresponds to the case in our example when the epidemic begins on island $P2$.

Further insight into the structure of systems (5.2.13) and (5.2.15) can be obtained by a change of variables. Excluding the exceptional case $a_{11} = a_{22}$, let

$$x_1 = y_1 - y_2 a_{12}/(a_{11} - a_{22}),$$

$$x_2 = y_2, \tag{5.2.17}$$

Model (5.2.13) maps into

$$\dot{y}_1 = a_{11}y_1 + v,$$

$$\dot{y}_2 = a_{22}y_2. \tag{5.2.18}$$

Thus (5.2.13) is equivalent to a collection of two non-interacting subsystems. Clearly y_2 is secure from the perturbing force $v(t)$. This is precisely the effect of the "not-putting-all-eggs-in-one-basket" kind.

Model (5.2.15) maps into

$$\dot{y}_1 = a_{11}y_1 + v(t)a_{12}/(a_{11} - a_{22}),$$

$$\dot{y}_2 = a_{22}y_2 + v(t). \tag{5.2.19}$$

We have decoupled the state equations, but if $a_{12} \neq 0$, the perturbing force acts on both subsystems. In fact, when (5.2.11) is satisfied there is no nonsingular linear transformation which maps (5.2.10) into another model with an isolated subsystem, i.e. a subsystem which does not interact with other subsystems and the perturbing force.

By using linear difference equation models, we can demonstrate that age structure in a population reduces the vulnerability of the population to a disturbance which perturbs only members of the population of certain ages and which acts during a period which is brief compared with the life span of the species.

5.3. STABILITY OF NONLINEAR MULTISPECIES MODELS

Suppose an ecosystem has m interacting species. Let N_i be the population of the ith species. A model of the system is

$$\dot{N}_i = N_i F_i(N_1, N_2, \ldots, N_m), \quad i = 1, 2, \ldots, m \tag{5.3.1}$$

where $F_1, F_2, \ldots, F_m$ are general nonlinear functions.

We shall examine the stability of the equilibrium $N^\star$ which satisfies the equations

$$F_i(N) = 0, \quad i = 1, 2, \ldots, m. \tag{5.3.2}$$

This equilibrium is feasible (positive) if $N_i^\star > 0$ for $i = 1, 2, \ldots, m$. By definition $N^\star$ is globally stable if every solution of (5.3.1) which begins in the positive orthant $\mathrm{R}_+^m = \{N \mid N_i > 0, i = 1, 2, \ldots, m\}$ remains in it for all finite values of t and converges to $N^\star$ as $t \to \infty$.

The most effective analytical method to establish the global or finite stability of $N^\star$ is the direct method of Liapunov. This requires the construction of a Liapunov function. One way to construct a Liapunov function for a multispecies model is to make a positive linear combination of Liapunov functions for single-species models.

Let $V_i(N_i)$ be a Liapunov function for the logistic model of a single-species population. A general expressinn for $V_i(N_i)$ is

$$V_i(N_i) = \int_{N_i^\star}^{N_i} [f_i(S_i)/g_i(S_i)] \, dS_i \tag{5.3.3}$$

where $f_i(S_i)$ is a continuous function such that $f_i(S_i) < 0$ for all $S_i \in (0, N_i^\star)$, $f_i(S_i) > 0$ for all $S_i \in (N_i^\star, \infty)$, and $g_i(S_i) > 0$ for all $S_i > 0$. Furthermore $f_i(S_i)$ and $g_i(S_i)$ are chosen so that $V_i(N_i) \to \infty$ as $N_i \to 0+$ or as $N_i \to \infty$.

Example 5.3.1. (a) If $f_1(S_1) = S_1 - N_1^\star$ and $g_1(S_1) = S_1$, then

$V_1(N_1) = N_1 - N_1^\star - N_1^\star \ln(N_1/N_1^\star)$.

(b) If $f_2(S_2) = S_2^2 - N_2^{\star 2}$ and $g_2(S_2) = S_2^2$, then

$V_2(N_2) = N_2 - 2N_2^\star + N_2^{\star 2}/N_2$.

Let $c_1, c_2, \ldots, c_m$ be positive constants. A candidate to act as a Liapunov function for (5.3.1) is

$$V(N) = c_1 V_1(N_1) + c_2 V_2(N_2) + \cdots + c_m V_m(N_m). \tag{5.3.4}$$

Along a solution of (5.3.1),

$$\dot{V}(N) = \sum_{i=1}^{m} c_i (\partial V_i / \partial N_i) N_i F_i(N). \tag{5.3.5}$$

By Theorem 3.3.5 the equilibrium $N^\star$ is globally stable if $\dot{V} \leqslant 0$ for all $N \in \mathbf{R}_+^m$, and $\dot{V}(N)$ does not vanish identically along a solution of (5.3.1) except for the trivial solution $N = N^\star$. The next theorem follows by using a particular class of Liapunov function.

Theorem 5.3.1. *Suppose model (5.3.1) has a positive equilibrium at $N^\star$. Let $f_i(s)$ and $g_i(s)$ be a pair of functions such that $V_i(N_i)$ of (5.3.3) is a Liapunov function for a single-species model with an equilibrium at $N_i^\star$. The equilibrium $N^\star$ of (5.3.3) is globally stable if there exist positive constants $c_1, c_2, \ldots, c_m$ such that the function*

$$W(N) = \sum_{i=1}^{m} c_i[f_i(N_i)/g_i(N_i)]N_iF_i(N) \leqslant 0 \tag{5.3.6}$$

for all $N \in \mathbf{R}_+^m$ and $W(N)$ does not vanish identically along a solution of (5.3.1) except for $N = N^\star$.

Proof. Let $V(N)$ be the function given in (5.3.4) where $V_i(N_i)$ is given by (5.3.3). Along a solution of (5.31.) we have $\dot{V}(N) = W(N)$. The theorem follows by LaSalle's (1976) extension of the direct method of Liapunov.

If $f_i(N_i) = N_i - N_i^\star$ and $g_i(N_i) = N_i$, we deduce that $N^\star$ is globally stable if there exist positive constants $c_1, c_2, \ldots, c_m$ such that

$$\dot{V}(N) = \sum_{i=1}^{m} c_i(N_i - N_i^\star)F_i(N) \leqslant 0 \tag{5.3.7}$$

for all $N \in \mathbf{R}_+^m$ and $\dot{V}(N)$ does not vanish identically along a solution of (5.3.1) except for the trivial solution $N = N^\star$.

Example 5.3.2. Following Hsu (1978 a) let us examine the prey--predator model

$$\dot{N}_1 = N_1[b(N_1) - N_2a_{12}(N_1)],$$
$$\dot{N}_2 = N_2[a_{21}(N_1)]$$

where $b(N_1)$, $a_{12}(N_1)$ and $a_{21}(N_1)$ are functions of N_1. We assume that it has a positive equilibrium at $N^\star = (N_1^\star, N_2^\star)$.

If $c_1 = 1$, $c_2 = 1$, $f_1(N_1) = a_{21}(N_1)$, $g_1(N_1) = N_1a_{12}(N_1)$, $f_2(N_2) = N_2 - N_2^\star$ and $g_2(N_2) = N_2$, the function $\dot{V}(N)$ of (5.3.6) becomes

$$\begin{aligned}\dot{V}(N) &= a_{21}(N_1)[b(N_1)/a_{12}(N_1) - N_2^\star] \\ &= a_{21}(N_1)[b(N_1)/a_{12}(N_1) - b(N_1^\star)/a_{12}(N_1^\star)].\end{aligned}$$

It follows that $N^\star$ is globally stable if $a_{21}(N_1)[b(N_1)/a_{12}(N_1) - b(N_1^\star)/a_{12}(N_1^\star)] < 0$ for all $N_1 > 0$ and $N_1 \neq N_1^\star$.

From experience, we find that the function (5.3.4) is a good candidate to act as a Liapunov function if model (5.3.1) is a collection of populations each of which is self-regulating and the sum total of the regulating intraspecific interactions dominates the destabilizing effects of the interspecific interactions, If a population in the community has unstable intraspecific intercations, it is worthwhile to subject (5.3.1) to a preliminary transformation. In this manner, Hsu (1978 c) was able to establish the global stability of Lotka—Volterra models in which not all the species are self-regulating.

If $\partial F_1/\partial N_1 > 0$ at the equilibrium, the N_1-population is said to be not self-regulating at the equilibrium. In this case it may be worthwhile to subject model (5.3.1) to the Poincare transformation

$$N_1 = 1/x_1, \qquad N_1 = x_i/x_1 \quad \text{for } i = 2, 3, \ldots, m. \tag{5.3.8}$$

Under this transformation, model (5.3.1) becomes

$$\dot{x}_1 = -x_1 F_1(1/x_1, x_2/x_1, \ldots, x_m/x_1),$$

$$\dot{x}_1 = x_i(F_i - F_1), \quad i = 2, 3, \ldots, m. \tag{5.3.9}$$

The positive equilibrium $N^\star$ is mapped into a positive equilibrium which is denoted by $x^\star$. Sometimes it is easy to construct a Liapunov function for the transformed model when it is difficult to do so for the original model. This combination of a preliminary transformation and a search for a Liapunov function may be used repeatedly until a suitable Liapunov function is constructed.

Using the Liapunov function

$$V(x) = \sum_{i=1}^{m} c_i[x_i - x_i^\star - x_i^\star \ln(x_i/x_i^\star)] \tag{5.3.10}$$

we deduce that $x^\star$ of (5.3.9) is globally stable if there exist positive constants $c_1, c_2, \ldots, c_m$ such that

$$\dot{V} = -c_1(x_1 - x_1^\star)F_1 + \sum_{i=2}^{m} c_i(x_i - x_i^\star)(F_i - F_1) \leqslant 0. \tag{5.3.11}$$

and $\dot{V}$ does not vanish identically along a solution, except for $x = x^\star$.

Example 5.3.3. Consider the prey—predator model

$$\dot{N}_1 = N_1(0.25 + N_1 - 5N_2),$$

$$\dot{N}_2 = N_2(-0.5 + 2N_1 - 6N_2).$$

It has an equilibrium at (1.0, 0.25).

Under the Poincare transformation $N_1 = 1/x_1$, $N_2 = x_2/x_1$, we get

$$\dot{x}_1 = (-1 - 0.25x_1 + 5x_2),$$

$\dot{x}_2 = x_2(1 - 0.75x_1 - x_2)/x_1$.

Using the Liapunov function (5.3.10) with $c_1 = 0.75$ and $c_2 = 5$, we get

$\dot{V} = -[0.1875(x_1 - 1)^2 + 5(x_2 - 0.25)^2]/x_1$.

Therefore the transformed model is globally stable. Hence the original model is globally stable.

5.4. STABILITY OF MULTISPECIES LOTKA—VOLTERRA MODELS

The Lotka—Volterra models form the simplest class of nonlinear population models. A highly nonlinear population can always be approximated by a Lotka—Volterra model in the neighbourhood of an equilibrium. Thus an analysis of a Lotka—Volterra model of a community may provide more information of the stability properties of the community than an eigenvalues analysis of a more nonlinear model of the system.

A Lotka—Volterra model for m interacting species is

$$\dot{N}_i = N_i\left[b_i + \sum_{j=1}^{m} a_{ij}N_j\right], \quad i = 1, 2, \ldots, m \tag{5.4.1}$$

where b_i, a_{ij} for $i, j = 1, 2, \ldots, m$ are constants. Each equation in (5.4.1) has two factors. The model could have up to a maximum of 2^m equilibria. Let N^* denote the equilibrium which satisfies the set of equations,

$$b_i + \sum_{j=1}^{m} a_{ij}N_j^* = 0, \quad i = 1, 2, \ldots, m. \tag{5.4.2}$$

By definition N^* is a positive or feasible equilibrium if $N_i^* > 0$ for $i = 1, 2, \ldots, m$.

Using eq. (5.4.2), model (5.4.1) becomes

$$\dot{N}_i = N_i \sum_{j=1}^{m} a_{ij}(N_j - N_j^*), \quad i = 1, 2, \ldots, m. \tag{5.4.3}$$

Let N^* be an equilibrium of (5.3.1) and $a_{ij} = \partial F_i/\partial N_j$ at N^*. It follows that (5.4.3) is a local approximation of (5.3.1).

The equilibrium N^* is locally stable if all the real parts of the eigenvalues of the matrix $(N_i^* a_{ij})$ are negative. Note that in general the stability properties of the matrix $(N_i^* a_{ij})$ are different from those of the matrix $A = (a_{ij})$. The next theorem describes some conditions for global stability.

Theorem 5.4.1. *The positive equilibrium N^* is globally stable if there exists a positive diagonal matrix C such that $CA + A^{\mathrm{T}}C$ is negative semi-definite and the function*

$$W(N) = (1/2)(N - N^\star)^T(CA + A^TC)(N - N^\star) \tag{5.4.4}$$

does not vanish identically along a solution of (5.4.1) except for $N = N^\star$.

Proof. Let $C = \text{diag}(c_1, c_2, \ldots, c_m)$ where $c_1, c_2, \ldots, c_m$ are positive constants, and

$$V(N) = \sum_{i=1}^{m} c_i[N_i - N_i^\star - N_i^\star \ln(N_i/N_i^\star)]. \tag{5.4.5}$$

Along solutions of (5.4.1) we have

$$\dot{V} = (1/2)(N - N^\star)^T(CA + A^TC)(N - N^\star) = W(N). \tag{5.4.6}$$

By assumption $W(N)$ does not vanish identically along a solution of (5.4.1) except for $N = N^\star$. Hence by LaSalle's extension of the direct method of Liapunov $N^\star$ is globally stable.

Example 5.4.1. Consider the prey—predator model

$$\dot{N}_1 = N_1[b_1 - a_{12}N_2],$$

$$\dot{N}_2 = N_2[-d + Ea_{12}N_1 - a_{22}N_2]$$

where b_1, a_{12}, a_{22}, E and d are positive constants. This model has a positive equilibrium at $N^\star = [(da_{12} + ba_{22})/Ea_{12}^2, b/a_{12}]$. The interaction matrix is

$$A = \begin{pmatrix} 0 & -a_{12} \\ Ea_{12} & -a_{22} \end{pmatrix}.$$

If $C = \text{diag}(E, 1)$, the matrix $CA + A^TC = \text{diag}(0, -2a_{22})$. Thus $CA + A^TC$ is negative semidefinite. The function $\dot{V}(N) = -(a_{22})(N_2 - b/a_{12})^2$ does not vanish identically along a solution of the model except for $N = N^\star$. Hence the equilibrium $N^\star$ is globally stable.

Similarly it can be shown that a longer food chain is globally stable if it has a positive equilibrium. Consider the model

$$\dot{N}_1 = N_1[b_1 - a_{12}N_2],$$

$$\dot{N}_2 = N_2[-d_2 + e_2a_{12}N_1 - a_{23}N_3],$$

$$\dot{N}_3 = N_3[-d_3 + e_3a_{23}N_2 - a_{33}N_3]$$

where all the parameters are positive. If the equilibrium at $N^\star$ where $N_2^\star = b_1/a_{12}$, $N_3^\star = (e_3a_{23}N_2^\star - d_3)/a_{33}$, $N_1^\star = (d_2 + a_{23}N_3^\star)/(e_2a_{12})$ is positive, then it is globally stable.

Next consider the case in which the species at the lowest trophic level is self-regulating. For a three-species food chain, we have

$$\dot{N}_1 = N_1[b_1 - a_{11}N_1 - a_{12}N_2],$$

$$\dot{N}_2 = N_2[-d_2 + e_2a_{12}N_1 - a_{23}N_3],$$

$$\dot{N}_3 = N_3[-d_3 + e_3a_{23}N_2].$$

It has an equilibrium at $N^\star$ where $N_2^\star = d_3/(e_3a_{23})$, $N_1^\star = (b_1 - a_{12}N_2^\star)/a_{11}$ and $N_3^\star = (e_2a_{12}N_1^\star - d_2)/a_{23}$. If we choose $c_1 = 1$, $c_2 = 1/e_2$ and $c_3 = 1/(e_2e_3)$, then the function $V(N)$ of (5.4.4) is a Liapunov function. We conclude that the equilibrium $N^\star$ is globally stable.

In general, for $m > 2$, it is difficult to construct a positive diagonal matrix C such that $CA + A^TC$ is negative semidefinite. The case when $m = 3$ has been studied by Cross (1978). We shall establish some sufficient conditions for a Lotka—Volterra model of a two-species system to be globally stable.

Theorem 5.4.2. *For* $m = 2$, *model (5.4.1) is globally stable if* (i) *it has a positive equilibrium at* $N^\star$, *and* (ii) $a_{11} < 0$, $a_{22} < 0$ *and* $\det A > 0$.

Proof. We shall show that condition (ii) implies that there exists a positive diagonal matrix C such that $CA + A^TC$ is negative definite. The matrix $CA + A^TC$ is negative definite if $c_1a_{11} < 0$, $c_2a_{22} < 0$ and

$$4c_1c_2a_{11}a_{22} - (c_1a_{12} + c_2a_{21})^2 > 0. \tag{5.4.7}$$

Clearly $c_1a_{11} < 0$ and $c_2a_{22} < 0$ if condition (ii) is satisfied.

For condition (5.4.7), there are three cases: (a) $a_{12}a_{21} = 0$, (b) $a_{12}a_{21} > 0$, and (c) $a_{12}a_{21} < 0$.

In case (a), either $a_{12} = 0$ or $a_{21} = 0$, or both are equal to zero. Suppose $a_{12} = 0$. Condition (5.4.6) reduces to

$$c_2(4c_1a_{11}a_{22} - c_2a_{21}^2) > 0. \tag{5.4.8}$$

Clearly it is possible to choose positive constants c_1 and c_2 so that (5.4.8) is satisfied. Similar arguments apply when $a_{21} = 0$, or a_{12} and a_{21} are both equal to zero. Biologically, case (a) contains interactions known as amensalism, commensalism and neutralism.

In case (b), $a_{12}a_{21} > 0$. This means either $a_{12} < 0$ and $a_{21} < 0$, or $a_{12} > 0$ and $a_{21} > 0$. These subcases are known as competition and mutualism respectively. Condition (5.4.7) can be rewritten as

$$4c_1c_2(a_{11}a_{22} - a_{12}a_{21}) - (c_1a_{12} - c_2a_{21})^2 > 0. \tag{5.4.9}$$

It is possible to choose positive numbers c_1 and c_2 so that

$$c_1a_{12} - c_2a_{21} = 0. \tag{5.4.10}$$

By assumption $\det A > 0$. Hence (5.4.9) is satisfied.

In case (c), $a_{12}a_{21} < 0$. This represents a prey—predator interaction. As a_{12} and a_{21} have opposite signs, it is possible to choose positive numbers c_1 and c_2 such that

$$c_1a_{12} + c_2a_{21} = 0. \tag{5.4.11}$$

It follows that (5.4.7) is satisfied. This completes the proof of the theorem.

We deduce from Theorem 5.4.2 that for two-species Lotka—Volterra models of competition or mutualism, local stability implies global stability.

For $m > 2$, there is a simple set of sufficient conditions for model (5.4.1) to be globally stable.

Theorem 5.4.3. *Model (5.4.1) is globally stable if it has a positive equilibrium at $N^\star$ and there exists a matrix G such that*

$$a_{ii} \leqslant G_{ii} \quad \text{and} \quad |a_{ij}| \leqslant G_{ij} \quad \textit{for } i \neq j \tag{5.4.12}$$

for i, j = 1, 2, . . . , m and all the leading principal minors of —G are positive.

Proof. The off-diagonal elements of the matrix G are nonnegative. If all the leading principal minors of $-G$ are positive, then G is an M-matrix. Araki and Kondo (1972) showed that an M-matrix has a positive diagonal matrix C such that $CG + G^TC$ is negative definite.

Using (5.4.6) and (5.4.12), we get

$$\dot{V} = \sum_{ij}^{m} c_i a_{ij}(N_i - N^\star)(N_j - N_j^\star) \leqslant (1/2)(N - N^\star)^T(CG + G^TC)(N - N^\star). \tag{5.4.13}$$

It follows that $\dot{V}(N)$ is negative definite. Hence $N^\star$ is globally stable.

Example 5.4.2. The three-species model

$$\dot{N}_1 = N_1(2.5 - 2N_1 - N_2 + N_3),$$

$$\dot{N}_2 = N_2(4.5 - N_1 - 3N_2 - N_3),$$

$$\dot{N}_3 = N_3(-1 + N_1 + N_2 - 2N_3)$$

has an equilibrium at (1, 1, 0.5).

A suitable G matrix is

$$\begin{pmatrix} -2 & 1 & 1 \\ 1 & -3 & 1 \\ 1 & 1 & -2 \end{pmatrix}.$$

All the leading principal minors of $-G$ are positive. It follows that the equilibrium (1, 1, 0.5) is globally stable.

Theorem 5.4.1 can only be applied to a Lotka—Volterra model which satisfies the condition $a_{ii} \leqslant 0$ for $i = 1, 2, \ldots, m$. But Hsu (1978 c) showed that it is possible to establish global stability in a Lotka—Volterra model which violates this condition by first subjecting the model to a Poincare transformation. We shall demonstrate this approach by applying it to the two species Lotka—Volterra model,

$$\dot{N}_1 = N_1(b_1 + a_{11}N_1 + a_{12}N_2),$$
$$\dot{N}_2 = N_2(b_2 + a_{21}N_1 + a_{22}N_2). \tag{5.4.14}$$

Theorem 5.4.4. *Model (5.4.1) is globally stable if* (i) *it has a positive equilibrium and* (ii) $b_1 > 0$, $a_{12} > a_{22}$ *and* $b_2a_{12} > b_1a_{22}$.

Proof. Under the Poincare transformation

$$N_1 = 1/x_1, \qquad N_2 = x_2/x_1 \tag{5.4.15}$$

model (5.4.14) becomes

$$\dot{x}_1 = x_1[-a_{11} - b_1x_1 - a_{12}x_2]/x_1,$$
$$\dot{x}_2 = x_2[(a_{21} - a_{11}) + (b_2 - b_1)x_1 + (a_{22} - a_{12})x_2]/x_1. \tag{5.4.16}$$

The equilibrium $N^\star$ is mapped into the positive equilibrium $x^\star = (1/N_1^\star, N_2^\star/N_1^\star)$.

For convenience, let

$$H = \begin{pmatrix} -b_1 & -a_{12} \\ b_2 - b_1 & -(a_{12} - a_{22}) \end{pmatrix}. \tag{5.4.17}$$

Using a Liapunov function of the form in (5.4.5), we deduce that $x^\star$ is globally stable if there exists a positive diagonal matrix C such that

$$\dot{V} = (1/2)(x - x^\star)^T(CH + H^TC)(x - x^\star)/x_1 < 0 \tag{5.4.18}$$

for all $x \in R_+^2$ and $x \neq x^\star$. Using arguments similar to those in the proof of Theorem 5.4.2, we deduce that $x^\star$ is globally stable if condition (ii) is satisfied.

Corollary 5.4.1. *Model (5.4.1) is globally stable if* (i) *it has a positive equilibrium, and* (ii) $b_2 > 0$, $a_{21} > a_{11}$ *and* $b_1a_{21} > b_2a_{11}$.

This corollary can be proved by first subjecting (5.4.1) to the Poincare transformation

$$N_1 = y_1/y_2, \qquad N_2 = 1/y_2.$$

The result can then be established by using arguments similar to those which are used to prove Theorem 5.4.4.

Hsu (1978 c) showed that the Poincare transformation can also be used to establish that model (5.4.1) has an unbounded solution. We shall examine how this approach can be used by applying it to model (5.4.14). If $a_{11} > 0$ and $a_{11} > a_{21}$, then there exists a solution of (5.4.16) with its initial state in the positive orthant which converges to $x = (0, 0)$ as $t \to \infty$. This implies that (5.4.14) has a solution such that $N_1(t) \to \infty$ as $t \to \infty$.

5.5. STABILITY OF A CLASS OF NONLINEAR MODELS

If a model of an ecosystem is more nonlinear than a Lotka—Volterra model, it is very difficult to use Theorem 5.3.1 to establish the global or finite stability of its equilibrium. For $m > 2$, it is difficult to verify that a nonlinear function, like $\dot{V}(N)$, is negative semidefinite in a given region. The condition that $\dot{V}(N)$ is negative semidefinite in a given region is equivalent to the condition that the global maximum of $\dot{V}(N)$ is equal to zero. Unfortunately the available techniques for global optimization (see Dixon and Szego, 1975) are limited to very special classes of functions.

In this section we shall examine an algebraic criterion for a nonlinear ecosystem model to be globally stable. Unfortunately it is a very conservative condition. Its main value is that it provides a deep insight into conditions for a complex ecosystem to be globally stable. It also identifies a class of highly nonlinear ecosystem models which have qualitatively the same dynamical behavior as a subset of Lotka—Volterra models.

A general model for m interacting species is

$$\dot{N}_1 = N_i F_i(N_1, N_2, \ldots, N_m), \quad i = 1, 2, \ldots, m. \tag{5.5.1}$$

Let $N^\star$ denote a positive equilibrium which satisfies the equations

$$F_i(N) = 0, \quad i = 1, 2, \ldots, m. \tag{5.5.2}$$

Let G be a constant matrix such that

$$\partial F_i/\partial N_i \leqslant G_{ii}, \quad i = 1, 2, \ldots, m, \tag{5.5.3}$$

$$|\partial F_i/\partial N_j| \leqslant G_{ij} \quad \text{for } i \neq j \tag{5.5.4}$$

for all $N \in \mathrm{R}_+^m = \{N \mid N_i > 0, i = 1, 2, \ldots, m\}$, the positive orthant.

Theorem 5.5.1. *The equilibrium $N^\star$ of (5.5.1) is globally stable if there exists a constant matrix G which satisfies (5.5.3) and (5.5.4), and all the leading principal minors of $-G$ are positive.*

Proof. If $c_1, c_2, \ldots, c_m$ are positive constants, let

$$V(N) = \sum_{i=1}^{m} c_i[N_i - N_i^\star - N_i^\star \ln(N_i/N_i^\star)]. \tag{5.5.5}$$

Along solutions of (5.5.1) we have

$$\dot{V}(N) = \sum_{i=1}^{m} c_i(N_i - N_i^\star)F_i(N). \tag{5.5.6}$$

Applying Taylor's Theorem to $\dot{V}(N)$, we get

$$\dot{V}(N) = \sum_{i,j=1}^{m} c_i(N_i - N_i^\star)(\partial F_i/\partial N_j)(N_j - N_j^\star) \tag{5.5.7}$$

in which the partial derivatives are computed at a set of points between N and $N^\star$.

Let $C = \text{diag}(c_1, c_2, \ldots, c_m)$. Eq. (5.5.3), (5.5.4) and (5.5.7) give

$$\dot{V} \leqslant \sum_{i=1}^{m} c_i G_{ii}(N_i - N_i^\star)^2 + \sum_{i=1}^{m} \sum_{j \neq i}^{m} c_i G_{ij}|N_i - N_i^\star||N_j - N_j^\star|$$

$$= (1/2)y^{\mathrm{T}}(CG + G^{\mathrm{T}}C)y, \tag{5.5.8}$$

where $y = (|N_i - N_i^\star|)$. Clearly $\dot{V}(N)$ is negative definite if $CG + G^{\mathrm{T}}C$ is negative definite.

The off-diagonal elements of G are nonnegative. If all the leading principal minors of $-G$ are positive, the matrix G is an M-matrix. For such a matrix, there exists a positive diagonal matrix C such that $CG + G^{\mathrm{T}}C$ is negative definite (Araki and Kondo, 1972). Hence $N^\star$ is globally stable.

Theorem 5.5.1 implies that an ecosystem model is globally stable if it is a collection of interacting stable subsystems such that the interactions between subsystems, form the total system's point of view, are weaker than the self-interactions of the subsystems. If an ecosystem model satisfies eqs. (5.5.3) and (5.5.4), the condition that all the leading principal minors of $-G$ are positive provides a criterion that the self-regulating intraspecific interactions are stronger than the interspecific interactions. The next example demonstrates that this is not a straightforward and intuitive concept.

Example 5.5.1. Consider two matrices G_1 and G_2 where

$$G_1 = \begin{pmatrix} -1 & 0.6 & 0.3 \\ 0.5 & -1 & 0.7 \\ 0.7 & 0.4 & -1 \end{pmatrix}, \qquad G_2 = \begin{pmatrix} -1 & 8 & 0.1 \\ 0.01 & -1 & 0.2 \\ 0.2 & 2 & -1 \end{pmatrix}.$$

Let the matrices be associated with two nonlinear ecosystem models via conditions (5.5.3) and (5.5.4). From (5.5.4) it is clear that the interspecific interactions can have positive or negative signs. Therefore the interspecific interactions could be any type of interactions.

In G_1 all the off-diagonal elements are numerically smaller than the diagonal elements. On the other hand, G_2 has some off-diagonal elements which are numerically larger than its diagonal ones. Thus, at first sight, the intraspecific interactions are stronger than the interspecific interactions for a multispecies system with matrix G_1, and vice-versa for a community with matrix G_2. But $\det(-G_1) < 0$, and therefore G_1, does not satisfy the given criterion for a system to have intraspecific interactions, which, from the total system's point of view, are stronger than its interspecific interactions. On the other hand, all the leading principal minors of $-G_2$ are positive; therefore, a community with matrix G_2 satisfies the given criterion for its intraspecific interactions to be stronger than its interspecific interactions.

We shall demonstrate that constant energy inputs to each population in an ecosystem model which satisfies Theorem 5.5.1 in the absence of these energy inputs, are stabilizing influences, If a model does not satisfy Theorem 5.5.1 this type of energy input may be destabilizing. It is known in system theory that a model with more than two components can be destabilized by making some of its diagonal elements more negative (Willems, 1971).

Let $I_1, I_2, \ldots, I_m$ be nonnegative constants. Consider the model

$$\dot{N}_i = N_i F_i(N) + I_i, \quad i = 1, 2, \ldots, m. \tag{5.5.9}$$

Suppose this model has a positive equilibrium at $N^\star$. By assumption there exists a constant matrix G such that conditions (5.5.3) and (5.5.4) are satisfied, and all the leading principal minors of $-G$ are positive. When $I_i \geqslant 0$ for $i = 1, 2, \ldots, m$, we have

$$\partial[F_i + I_i/N_i]/\partial N_i \leqslant G_{ii} - I_i/N_i^2 \leqslant G_{ii}. \tag{5.5.10}$$

It follows that if model (5.5.9) satisfies Theorem 5.5.1 when $I_i = 0$ for $i = 1, 2, \ldots, m$, it will also do so when $I_i \geqslant 0$ for $i = 1, 2, \ldots, m$, provided the latter has a positive equilibrium. Hence, constant energy inputs into an ecosystem model which satisfies Theorem 5.5.1 in the absence of the energy inputs, are stabilizing influences.

A population model can be destabilized by a constant fraction harvesting, or by a constant quota harvesting, or by enrichment (Goh, 1969/1970; Rosenzweig, 1971). We shall now identify a class of multispecies models which remain stable under constant fraction harvestings of the component species, provided the harvested community has a positive equilibrium with all the species.

Suppose the unexploited system is described by eq. (5.5.1). Let E_i be the constant fraction harvesting which is applied to the ith species in the community. Thus $E_1, E_2, \ldots, E_m$ are nonnegative constants. A model of the exploited system is

$$\dot{N}_i = N_i[F_i(N) - E_i], \quad i = 1, 2, \ldots, m. \tag{5.5.11}$$

We have

$$\partial(F_i - E_i)/\partial N_j = \partial F_i/\partial N_i \tag{5.5.12}$$

for $i, j = 1, 2, \ldots, m$. Hence if model (5.5.1) satisfies Theorem 5.5.1 so does model (5.5.11), provided that the latter has a positive equilibrium.

5.6. CONNECTIVE STABILITY

On an evolutionary time scale the character of the interactions between two species can change, e.g. from a competitive to a commensal interaction or vice-versa. On the other hand, during a change in seasons, a predator can switch from feeding on one species to another, Siljak (1978) has recently developed a new stability concept which may be useful for studying the consequences of such structural changes in a multispecies community. The new concept of stability is called connective stability. If a system remains stable in spite of certain structural changes in the interactions, then it is said to be connectively stable.

However, it is necessary to modify Siljak's results on connective stability before they can be applied to ecological models. In an ecosystem model the state variables must be nonnegative. This restricts the set of admissible structural perturbations to those which do not cause the equilibrium of an ecosystem model to become nonpositive. Siljak (1978) studied the connective stability of the model

$$\dot{x}_i = \sum_{j=1}^{m} A_{ij}(t, x)x_j, \quad i = 1, 2, \ldots, m, \tag{5.6.1}$$

where (A_{ij}) is a matrix of nonlinear functions. This model tacitly assumes that structural perturbations which cause interactions between species to be connected or disconnected do not shift the equilibrium from the origin. This assumption is generally not satisfied in a nonlinear model of an ecosystem.

Consider the prey—predator model,

$$\begin{aligned} \dot{N}_1 &= N_1[1.8 - N_1 - 0.8N_2], \\ \dot{N}_2 &= N_2[-0.1 + 1.1c_{21}N_1 - N_2]. \end{aligned} \tag{5.6.2}$$

When $c_{21} = 1$, the model has a positive equilibrium at (1, 1). It follows that the linearized dynamics in the neighbourhood of (1, 1) is

$$\begin{aligned} \dot{x}_1 &= -x_1 - 0.8x_2. \\ \dot{x}_2 &= 1.1x_1 - x_2. \end{aligned}$$

This linear system satisfies Siljak's conditions for connective stability.

When $c_{21} = 1$, eq. (5.6.2) reduces to

$$\begin{aligned} \dot{N}_1 &= N_1(1.8 - N_1 - 0.8N_2), \\ \dot{N}_2 &= N_2(-0.1 - N_2). \end{aligned} \tag{5.6.3}$$

The model has a nontrivial equilibrium at (1.88, —0.1); and three other equilibria at (0, 0), (0, —0.1) and (1.8, 0). It can be shown analytically that all the solutions which begin in the positive orthant tend to the equilibrium (1.8, 0) as t tends to infinity. This means that the perturbed model (5.6.3) does not represent a viable prey—predator system.

Siljak's pioneering analysis of connective stability in nonlinear systems can be modified to apply to ecosystem models. For the sake of simplicity, we shall use Lotka—Volterra models in this discussion. This analysis can be easily extended to the class of nonlinear ecosystem models which are analyzed in Section 5.5.

Consider the Lotka—Volterra model

$$\dot{N}_i = N_i \left[b_i(s) + \sum_{j=1}^{m} c_{ij}(s) G_{ij} N_j \right], \quad i = 1, 2, \ldots, m, \tag{5.6.4}$$

where $b_i(s)$ and $c_{ij}(s)$ are functions of s. Let $c_{ii} = 1$ for $i = 1, 2, \ldots, m$ and $-1 \leqslant c_{ij}(s) \leqslant 1$ for $i \neq j$. Let $G = (G_{ij})$ be a constant matrix such that $G_{ii} < 0$ for $i = 1, 2, \ldots, m$ and $G_{ij} > 0$ for $i \neq j$. The assumption that c_{ij} can be positive or negative when $i \neq j$ implies that eq. (5.6.4) is a model for a community with any type of interspecific interactions.

The nontrivial equilibrium of eq. (5.6.4) is

$$N^\star(s) = -E^{-1} b \tag{5.6.5}$$

where $b = (b_i)$ and $E = (E_{ij}) = (c_{ij} G_{ij})$.

We shall consider two cases. In the first case, the parameters b_i and c_{ij} are permitted to change impulsively but infrequently relative to the time scale of model (5.6.4). In the second case, let δ be a small number and $s = \delta t$. This means that $b_i(s)$ and $c_{ij}(s)$ are slowly varying functions of t. Therefore in both cases the question as to whether a new equilibrium $N^\star(s)$, which is created by structural perturbations and environmental changes, in positive or not, is separated from the question as to whether or not the newly created equilibrium is stable.

Goh and Jennings (1977) showed that the subset of Lotka—Volterra models, where each model has a positive equilibrium, is a small fraction in a set of randomly assembled Lotka—Volterra models. Furthermore, this fraction decreases rapidly with the number of species in a community. These results imply that the requirement that the nontrivial equilibrium of a Lotka—Volterra model must be positive imposes a very severe constraint on the values of its parameters. This in turn limits the class of structural perturbations which will not destroy the viability of a complex ecosystem.

The next theorem describes conditions under which model (5.6.4) is connectively stable in a restricted sense.

Theorem 5.6.1. *If model (5.6.4) is subjected to the above restricted class of structural perturbations such that the equilibrium $N^\star(s)$ remains positive, and if all the leading principal minors of $-G$ are positive, then $N^\star(s)$ remains globally stable.*

Proof. Eq. (5.6.4) and (5.6.5) give

$$\dot{N}_i = N_i \left[\sum_{j=1}^{m} c_{ij}(s) G_{ij} (N_j - N_j^\star) \right], i = 1, 2, \ldots, m \tag{5.6.6}$$

A Liapunov function for this system is

$$V(N) = \sum_{i=1}^{m} d_i[N_i - N_i^\star - N_i^\star \ln(N_i/N_i^\star)] \tag{5.6.7}$$

where $d_1, d_2, \ldots, d_m$ are positive constants and $N_i^\star$ is a function of s.

By assumption, $c_{ii} = 1$ for $i = 1, 2, \ldots, m$ and $|c_{ij}| \leqslant 1$ for $i \neq j$. It follows that along solutions of the model,

$$\begin{aligned} \dot{V}(N) &= \sum_{i=1}^{m} \sum_{j=1}^{m} d_i c_{ij} G_{ij}(N_i - N_i^\star)(N_j - N_j^\star) \\ &\leqslant \sum_{i=1}^{m} d_i G_{ii}(N_i - N_i^\star)^2 + \sum_{i=1}^{m} \sum_{j=i}^{m} d_i G_{ij}|N_i - N_i^\star|\,|N_j - N_j^\star| \\ &= (1/2)y^{\mathrm{T}}(DG + G^{\mathrm{T}}D)y \end{aligned} \tag{5.6.8}$$

where $D = \mathrm{diag}(d_1, d_2, \ldots, d_m)$ and $y = (|N_i - N_i^\star|)$. The diagonal elements of G are negative and its off-diagonal elements are nonnegative. If all the leading principal minors of $-G$ are positive, then G is an M-matrix. Therefore there exists a positive diagonal matrix D such that $DG + G^{\mathrm{T}}D$ is negative definite (Araki and Kondo, 1972). Hence $N^\star$ is globally stable in spite of certain changes in the parameters $\{b_i(s), c_{ij}(s)\}$.

Theorem 5.6.1 describes the conditions under which model (5.6.4) can track its equilibrium as its parameters vary because of certain environmental and structural perturbations. If model (5.6.4) satisfies the conditions of the theorem, then each of its submodels, when one or more species is absent, will also satisfy the theorem provided that it has a positive equilibrium. In other words, a perturbation which causes the extinction of one or more species in an ecosystem model which satisfies Theorem 5.6.1 will create a submodel which satisfies the same theorem, provided the reduced model has a positive equilibrium.

5.7. NONVULNERABILITY OF NONLINEAR MULTISPECIES MODELS

In the real world ecosystems are continually disturbed by unpredictable forces due to changes in weather, migrating species and diseases. Global stability in a model does not mean that an ecosystem model can persist in the face of continual and unpredictable disturbances. A way to establish an ecosystem's ability to do so is to use the theory of stochastic differential equations (May, 1974; Ludwig, 1975). Unfortunately it is very difficult to solve the Fokker—Plank equation for the full nonlinear stochastic model except for models with one or two variables. An alternative approach is to use Liapunov functions and the concept of nonvulnerability (Goh, 1976 a). We shall discuss how this approach can be applied to a class of nonlinear ecosystem models.

Let $u_i(t)$ be the magnitude of disturbance which acts on the growth rate of the ith species. Suppose a model of the ecosystem which is continually perturbed is

$$\dot{N}_i = N_i H_i(N, u), \quad i = 1, 2, \ldots, m \tag{5.7.1}$$

where $H_i(N, u)$ is a function of $N_1, N_2, \ldots, N_m, u_1, u_2, \ldots, u_m$. If the disturbances act in a density-independent manner, then model (5.7.1) reduces the form

$$\dot{N}_i = N_i F_i(N) + u_i(t) N_i, \quad i = 1, 2, \ldots, m. \tag{5.7.2}$$

Suppose $u_1(t), u_2(t), \ldots, u_m(t)$ are piecewise continuous functions which satisfy the a priori constraints,

$$a_i \leqslant u_i(t) \leqslant b_i, \quad i = 1, 2, \ldots, m \tag{5.7.3}$$

where $\{a_i, b_i\}$ is a set of constants. For convenience, we denote the set of admissible functions $\{u_i(t)\}$ by U.

Let $S(0)$ denote a set of desirable states of the ecosystem model in (5.7.1). Let Z denote a set of undesirable states. The system is said to be vulnerable relative to the sets U, $S(0)$ and Z during the time interval $[0, T]$, if there exists an admissible vector function $u(t)$ which drives the system from an initial state in $S(0)$ to the undesirable set Z during the time interval $[0, T]$. If there is no admissible disturbance $u(t)$ which can drive the system from the set $S(0)$ to Z during the time interval $[0, T]$, than the system is said to be nonvulnerable relative to the sets U, $S(0)$ and Z during the time interval $[0, T]$.

A way to test whether or not a system is vulnerable is to solve the following optimal control problem:

System: $\dot{N}_i = N_i H_i(N, u), \quad i = 1, 2, \ldots, m.$ (5.7.4)

Initially: $N(0) \in \partial S(0).$ (5.7.5)

Constraint: $u \in U.$ (5.7.6)

Terminally: $N(t_1) \in \partial Z.$ (5.7.7)

Objective: Minimize (t_1). (5.7.8)

Here ∂S and ∂Z are the boundaries of the sets $S(0)$ and Z. We assume that the set $S(0)$ and the complement of the set Z are simply connected regions. If this optimal control problem has a solution where $\min(t_1)$ is less than T, then model (5.7.1) is vulnerable; where $\min(t_1)$ is greater than T, the system is nonvulnerable. If $m \leqslant 3$ it is relatively easy to solve numerically this optimal control problem; as m increases, it rapidly becomes very difficult to do so.

Another method of establishing that an ecosystem model is nonvulnerable is to carry out a conservative analysis using Liapunov-like function. Let $V(N)$ denote a Liapunov function for (5.7.1) when $u_i(t) = 0$, for $i = 1, 2, \ldots, m$. In this analysis $V(N)$ is called a Liapunov-like function because it is not used to establish asymptotic stability. It is convenient to define the sets $S(0)$ and Z in terms of the level surfaces of $V(N)$. Let s and z be positive constants and let $S(0) = \{N \mid V(N) \leqslant s\}$ and $Z = \{N \mid V(N) \geqslant z\}$.

Along solutions of eq. (5.7.1), we have

$$\dot{V}(N, u) = \sum_{i=1}^{m} (\partial V/\partial N_i) N_i H_i(N, u). \tag{5.7.9}$$

The next theorem describes some sufficient conditions for model (5.7.1) to be nonvulnerable.

Theorem 5.7.1. *Model (5.7.1) is nonvulnerable relative to the set U, $S(0)$ and Z during the time interval $[0, T]$, if there exists a positive number p such that $s \leqslant p < z$ and the global maximum of $\dot{V}(N, u)$ for all $u \in U$ and $N \in \{N \mid V(N) = p\}$ is negative. Here T could be any positive number.*

Proof. The conditions in the theorem imply that $\dot{V}(N, u)$ is negative on the hypersurface $\{N \mid V(N) = p\}$ for all admissible disturbances. It follows that all solutions of (5.7.1) which begin in the set $\{N \mid V(N) \leqslant p\}$ remain in it indefinitely. Therefore the system cannot be driven from the set $P = \{N \mid V(N) \leqslant p\}$ to the set Z because $p < z$. But the set $S(0)$ is a subset of the set P. Hence the system is nonvulnerable relative to the sets U, $S(0)$ and Z during the time interval $[0, T]$.

For model (5.7.2) we can establish more refined conditions for it to be nonvulnerable. Let $C = \text{diag}(c_1, c_2, \ldots, c_m)$ be a positive diagonal matrix and

$$V(N) = \sum_{i=1}^{m} c_i [N_1 - N_i^\star - N_i^\star \ln(N_i/N_i^\star)] \tag{5.7.10}$$

where $N_i^\star > 0$ for $i = 1, 2, \ldots, m$. Here $N^\star = (N_i^\star)$ need not be an equilibrium of (5.7.2) when $u_i = 0$ for $i = 1, 2, \ldots, m$. Along solutions of (5.7.2) we have

$$\dot{V}(N, u) = (N - N^\star)^{\mathrm{T}} CF(N) + (N - N^\star)^{\mathrm{T}} Cu. \tag{5.7.11}$$

Corollary 5.7.1. *Model (5.7.2) is nonvulnerable relative to the sets U, $S(0)$ and Z during the time interval $[0, T]$, where T is any positive number, if there exist positive constants $c_1, c_2, \ldots, c_m, p$ such that $s \leqslant p < z$ and the global maximum of $\dot{V}(N, u)$ of (5.7.11) for all $u \in U$ and $N \in \{N \mid V(N) = p\}$ is negative.*

In practice it is very difficult to use Theorem 5.7.1 because global optimization algorithms are unreliable (Dixon and Szego, 1975). For this reason we shall now discuss a more conservative but reliable method, This is an extension of the algebraic method which was developed for Lotka—Volterra models by Goh (1976 a).

Suppose there exists a constant matrix G such that model (5.7.2) satisfies the inequalities

$$\partial F_i(N)/\partial N_i \leqslant G_{ii}, \quad i = 1, 2, \ldots, m, \tag{5.7.12}$$

$$|\partial F_i(N)/\partial N_j| \leqslant G_{ij}, \quad i \neq j \tag{5.7.13}$$

for all $N \in \mathbf{R}_+^m$. The off-diagonal elements of G are nonnegative. Therefore, if all the leading principal minors of $-G$ are positive, then G is an M-matrix. In this case there exists a positive diagonal matrix C such that $CG + G^{\mathrm{T}}C$ is negative definite. We shall normalize the elements of C by setting $c_1 = 1$. Let $c^\star$ be the largest element of c and $\lambda_{\min}$ be the smallest eigenvalues of the matrix $-(CG + G^{\mathrm{T}}C)/2$. For the sake of simplicity, we shall consider the special case in which the disturbance functions satisfy the constraints,

$$-b \leqslant u_i(t) \leqslant b, \quad i = 1, 2, \ldots, m. \tag{5.7.14}$$

Let $r^\star$ be the minimum distance from $N^\star$ to any of the coordinate axis hyperplanes. Thus $r^\star$ is the smallest number in the set $\{N_1^\star, N_2^\star, \ldots, N_m^\star\}$. Let r_0 be the smallest positive number which satisfies the inequality

$$\lambda_{\min} \geqslant [bc^\star m^{1/2}]/r \tag{5.7.15}$$

Consider the hypersphere

$$(N - N^\star)^{\mathrm{T}}(N - N^\star) = r^2 \tag{5.7.16}$$

where r satisfies the inequality $r_0 < r < r^\star$. Let $N^\star$ be a positive equilibrium of (5.7.2) when $u = 0$. Let p be the maximum value of $V(N)$ for all N on the hypersphere in (5.7.16). Let s and z be positive constants such that $s \leqslant p < z$ and let $S(0) = \{N \mid V(N) \leqslant s\}$ and $Z = \{N \mid V(N) > z\}$.

Theorem 5.7.2. *The ecosystem model (5.7.2) is nonvulnerable relative to the sets U, $S(0)$ and Z during the time interval $[0, T]$ where T is any positive number if model (5.7.2) satisfies (5.7.12) and (5.7.13) and there exists a positive diagonal matrix C such that inequality (5.7.15) is satisfied.*

Proof. By assumption, model (5.7.2) satisfies (5.7.12) and (5.7.13). Hence, along solutions of the model, we have

$$\begin{aligned} \dot{V}(N, u) &= (N - N^\star)^{\mathrm{T}} CF(N) + (N - N^\star)^{\mathrm{T}} Cu \\ &\leqslant (1/2) y^{\mathrm{T}}(CG + G^{\mathrm{T}}C)y + (N - N^\star)^{\mathrm{T}} Cu \end{aligned} \tag{5.7.17}$$

where $y = (|N_i - N_i^\star|)$.

The product $(N - N^\star)^{\mathrm{T}} Cu$ is an inner product which is induced by the positive diagonal matrix C. Let N be a point on the hypersphere (5.7.16). By the Cauchy—Schwarz inequality, we get

$$(N-N^{\star})^{\mathrm{T}} Cu \leqslant [(N-N^{\star})^{\mathrm{T}} C(N-N^{\star})]^{1/2} [u^{\mathrm{T}}Cu]^{1/2}$$
$$\leqslant c^{\star}r(u^{\mathrm{T}}u)^{1/2}$$
$$\leqslant c^{\star}rbm^{1/2}. \tag{5.7.18}$$

Eqs. (5.7.17) and (5.7.18) imply

$$\dot{V}(N, u) \leqslant \lambda_{\min} r^2 + rbc^{\star}m^{1/2} \tag{5.7.19}$$

Inequality (5.7.16) and $r_0 < r < r^{\star}$ imply that $\dot{V}(N, u)$ is negative on the hypersphere (5.7.16). Outside this hypersphere, the quadratic term involving $(N-N^{\star})$ in (5.7.17) dominates the linear term in $(N-N^{\star})$. Therefore, on and outside the hypersphere, $\dot{V}(N, u)$ is negative. Hence $\dot{V}(N, u)$ is negative on the level set $\{N \mid V(N) = p\}$. This implies that a solution of model (5.7.2) which begins in the set $S(0)$ cannot be driven by the set of admissible disturbances to the set Z. Hence the ecosystem model (5.7.2) is nonvulnerable relative to the sets U, $S(0)$ and Z during the time interval $[0, T]$ where T is any positive number.

The following example demonstrates how Theorems 5.7.1 and 5.7.2 may be used.

Example 5.7.1. Let N_1 be a prey population and N_2 be the population of its predator. Suppose the prey—predator system which is continually perturbed can be described by the model

$$\dot{N}_1 = N_1[22 - 0.1N_1 - 2.1N_2] + u_1(t)N_1,$$
$$\dot{N}_2 = N_2[-5 + 0.7N_1 - 0.2N_2] + u_2(t)N_2 \tag{5.7.20}$$

where $u_1(t)$ and $u_2(t)$ are piecewise continuous disturbances. Suppose $u_1(t)$ and $u_2(t)$ satisfy the constraints

$$-0.11 \leqslant u_i(t) \leqslant 0.11, \quad i = 1, 2, \tag{5.7.21}$$

In the absence of the disturbances, $u_1 = 0$, $u_2 = 0$, the model has a positive equilibrium at $N^{\star} = (10, 10)$. By trial and error, we choose $c_1 = 1.0$ and $c_2 = 2.94$ for the $V(N)$ function in (5.7.10). The level set of this $V(N)$ function which passes through the point (8.37, 10) is $V(N) = 0.1493$. The global maximum of $\dot{V}(N, u)$ for all admissible functions $u_1(t)$ and $u_2(t)$ and for all $N \in \{N \mid V(N) \geqslant 0.1493, N_1 > 0, N_2 > 0\}$ is equal to -0.0035 when $N_1 = 9.511$, $N_2 = 10.428$, $u_1 = -0.11$ and $u_2 = 0.11$.

Let p be any positive number larger than 0.1493, and s and z be positive numbers such that $s \leqslant p < z$. Let $S(0) = \{N \mid V(N) \leqslant s\}$ and $Z = \{N \mid V(N) \geqslant z\}$. By Theorem 5.7.1 model (5.7.20) is nonvulnerable relative to $S(0)$, Z and the admissible set of disturbances, during the time interval $[0, T]$ where T is any positive number.

We can use Theorem 5.7.2 to establish that (5.7.20) is nonvulnerable relative to different sets $S(0)$ and Z. Allowing $c_1 = 1$ and $c_2 = 3$, we get $\lambda_{min} = 0.1$. In this case $V(N)$ is slightly different from before. Let $p \geqslant 5.79$, and let s and z be positive numbers such that $s \leqslant p < z$. By Theorem 5.7.2 model (5.7.20) is nonvulnerable relative to the sets $S(0) = \{N \mid V(N) \leqslant s\}$ and $Z = \{N \mid V(N) > \geqslant z\}$ during the time interval $[0, T]$ where T is any positive number. The set of admissible disturbances is the same as before. We conclude that Theorem 5.7.1 is more effective than Theorem 5.7.2 provided we can compute the global maximum of $\dot{V}(N, u)$. Unfortunately, global optimization algorithms are at present unreliable.

5.8. SECTOR STABILITY

By definition, a nonnegative equilibrium of an ecosystem model is sector stable, if every solution of the system, which begins in a nonnegative neighbourhood, remains in the same or larger nonnegative neighbourhood for all finite values of t, and converges to the equilibrium as t tends to infinity. We shall demonstrate that the concept of sector stability is important in the analysis of a complex ecosystem model. A nontrivial mathematical criterion is given for the total system property which states that the self-regulating intraspecific interactions are stronger than the interspecific interactions. This total system property is sufficient for an ecosystem model to be globally sector stable.

The question as to whether a nonnegative equilibrium of an ecosystem model is "stable" occurs in the study of some interesting processes in ecology, e.g., succession and extinction. The concept of asymptotic stability cannot be directly used to study the behavior of this type of equilibrium, because it is defined in terms of open neighbourhoods. But for a nonnegative equilibrium to be "stable", we require only that every trajectory which begins in a nonnegative neighbourhood, remains in it or a larger nonnegative neighbourhood for all finite values of t, and converges to the equilibrium as $t \to \infty$. If a nonnegative equilibrium has this property, it is said to be *sector stable.*

In a well-known experiment, Paine (1966) found that the removal of a predator starfish from an area resulted in the reduction of a 15-species community to an 8-species community. This experiment gives rise to an interesting question as to whether the composition of the 8-species community is sensitive to the densities of the 15 species at the time when the starfish population was removed. The concept of sector stability can provide the answer to this question if an adequate model of the 15 species is available.

Simberloff (1976) carried out a large number of censuses of arboreal arthropod populations on eight islands during a 3-year period. He found that members of a group of 22 species out of a total of 254 species were either present at all the censuses or were absent only once. This suggests that the presence of the other species on an island was maintained by continual migrations. If this is the case, the concept of sector stability may provide some useful insight into the composition of the species in the communities on the islands if models of the communities are available.

A way to model a spatially heterogenous community is to use a set of differential equations (Levin, 1974). One equation is used to describe the dynamics of a subpopulation in a patch. The interactions between the equations describe the migrations between different patches and the interactions between different species in a patch. A species in this type of environment can persist even if it is extinct in some of the patches for most of the time. This calls for the analysis of a nonnegative equilibrium to determine whether or not it is sector stable.

Roberts (1974), Gilpin (1975 b) and Goh and Jennings (1977) have shown that only a small fraction of randomly assembled Lotka—Volterra models have a (feasible) positive equilibrium. Therefore the requirement that an ecosystem model should be positive imposes a very severe constraint on the parameters of a complex ecosystem model. On the other hand, a given nondegenerate Lotka—Volterra model for m species has 2^m equilibria. Usually a significant fraction of these equilibria are nonnegative. The appropriate stability concept for studying the convergence properties of a nonnegative equilibrium is sector stability.

The convergence of trajectories of a multispecies model, which begin in the positive orthant, to a nonnegative equilibrium has been studied by Leung (1976), Goh (1978 a), Hsu (1978 b) and Case and Casten (1979).

Suppose a model of a multispecies system is

$$\dot{N}_i = N_i F_i(N_1, N_2, \ldots, N_m), \quad i = 1, 2, \ldots, m \tag{5.8.1}$$

where $F_1(N), F_2(N), \ldots, F_m(N)$ have continuous partial derivatives everywhere in the state space. This model has an equilibrium at $N^\star$ if $\dot{N}_i = 0$ for $i = 1, 2, \ldots, m$ at $N^\star$. An equilibrium of (5.8.1) is obtained by setting $N_i = 0$ or $F_i(N) = 0$ for $i = 1, 2, \ldots, m$ and solving the resultant set of equations. In a nondegenerate Lotka—Volterra model, this procedure leads to 2^m equilibria.

Let $M = \{1, 2, \ldots, m\}$, P be a proper subset of M, and $Q = M - P$. An equilibrium $N^\star$ of (5.8.1) is nonnegative if $N_i^\star > 0$ for all $i \in P$, and $N_i^\star = 0$ for all $i \in Q$. Let I be a subset of M, $V(N) = \|N\|$ be a norm of N and ϵ and δ be positive numbers. Let

$$R(\epsilon, I) = \{N \mid V(N) < \epsilon, N_i > 0 \text{ for } i \in I \text{ and } N_i \geqslant 0 \text{ for } i \in M - I\}, \tag{5.8.2}$$

$$S(\epsilon, I) = \{N \mid V(N) < \epsilon, N_i > 0 \text{ for } i \in I \text{ and } N_i = 0 \text{ for } i \in M - I\}. \tag{5.8.3}$$

Definition. The nonnegative equilibrium $N^\star$ is *sector stable*, if, for each small positive number ϵ, there exists a corresponding positive number $\delta(\epsilon)$, such that every solution of (5.8.1) which begins in the set $R(\epsilon, P)$, remains in $R(\delta(\epsilon), P)$ for all finite values of t and converges to $N^\star$ as $t \to \infty$.

This definition is similar to that for asymptotic stability. The difference is that in the case of sector stability, the trajectories are restricted to a nonnegative part of an open neighbourhood of the equilibirum.

We shall establish some sufficient conditions for a nonnegative equilibrium to be locally stable sector. Let the matrix

$$K = (N_i^\star \partial F_i/\partial N_j) \tag{5.8.4}$$

where $i, j \in P$ and the partial derivative $\partial F_i/\partial N_j$ is computed at $N^\star$.

Theorem 5.8.1. *The nonnegative equilibrium $N^\star$ of (5.8.1) is locally sector stable if all the eigenvalues of K have negative real parts, and*

$$F_i(N^\star) < 0 \quad \textit{for all } i \in Q. \tag{5.8.5}$$

Proof. Without loss of generality, we relable the index set so that $P = \{1, 2, \ldots, n\}$ where n is a positive integer. Let $x_i = N_i - N_i^\star$ for $i = 1, 2, \ldots, m$. The linearized dynamics of (5.8.1) is

$$\dot{x} = \begin{pmatrix} N_i^\star \partial F_i/\partial N_j & N_i^\star \partial F_i/\partial N_k \\ 0 & \delta_{hk} F_k \end{pmatrix} x \tag{5.8.6}$$

where $i, j = 1, 2, \ldots, n; h, k = n+1, n+2, \ldots, m; \delta_{hk} = 1$ if $h = k$ and $\delta_{hk} = 0$ if $h \neq k$.

The matrix in (5.8.6) is a block upper triangular matrix. All its eigenvalues have negative real parts because the submatrix K is stable and condition (5.8.5) is satisfied. Therefore the equilibrium of (5.8.6) is asymptotically stable. It follows that there exists a Liapunov function $V(x) = x^{\mathrm{T}}Bx$ such that $\dot{V}(x) < 0$ along all nontrivial solutions of (5.8.6).

The proof is completed by showing that every solution of (5.8.1) which begins in the set $\mathrm{R}(\epsilon, P)$ remains in the set $R(\delta(\epsilon), P)$ for all finite values of t,

Consider the solutions of (5.8.1) which begin in the set $S(\epsilon, M)$. By assumption, the functions $F_1(N), F_2(N), \ldots, F_m(N)$ have continuous partial derivatives everywhere in the state space. Therefore the standard local existence and uniqueness theorem for differential equations is applicable to (5.8.1). By inspection, any solution of (5.8.1) which begins on a coordinate axis hyperplane remains in it for all finite values of t. It follows that all the solutions of (5.8.1) which begin in $S(\epsilon, M)$ do not intersect a coordinate axis hyperplane for all finite values of t. This invariance property and the condition $\dot{V}(x) < 0$ imply that a solution of (5.8.1) which begins in the set $S(\epsilon, M)$ will remain in the set $S(\delta(\epsilon), M)$ for all finite values of t and converge to $N^\star$ as $t \to \infty$.

Let I be equal to P or let it be a superset of P. The above line of reasoning can be used to show that there exists $\delta(\epsilon)$, such that every solution of (5.8.1) which begins in the set $S(\epsilon, I)$ remains in $S(\delta(\epsilon), I)$ for all finite values of t and tends to $N^\star$ as $t \to \infty$. The set $R(\epsilon, P)$ is the union of the sets $S(\epsilon, I)$ where I is a superset of P. It follows that $N^\star$ is locally sector stable.

The most powerful analytical method for establishing finite and global stability is the direct method of Liapunov. We shall modify this method so that it can be used to establish finite and global sector stability of a nonnegative equilibrium.

Let $c_1, c_2, \ldots, c_m$ be positive constants. Consider the function

$$V(N) = \sum_{i \in P} c_i[N_i - N_i^\star - N_i^\star \ln(N_i/N_i^\star)] + \sum_{i \in Q} c_i\,|N_i|. \tag{5.8.7}$$

The level sets of $V(N)$ are closed hypersurfaces. For $i \in P$, the function $V(N) \to \infty$ as $N_i \to \infty$ or as $N_i \to 0+$. For $i \in Q$, the function $V(N) \to \infty$ as $|N_i| \to \infty$.

Let d be a finite positive number, and let $V(N)$ in (5.8.2) and (5.8.3) be the function in (5.8.7). By definition, the nonnegative equilibrium $N^\star$ is sector stable relative to the set $R(d, P)$, if every solution of (5.8.1) which begins in the set $R(d, P)$ remains in it for all finite values of t and converges to $N^\star$ as $t \to \infty$.

Theorem 5.8.2. *The nonnegative equilibrium $N^\star$ of (5.8.1) is sector stable relative to the set $R(d, P)$, if there exist positive constants $c_1, c_2, \ldots, c_m$ such that the function*

$$W(N) = \sum_{i=1}^{m} c_i(N_i - N_i^\star)F_i(N) \leqslant 0 \tag{5.8.8}$$

at every point in $R(d, P)$, and $W(N)$ does not vanish identically along a solution of (5.8.1), except for $N = N^\star$.

Proof. In the set $S(d, M)$, the total time derivative of $V(N)$ along solutions of (5.8.1) is

$$\dot{V}(N) = \sum_{i \in P} c_i(N_i - N_i^\star)F_i(N) + \sum_{i \in Q} c_i N_i F_i(N). \tag{5.8.9}$$

But $N_i^\star = 0$ for $i \in Q$. Therefore $N_i = N_i - N_i^\star$ for all $i \in Q$. Hence $\dot{V}(N) = W(N)$ for all $N \in S(d, M)$.

The standard local existence and uniqueness theorem for differential equations implies that every solution of model (5.8.1), which begins in the set $S(d, M)$, does not intersect any of the coordinate axis hyperplane $\{N \mid N_i = 0\}$ where $i = 1, 2, \ldots, m$. This behavior of the solutions of (5.8.1), and the conditions that (i) $\dot{V}(N) \leqslant 0$ and (ii) $\dot{V}(N)$ does not vanish identically along a nontrivial solution in the set $S(d, M)$, imply that every solution of (5.8.1) which begins in $S(d, M)$ remains in it for all finite values of t and converges to $N^\star$ as $t \to \infty$. A similar argument applies to $S(d, I)$ where I is any proper superset of P. In the reduced state space, where $N_i \equiv 0$ for all $i \in Q$, the equilibrium $N^\star$ is a positive equilibrium. The conditions of this theorem are sufficient for $N^\star$ to be stable relative to the set $S(d, P)$.

The set $R(d, P)$ is the union of all the sets $S(d, I)$ where I is any proper superset of P and of the set $S(d, P)$. We conclude that $N^\star$ is sector stable relative to the set $R(d, P)$.

Let the limit of $R(d, P)$ as $d \to \infty$ be denoted by $R(\infty, P)$. Hence

$$R(\infty, P) = \{N \mid N_i > 0 \text{ for } i \in P \text{ and } N_i \geqslant 0 \text{ for } i \in Q\}. \tag{5.8.10}$$

By definition, the nonnegative equilibrium $N^\star$ is globally sector stable if it is sector stable relative to the set $R(\infty, P)$. The next corollary follows immediately from the definition of global sector stability and Theorem 5.8.2.

Corollary 5.8.1. *The nonnegative equilibrium $N^\star$ of (5.8.1) is globally sector stable if there exist positive constants $c_1, c_2, \ldots, c_m$, such that at every point in $R(\infty, P)$, the function $W(N)$ of (5.8.8) is nonpositive and does not vanish identically along a solution of (5.8.1), except for $N = N^\star$.*

We shall apply Corollary 5.8.1 to a nonnegative equilibrium of the general Lotka—Volterra model

$$\dot{N}_i = N_i \left[b_i + \sum_{j=1}^{m} a_{ij}N_j\right], \quad i = 1, 2, \ldots, m. \tag{5.8.11}$$

Let P be a subset of $M = \{1, 2, \ldots, m\}$ and $Q = M - P$, and let $A = (a_{ij})$. Suppose $N^\star$ is an equilibrium of (5.8.11), such that $N_i^\star > 0$ for all $i \in P$, and $N_i^\star = 0$ for all $i \in Q$. Corollary 5.8.1 implies that $N^\star$ is globally sector stable if: (i) there exists a positive diagonal matrix C such that $CA + A^TC$ is negative semidefinite; (ii) for all $i \in Q$,

$$F_i(N^\star) = b_i + \sum_{j=1}^{m} a_{ij}N_j^\star \leqslant 0;$$

and (iii) the function

$$\dot{V}(N) = (1/2)\,(N - N^\star)^T(CA + A^TC)(N - N^\star) + \sum_{i \in Q} c_iN_iF_i(N^\star)$$

does not vanish identically along a solution of (5.8.11), except for $N = N^\star$.

In practice it is very difficult to verify condition (5.8.8) when $m > 2$ and when $\{F_i(N)\}$ is a set of nonlinear functions. The next theorem provides a conservative method for using Corollary 5.8.1.

Theorem 5.8.3. *Suppose there exists a constant matrix G such that*

$$\partial F_i(N)/\partial N_i \leqslant G_{ii}, \quad i = 1, 2, \ldots, m, \tag{5.8.12}$$

$$|\partial F_i(N)/\partial N_j| \leqslant G_{ij}, \quad i \neq j \tag{5.8.13}$$

for all N in an open set containing $R(\infty, P)$. If all the leading principal minors of $-G$ are positive, and

$$F_i(N^\star) \leqslant 0 \quad \text{for all } i \in Q, \tag{5.8.14}$$

then the nonnegative equilibrium $N^\star$ is globally sector stable.

Proof. Applying the mean value theorem to each of the functions $F_1(N)$, $F_2(N), \ldots, F_m(N)$, eq. (5.8.9) gives

$$\dot{V}(N) = \sum_{i=1}^{m} \sum_{j=1}^{m} c_i(N_i - N_i^\star)(\partial F_i/\partial N_j)(N_j - N_j^\star) + \sum_{i \in Q} c_i N_i F_i(N^\star). \tag{5.8.15}$$

For each value of i, the vector $(\partial F_i/\partial N_j)$ is computed at a point between N and $N^\star$.

Eq. (5.8.12), (5.8.13) and (5.8.15) give

$$\dot{V}(N) \leqslant (1/2)Y^{\mathrm{T}}(CG + G^{\mathrm{T}}C)Y + \sum_{i \in Q} c_i N_i F_i(N^\star) \tag{5.8.16}$$

where $Y = (|N_i - N_i^\star|)$ and $C = \mathrm{diag}(c_1, c_2, \ldots, c_m)$.

The off-diagonal elements of G are nonnegative, and by assumption, all the leading principal minors of $-G$ are positive. Hence G is an M-matrix. It follows that there exists a positive diagonal matrix C such that $CG + G^{\mathrm{T}}C$ is negative definite (see Araki and Kondo, 1972). This property of G and (5.8.14) imply that $\dot{V}(N)$ is negative definite in $R(\infty, P)$. Hence $N^\star$ is globally sector stable.

Example 5.8.1. Consider the following model for three competing species,

$$\dot{N}_1 = N_1(11.7 - 4N_1 - 0.2N_2 - 0.1N_3),$$
$$\dot{N}_2 = N_2(1.2 - 0.8N_1 - N_2 - 0.2N_2),$$
$$\dot{N}_3 = N_3(3 - 2N_1 - N_2 - 2N_3). \tag{5.8.17}$$

It has 2^3 equilibria at $(3, -1, -1)$, $(0, 1, 1)$, $(0, 6/5, 0)$, $(0, 0, 3/2)$, $(0, 0, 0)$, $(231/78, 0, -114/78)$, $(117/40, 0, 0)$ and $(573/192, -57/48, 0)$. Thus there are four nonnegative equilibria.

Model (5.8.17) is a special case of (5.8.11). Let $G_{ii} = a_{ii}$ for $i = 1, 2, 3$ and $G_{ij} = |a_{ij}|$ for $i \neq j$. All the leading principal minors of the matrix

$$-G = \begin{pmatrix} 4 & -0.2 & -0.1 \\ -0.8 & 1 & -0.2 \\ -2 & -1 & 2 \end{pmatrix}$$

are positive.

At the equilibrium, $N^\star = (117/40, 0, 0)$, we have

$$F_1(N^\star) = 0, \qquad F_2(N^\star) = -1.14, \qquad F_3(N^\star) = -2.85.$$

Hence condition (5.8.14) is satisfied. By Theorem 5.8.3 the equilibrium (117/40, 0, 0) is globally sector stable. This means that every solution which begins in the set $\{N \mid N_1 > 0, N_2 \geqslant 0, N_3 \geqslant 0\}$ remains in it and converges to the equilibrium (117/40, 0, 0) as $t \to \infty$.

If $N_1 = 0$ the reduced system has an equilibrium at (0, 1, 1). This equilibrium is globally stable relative to the positive quadrant of the (N_2, N_3)-space. This means that the N_2 and N_3 species can coexist in the absence of the N_1 species. But if the N_1 species invade the region occupied by the N_2 and N_3 species, it will drive the N_2 and N_3 species to extinction. This process may be used to model the replacement of some species during a succession.

From studies of island communities, it is clear that the presence of many species on islands is maintained by continual immigrations. Let us suppose that the three species immigrate at constant rates I_1, I_2 and I_3. In this case, model (5.8.17) is replaced by

$$\dot{N}_1 = N_1[11.7 - 4N_1 - 0.2N_2 - 0.1N_3] + I_1,$$
$$\dot{N}_2 = N_2[1.2 - 0.8N_1 - N_2 - 0.2N_3] + I_2,$$
$$\dot{N}_3 = N_3[3 - 2N_1 - N_2 - 2N_3] + I_3. \tag{5.8.18}$$

If N_1 is small, and N_2 and N_3 are finite, then $\dot{N}_1 > 0$. On the other hand, if $N_1 \gg N_2$ and $N_1 \gg N_3$, and N_1 is large, then $\dot{N}_1 < 0$. Similar comments apply to $\dot{N}_2$ and $\dot{N}_3$. If (5.8.18) has a positive equilibrium at $N^\star$, then by Theorem 5.5.1 the equilibrium $N^\star$ is globally stable. Thus in this case, continuous immigrations at constant rates provide a stabilizing influence.

Example 5.8.2. A model of a chemostat with n species utilizing one limiting substrate is

$$\dot{S} = S\left[(S^0/S - 1)D - \sum_{i=1}^{n} k_iX_i/(a_i + S)\right], \tag{5.8.19}$$

$$\dot{X}_i = X_i[m_iS/(a_i + S) - D_i],\ i = 1, 2, \ldots, m, \tag{5.8.20}$$

where S^0, D, k_i, a_i, m_i and D_i are positive constants. The limiting behavior of this model has been studied by Hsu (1978 b).

For convenience, let $h_i = a_iD_i/(m_i - D_i)$ for $i = 1, 2, \ldots, m$. If $0 < h_1 < h_2 < \cdots < h_n$ and $h_1 < S^0$, then the equilibrium $(S^\star, X^\star) = (h_1, X_1^\star, 0, 0, \ldots, 0)$ is globally sector stable. For $i = 1$, eq. (5.8.20) gives $S^\star = h_1$ and eq. (5.8.19) gives $X_1^\star = D(a_1 + h_1)(S^0 - h_1)/h_1$ if $X_2^\star = X_3^\star = \cdots X_n^\star = 0$.

By Theorem 5.8.2 $(S^\star, X^\star)$ is globally stable if there exist positive constants $c_0, c_1, c_2, \ldots, c_n$ such that

$$W(S, X) = c_0(S - h_1) \left[(S^0/S - 1)D - \sum_{i=1}^{n} k_i X_i/(a_i + S)\right]$$

$$+ \sum_{i=1}^{n} c_i(X_i - X_i^\star)[m_i S/(a_i + S) - D_i]$$

is negative semidefinite in the set $\{(S,X) \mid S > 0, X_1 > 0, X_i \geqslant 0, i = 2, 3, \ldots$ $\ldots, n\}$ and it does not vanish identically along a nontrivial solution. Let $c_0 = 1$, $c_i = k_i/(m_i - D_i)$ for $i = 1, 2, \ldots, m$. After some manipulations, we get

$$W = -\frac{(S - h_1)^2 D(h_1 S + a_1 S^0)}{h_1 S(a_1 + S)} + \sum_{i=2}^{n} k_i(h_1 - h_i) \frac{X_i}{(a_i + X_i)}.$$

It follows that $W(S, X)$ is negative semidefinite. Moreover it can be shown that $W(S, X)$ does not vanish identically along a nontrivial solution. Hence $(S^\star, X^\star)$ is globally sector stable.

In a complex ecosystem model, the condition that an equilibrium should be positive imposes a very severe constraint on the set of parameters in the model. In a set of randomly assembled Lotka—Volterra models for m species, Goh and Jennings (1977) showed that the fraction of models with a positive equilibrium is approximately equal to 2^{-m}. On the other hand, a given non-degenerate Lotka—Volterra model has 2^m equilibria; and Gilpin and Case (1976) showed that usually a significant fraction of these equilibria are non-negative (partially feasible). This observation suggests that the sector stability of a nonnegative equilibrium is important in the analysis of a complex ecosystem model.

In a complex ecosystem model, we can expect that there is a nonnegative equilibrium which is globally stable if the sum total of the self-regulating intraspecific interactions is stronger than the sum total of the interspecific interactions. The complex ecosystem model (5.8.1) has this type of total system property if there exists a constant matrix G such that conditions (5.8.12) and (5.8.13) are satisfied and all the leading principal minors of $-G$ are positive. In such a model the species composition at steady state is independent of the initial densities of the species, provided that the initial set of species includes all the species at the steady state. Such an ecosystem model can withstand a wide range of large perturbations of the initial state.

5.9. POPULATION MODELS WITH TIME DELAYS

In this section we shall first examine single-species models with time delays. We shall then make a general stability analysis of multi-species models with continuous time delays.

The growth rate of a population in the real world does not respond instantaneously to changes in the population level. For example, the growth rate of a mammal population could have time delay because of a gestation period or because of the time it takes a female animal to mature. Population models with time delays have been studied by Volterra (1931), Wangersky and Cunningham (1957), May (1974), MacDonald (1976), Cushing (1977), Wörz-Busekros (1978), Brauer (1977) and others.

A well-known population model with a time delay is

$$\dot{N}(t) = rN(t)[1 - N(t-T)/K] \tag{5.9.1}$$

where r, K and T are positive constants. It has been suggested that this model describes a herbivore population grazing on vegetation which takes T units of time to recover. In this case the term $1 - N(t-T)/K$ is a measure of the abundance of the vegetation at time t. It is more plausible that the level of vegetation at time t depends on the total past history of the herbivore population. If so, model (5.9.1) can be replaced by either one of the following models,

$$\dot{N}(t) = N(t)\left[r - cN(t) - w \int_{-\infty}^{t} \exp(-a(t-s))N(s)ds\right], \tag{5.9.2}$$

$$\dot{N}(t) = N(t)\left[r - cN(t) - w \int_{-\infty}^{t} (t-s)\exp(-a(t-s))N(s)ds\right], \tag{5.9.3}$$

where r, c, w and a are positive constants.

Model (5.9.2) can be converted into a system of differential equations. Let

$$y(t) = \int_{-\infty}^{t} \exp(-a(t-s))N(s)\, ds. \tag{5.9.4}$$

Using (5.9.2) and (5.9.4), we get

$$\dot{N} = N[r - cN - wy],$$

$$\dot{y} = -ay + N. \tag{5.9.5}$$

The y variable has no direct biological meaning. But $x = r - wy$ is a measure of the abundance of the vegetation. In place of (5.9.5) we have

$$\dot{N} = N(x - cN),$$

$$\dot{x} = ar - ax - wN. \tag{5.9.6}$$

Thus the vegetation is replenished at a constant rate and it has a constant per capita natural death rate. This model is applicable to a vegetation whose growth rate is limited by available growing sites and which has a large seed bank in the ground. Furthermore, the vegetation is harvested by the herbivore at a rate which depends on the herbivore population but not on the vegetation level.

Model (5.9.3) can also be converted into a system of differential equations.

Let

$$y_1(t) = \int_{-\infty}^{t} (t-s)\exp(-a(t-s))N(s)\,\mathrm{d}s, \tag{5.9.7}$$

$$y_2(t) = \int_{-\infty}^{t} \exp(-a(t-s))N(s)\,\mathrm{d}s. \tag{5.9.8}$$

Using (5.9.3), (5.9.7) and (5.9.8), we get

$$\begin{aligned} \dot{N} &= N(r - cN - wy_1), \\ \dot{y}_1 &= -ay_1 + y_2, \\ \dot{y}_2 &= -ay_2 + N. \end{aligned} \tag{5.9.10}$$

It is not clear how a biological explanation can be provided for the differential equations for y_1 and y_2.

From eqs. (5.9.4), (5.9.7) and (5.9.8), we deduce that the variables $y(t)$, $y_1(t)$ and $y_2(t)$ are positive for all values of t. This property of the y, y_1, y_2 variables plays a role in an analysis of the stability of models (5.9.5) and (5.9.10).

Model (5.9.5) has a positive equilibrium at $(N^\star, y^\star) = (ar/(ac+w), r/(ac+w))$. We shall show that $(N^\star, y^\star)$ is globally stable. Model (5.9.5) may be rewritten in the form

$$\begin{aligned} \dot{N} &= N[-c(N-N^\star) - w(y-y^\star)], \\ \dot{y} &= -a(y-y^\star) + (N-N^\star). \end{aligned} \tag{5.9.11}$$

Consider the function

$$V(N, y) = N - N^\star - N^\star \ln(N/N^\star) + (w/2)(y-y^\star)^2. \tag{5.9.12}$$

Along solutions of (5.9.11) we get

$$\dot{V} = -c(N-N^\star)^2 - aw(y-y^\star)^2. \tag{5.9.13}$$

The function $V(N, y)$ does not tend to infinity as $y \to 0+$. Thus it is necessary to check that every admissible solution $y(t)$ of (5.9.11) is positive for all $t > 0$. From (5.9.4) we find that this is indeed the case. More precisely, we deduce from (5.9.5) that $y(t) > 0$ for all $t \geqslant 0$ if $y(0) > 0$ and $N(t) \geqslant 0$. From (5.9.4) we have $y(0) > 0$. Hence $y(t) > 0$ for all $t > 0$. Therefore we conclude that every solution of (5.9.11) which begins in the positive quadrant of the (N, y)-space remains in it for all positive values of t and converges to $(N^\star, y^\star)$ as $t \to \infty$.

There is another reason for time delay to occur in the growth rate of a single-species population. It takes time for an animal or plant to mature. Let $N(t)$ denote the number of mature females in a population with a constant sex ratio. Suppose a female animal takes T units of time to mature. A model of this type of population is of the form

$$\dot{N}(t) = -dN(t) + F[N(t-T)], \qquad (5.9.14)$$

where d, a positive constant, is the per capita death rate of adult females and $F[N(t-T)]$ is the rate at which maturing females are recruited into the mature group of females.

Let us assume that model (5.9.14) has a positive equilibrium at $N^\star$ which is a solution of the equation

$$dN^\star = F(N^\star). \qquad (5.9.15)$$

Let $x(t) = N(t) - N^\star$. The linearized dynamics is

$$\dot{x}(t) = -dx(t) + px(t-T) \qquad (5.9.16)$$

where $p = F'(N^\star)$, the derivative of $F(N)$ at $N^\star$.

Brauer (1978) has given necessary and sufficient conditions for (5.9.16) to be stable. He has also established some interesting results on the manner in which the characteristic return time of this equation depends on the parameters d, p and T. We shall apply Brauer's results to (5.9.14).

The eigenvalue equation for (5.9.16) is

$$[p\exp(-\lambda T) - \lambda] - d = 0. \qquad (5.9.17)$$

Let λ_m be the largest real part of the solutions of (5.9.17). If λ_m is negative, then (5.9.16) is stable and the characteristic return time, or settling time, is $(-1/\lambda_m)$. For convenience, let $u = -\lambda_m$. If $p < 0$, the equation which determines u depends on $T^\star$ which is the solution of

$$pT^\star\exp(dT^\star + 1) + 1 = 0. \qquad (5.9.18)$$

If $p < 0$ and $0 \leqslant T < T^\star$, the characteristic return time of (5.9.16) is a decreasing function of T, and u is the solution of the equation

$$(-d + u)T\exp -(d + u)T + pT\exp(dT) = 0. \qquad (5.9.19)$$

If $p < 0$ and $T > T^\star$, then the characteristic return time is an increasing function of T, and u is a solution of

$$[u/\sin(u)]\exp[-u/\tan(u)] + pT\exp(dT) = 0. \qquad (5.9.20)$$

Example 5.9.1. Goh and Agnew (1979) fitted the model

$$N(t+1) = sN(t) + aN(t-8)\exp[-bN(t-8)] \qquad (5.9.21)$$

to the data for the southern hemisphere stock of the fin whale which has been given by Allen (1973). The parameter values obtained are $s = 0.96$, $a = 0.12$ and $1/b = 3 \times 10^5$.

A differential equation which approximates this difference equation is

$$\dot{N}(t) = -(1-s)N(t) + aN(t-8)\exp[-bN(t-8)]. \qquad (5.9.22)$$

This model has an equilibrium at $N^\star = (3\ln 3) \times 10^5$. The linearized equation for (5.9.22) is (5.9.16) where $d = 0.04$, $p = -0.003944$. For these parameter values, (5.9.18) has solution $T^{\star} = 29.1$. Since $T = 8 < 29.1$, u is the solution of (5.9.19). We get $u = 0.625$. Thus the characteristic return time is 1.6. In view of the time delay of 8 years, this value of the characteristic return time appears small. The reason for this is the stabilizing term $-(1 - s)N(t)$ in (5.9.22).

We shall now make a general analysis of a class of nonlinear integro-differential equations. This discussion generalizes the results of Wörz-Busekros (1978) for Lotka—Volterra equations with continuous time delays. Consider equations of the form

$$\dot{N}_i = N_i F_i(N) + \sum_{h,k=1}^{p} N_i H_{ih} \int_{-\infty}^{t} E_{hk}(t, s) G_k(N(s))\, ds \qquad (5.9.23)$$

where $F_i(N)$, $G_k(N)$ are nonlinear functions of N; (H_{ih}) is a constant matrix; and (E_{hk}) is a matrix whose elements are some functions of t and s.

Some equations of the form (5.9.23) can be reduced to a system of differential equations. This reduction can be usually carried out in different ways. In the transformation of (5.9.23) to a system of differential equations, it is useful to note the fact that a general solution of the autonomous vector equation,

$$\dot{x}(t) = Ax(t) + Bu(t), \qquad (5.9.24)$$

where A and B are constant matrices, is of the form

$$x(t) = M(t, t_0)x(t_0) + \int_{-\infty}^{t} M(t, s)Bu(s)\, ds. \qquad (5.9.25)$$

Here $M(t, s)$ is the transition matrix which has many interesting properties (see Willems, 1970).

Due to the fact that (5.9.23) can be reduced to a system of differential equations in a nonunique manner, it is convenient for us to examine the differential equation system of the form

$$\dot{N}_i = N_i F_i(N) + N_i \sum_{s=1}^{m} Q_{is} y_s, \quad i = 1, 2, \ldots, m,$$

$$\dot{y}_r = \sum_{s=1}^{n} A_{rs} y_s + \sum_{i=1}^{m} B_{ri} G_i(N), \quad r = 1, 2, \ldots, n \qquad (5.9.26)$$

instead of (5.9.23). In model (5.9.26), (A_{rs}), (B_{ri}) and (Q_{is}) are constant matrices. In matrix notation, model (5.9.26) is

$$\dot{N} = \mathrm{diag}(N_1, N_2, \ldots, N_m)[F(N) + Qy],$$

$$\dot{y} = Ay + BG(N). \qquad (5.9.27)$$

Generally, the variables $y_1, y_2, \ldots, y_n$ can take positive and negative values. But in some cases (e.g. in (5.9.5) and (5.9.10)) these variables take only positive values. Let $\mathbf{R}_+^m \times \mathbf{R}^n$ denote the region in which the N_i variables are restricted to positive values and the y_s variables are unrestricted in sign.

Let $(N^\star, y^\star)$ denote an equilibrium of (5.9.27) which satisfies the equations

$$F(N^\star) + Qy^\star = 0,$$
$$Ay^\star + BG(N^\star) = 0. \tag{5.9.28}$$

Suppose $(N^\star, y^\star)$ is inside the region $\mathbf{R}_+^m \times \mathbf{R}^n$. Eqs. (5.9.27) and (5.9.28) give

$$\dot{N} = \mathrm{diag}(N_1, N_2, \ldots, N_m)[F(N) - F(N^\star) + Q(y - y^\star)],$$
$$\dot{y} = A(y - y^\star) + B[G(N) - G(N^\star)]. \tag{5.9.29}$$

By definition $(N^\star, y^\star)$ is globally asymptotically stable if every solution of (5.9.27) which begins in the region $\mathbf{R}_+^m \times \mathbf{R}^n$ remains in it for all finite values of t and converges to $(N^\star, y^\star)$ as $t \to \infty$.

Suppose the homogenous system

$$\dot{y}_r = \sum_{s=1}^{n} A_{rs} y_s, \quad r = 1, 2, \ldots, n \tag{5.9.30}$$

is stable. In this case there exists a positive definite symmetric matrix P such that $PA + A^{\mathrm{T}}P$ is negative definite.

Theorem 5.9.1. *The equilibrium $(N^\star, y^\star)$ is globally asymptotically stable if there exists a positive diagonal matrix C and a positive definite matrix P such that*

$$W(N, y) = (N - N^\star)^{\mathrm{T}} C[F(N) - F(N^\star) + Q(y - y^\star)]$$
$$\qquad + (y - y^\star)^{\mathrm{T}} P[A(y - y^\star) + BG(N) - BG(N^\star)] \leqslant 0 \tag{5.9.31}$$

for all $(N, y) \in \mathbf{R}_+^m \times \mathbf{R}^n$ and $W(N, y)$ does not vanish identically along a solution of (5.9.27) in $\mathbf{R}_+^m \times \mathbf{R}^n$, except for $(N, y) = (N^\star, y^\star)$.

Proof. We shall show that the function

$$V(N, y) = \sum_{i=1}^{m} c_i[N_i - N_i^\star - N_i^\star \ln(N_i/N_i^\star)] + (1/2)(y - y^\star)^{\mathrm{T}} P(y - y^\star) \tag{5.9.32}$$

where $C = \mathrm{diag}(c_1, c_2, \ldots, c_m)$, is a Liapunov function.

Along solutions of (5.9.27) we have

$$\dot{V}(N, y) = W(N, y).$$

Hence $V(N, y)$ is a Liapunov function. Using LaSalle's extension of the direct method of Liapunov (see LaSalle, 1976) and the fact that $(N^\star, y^\star)$ is the only invariant point in $\mathbf{R}_+^m \times \mathbf{R}^n$, we conclude that $(N^\star, y^\star)$ is globally asymptotically stable.

Let L be an $m \times m$ constant matrix and b be a constant vector. Suppose $F(N) = b + LN$ and $G = N$. Eq (5.9.27) becomes

$$\dot{N} = \mathrm{diag}(N_1, N_2, \ldots, N_m)[L(N - N^\star) + Q(y - y^\star)],$$

$$\dot{y} = A(y - y^\star) + B(N - N^\star). \tag{5.9.33}$$

From (5.9.31) and (5.9.33) we get

$$W(N, y) = (1/2)Z^{\mathrm{T}}RZ \tag{5.9.34}$$

where

$$Z = \begin{pmatrix} N - N^\star \\ y - y^\star \end{pmatrix}, \qquad R = \begin{pmatrix} CL + L^{\mathrm{T}}C & CQ + B^{\mathrm{T}}P \\ Q^{\mathrm{T}}C + PB & PA + A^{\mathrm{T}}P \end{pmatrix}. \tag{5.9.35}$$

Therefore the equilibrium $(N^\star, y^\star)$ of (5.9.33) is globally asymptotically stable if: (i) $N^\star$ is a positive vector; (ii) there exists a positive diagonal matrix C and a positive definite symmetric matrix P such that R is negative semi-definite; and (iii) $W(N, y)$ of (5.9.34) does not vanish identically along an admissible solution of (5.9.33), except for $N = N^\star$.

We note that

$$R = \begin{pmatrix} C & 0 \\ 0 & P \end{pmatrix} \begin{pmatrix} L & Q \\ B & A \end{pmatrix} + \begin{pmatrix} L^{\mathrm{T}} & B^{\mathrm{T}} \\ Q^{\mathrm{T}} & A^{\mathrm{T}} \end{pmatrix} \begin{pmatrix} C & 0 \\ 0 & P \end{pmatrix}. \tag{5.9.36}$$

For convenience, let

$$S = \begin{pmatrix} L & Q \\ B & A \end{pmatrix}. \tag{5.9.37}$$

A sufficient condition for R to be negative definite is that there exists a positive diagonal matrix D such that $DS + S^{\mathrm{T}}D$ is negative definite. Let $G_{ii} = S_{ii}$ for $i = 1, 2, \ldots, m + n$ and $G_{ij} = |S_{ij}|$ for $i \neq j$. If all the leading principal minors of $-G$ are positive, then G is an M-matrix. Then there exists a positive diagonal matrix D such that $DG + G^{\mathrm{T}}D$ is negative definite. For this matrix D, the matrix $DS + S^{\mathrm{T}}D$ is negative definite.

Example 5.9.2. Model (5.9.10) has a positive equilibrium at $(N^\star, y_1^\star, y_2^\star)$ where $y_1^\star = r/(w + ca^2)$, $y_2^\star = ay_1^\star$, $N^\star = a^2y_1^\star$. For this model,

$$S = \begin{pmatrix} -c & -w & 0 \\ 0 & -a & 1 \\ 1 & 0 & -a \end{pmatrix}. \tag{5.9.38}$$

The parameters c and w are positive. Consider the matrix

$$G = \begin{pmatrix} -c & w & 0 \\ 0 & -a & 1 \\ 1 & 0 & -a \end{pmatrix}. \quad (5.9.39)$$

All the leading principal minors of $-G$ are positive if $a^2 c > w$. Hence if $a^2 c > w$. the equilibrium $(N^\star, y_1^\star, y_2^\star)$ is globally asymptotically stable.

5.10. SPATIAL HETEROGENEITY IN POPULATION DYNAMICS

In the real world spatial heterogeneity plays a crucial role in the persistence and stability of populations. A classic example which demonstrated the importance of spatial heterogeneity was carried out by Huffaker (1958) using mites. The role of spatial heterogeneity in the dynamics of interacting populations has been the subject of much recent theoretical and experimental research; this research has been reviewed by Levin (1976 a).

We can distinguish three different mathematical approaches to studying the effects of spatial heterogeneity on population dynamics. In the first approach the changes in the number of patches occupied by different species are studied. This approach has been developed by Levins and Culver (1971), Horn and MacArthur (1972), Slatkin (1974), Hastings (1977), Caswell (1978) and others. In the second approach, m species which are distributed over an interconnected network of n patches are considered. The dynamics of a typical patch and the effects of migrations on a patch are studied. This approach has been developed by Othmer and Scriven (1971), Karlin and McGregor (1972), Levin (1974) Christiansen and Fenchel (1977) and others. In the third approach the role of space and dispersal in population dynamics is studied by using partial differential equations to model the population dynamics. This approach was pioneered by Skellam (1951) and developed by Comins and Blatt (1974), Hadeler et al. (1974), Steele (1974), Rosen (1975), Levin (1976 b), Gopalsamy (1977), Jorne and Carmi (1977), Hastings (1978) and others.

We shall examine some aspects of these three approaches to understand the role of spatial heterogeneity in population dynamics.

Suppose a prey and its predator are distributed over a large number of patches in a region. Let X denote the fraction of patches which contains preys only, and Y denote the fraction of patches which contains preys and predators. It follows that the number of empty patches is $1 - X - Y$. Note that X and Y must satisfy the constraints

$$1 \geqslant X \geqslant 0, \qquad 1 - X \geqslant Y \geqslant 0. \quad (5.10.1)$$

Under the assumption that the Y type of patches decay at a constant rate, Hasting (1977) proposed that a model of the dynamics of the patches is

$$\dot{X} = a(1 - X - Y)(X + Y) - bXY,$$

$$\dot{Y} = bXY - Y. \quad (5.10.2)$$

In this model it is assumed that an empty patch can be invaded only by the prey to create an X type of patch. The invasion rate of the predator is proportional to the number of patches it currently occupies. Similarly the invasion rate of the prey is proportional to the number of patches it occupies. Hastings (1977) also studied a model which assumes that the predator eliminates the prey in a Y type of patch in some set time. This leads to a delay differential equation with a more complex behavior.

Using the Poincare—Bendixon Theorem, Hastings (1977) showed that (5.10.2) is globally stable if it has a positive and admissible equilibrium which, for convenience, is denoted by $(X^\star, Y^\star)$. This means that every solution of (5.10.2) which begins in the region $\{(X, Y) \mid 1 > X > 0, 1 - X > Y > 0\}$ remains in it for all finite values of t and converges to $(X^\star, Y^\star)$ as $t \to \infty$. Model (5.10.2) has a positive and admissible equilibrium if $b > 1$. This brief discussion demonstrates the type of results which can be obtained in this approach.

In the second approach a community of m species is assumed to be distributed over an interconnected network of n patches. Let N_i^s be the population of the ith species in the sth patch and M_i^s be the net immigration rate of the ith species into the sth patch. A general model of this network of patches is

$$\dot{N}_i^s = N_i^s F_i^s \, (N_1^s, N_2^s, \ldots, N_m^s) + M_i^s(N_i^1, N_i^2, \ldots, N_i^n) \qquad (5.10.3)$$

where $i = 1, 2, \ldots, m$ and $s = 1, 2, \ldots, n$.

When n is large it is very difficult to analyse model (5.10.3). To get some insight into the behavior of the model, we first examine the case in which there is no migration between any patch, i.e. $M_i^s = 0$ for $i = 1, 2, \ldots, m$ and $s = 1, 2, \ldots, n$. In this case it is likely that the model for each patch has one or more nonnegative equilibria which are locally sector stable. For global sector stability, the equations for each patch must satisfy Corollary 5.8.1. If the parameters or the dynamical equations are different in distinct patches, then different combinations of species will persist in different patches.

Example 5.10.1. Let N_1 and N_2 be the populations of two competing species in a patch. Suppose a model of the competition is

$$\dot{N}_1 = N_1[2 - N_1 - N_2],$$

$$\dot{N}_2 = N_2[b - N_1 - 2N_2] \qquad (5.10.4)$$

where b is a positive parameter.

This model has an equilibrium at $(4 - b, b - 2)$. It is positive if $4 > b > 2$. If $b > 4$, the nonnegative equilibrium at $(0, b/2)$ satisfies Theorem 5.8.3; therefore, it is globally sector stable. This means that every solution of (5.10.4) which begins in the set $\{N \mid N_1 \geqslant 0, N_2 > 0\}$ remains in it and converges to $(0, b/2)$. Similarly if $b < 2$, the equilibrium $(2, 0)$ is globally sector stable.

Thus if b varies from patch to patch in a region and there is no migration between the patches, then we can expect that there exist three types of patches in the region; some contain two coexisting species and others contain N_1 or N_2 species.

Christiansen and Fenchel (1977) showed that a competition between two species can be destabilized by one species immigrating at a constant rate. The immigration causes the system to lose its positive equilibrium.

Example 5.10.2. For $I = 0$ the model

$$\dot{N}_1 = N_1[2 - N_1 - N_2],$$

$$\dot{N}_2 = N_2[3 - N_1 - 2N_2] + I \tag{5.10.5}$$

has a globally stable equilibrium at (1, 1).

If $I > 0$ it has an equilibrium at $N^\star$ where $N_1^\star = 1.5 - (I + 0.25)^{1/2}$ and $N_2^\star = 0.5 + (I + 0.25)^{1/2}$. This equilibrium loses its positivity if $I > 2$.

Immigration into a patch at constant rates is a stabilizing influence in a class of models if they generate a positive equilibrium. Model (5.8.18) illustrates this behavior. We shall establish this result in a general manner. Let a model of a typical patch be

$$\dot{N}_i = N_i F_i(N_1, N_2, \ldots, N_m) + I_i \tag{5.10.6}$$

where I_i is a nonnegative constant and $i = 1, 2, \ldots, m$. If $I_i > 0$, then $\dot{N}_i > 0$ when $N_i = 0$. Let (5.10.6) have a nonnegative equilibrium which satisfies Theorem 5.8.3 if $I_i = 0$ for $i = 1, 2, \ldots, m$. Suppose, for $I_i \geqslant 0$ where $i = 1, 2, \ldots, m$, model (5.10.6) has a positive equilibrium at $N^\star$. In this case the immigration at constant rates is a stabilizing influence. We have

$$\partial(F_i + I_i/N_i)/\partial N_i = \partial F_i/\partial N_i - I_i/N_i^2 \leqslant G_{ii}, \tag{5.10.7}$$

$$\partial(F_i + I_i/N_i)/\partial N_j = \partial F_i/\partial N_j \quad \text{if } i \neq j. \tag{5.10.8}$$

By Theorems 5.5.1 and 5.8.3 $N^\star$ of (5.10.6) is globally stable. Therefore, in this case, immigration at constant rates is a stabilizing influence.

We shall now examine the consequences of migration between patches. In the simplest case the net exchange from the sth patch to the rth patch for the ith species is proportional to $N_i^s - N_i^r$. We have

$$\dot{N}_i^s = N_i^s F_i^s(N^s) + \sum_{r=1}^{n} D_i^{sr}(N_i^s - N_i^r), \quad i = 1, 2, \ldots, m \tag{5.10.9}$$

where D_i^{sr} is a nonegative diffusion coefficient. This class of models was studied by Levin (1974, 1976 b) and Hastings (1978).

Suppose (5.10.9) has a nonnegative equilibrium. It may not be possible to rewrite it in the same form as model (5.8.1). Therefore it may not be possible to use Theorem 5.8.1 to establish that a nonnegative equilibrium of (5.10.9) is locally sector stable. We shall now consider Levin's (1976 b)

Theorem for local stability which is more general than Theorem 5.8.1.
Suppose the model

$$\dot{Y}_i = G_i(Y_1, Y_2, \ldots, Y_{p+q}), \quad i = 1, 2, \ldots, p + q \tag{5.10.10}$$

has a nonnegative equilibrium at $Y^\star = (Y_1^\star, Y_2^\star, \ldots, Y_p^\star, 0, 0, \ldots, 0)$ where $Y_i^\star > 0$ for $i = 1, 2, \ldots, p$.

Theorem 5.10.1. *The nonnegative equilibrium $Y^\star$ of (5.10.10) is locally sector stable if:* (i) *at $Y^\star$ all the eigenvalues of the matrix* $(\partial G_i(Y)/\partial Y_j)$, *where $i, j = 1, 2, \ldots, p + q$, have negative real parts*; (ii) $\partial G_r(Y)/\partial Y_i = 0$ *for $i = 1, 2, \ldots, p$, $r = p + 1, p + 2, \ldots, p + q$; and* (iii) $\partial G_r(Y)/\partial Y_s \geqslant 0$ *for $r \neq s$ and $r, s = p + 1, p + 2, \ldots, p + q$.*

Proof. This proof is adapted from that given by Levin (1976 b). The linearized dynamics of (5.10.10) is

$$\dot{y}_i = \sum_{j=1}^{p+q} \frac{\partial G_i}{\partial Y_j}(Y^\star) y_j, \quad i = 1, 2, \ldots, p + q. \tag{5.10.11}$$

Condition (i) implies that $Y^\star$ is asymptotically stable. Hence solutions of (5.10.10) which begin in an open neighbourhood of $Y^\star$ will remain close to $Y^\star$.

The variables $y_1, y_2, \ldots, y_p$ can be positive or negative because $Y_i^\star > 0$ for $i = 1, 2, \ldots, p$. But the variables $y_s \geqslant 0$ for $s = p + 1, p + 2, \ldots, p + q$ because $Y_s^\star = 0$. If conditions (i) to (iii) are satisfied, then solutions of (5.10.10) which begin in an admissible nonnegative neighbourhood of $Y^\star$ will remain in the same or larger admissible nonnegative neighbourhood. Hence $Y^\star$ is locally sector stable.

Before we consider some examples we shall make some general observations on competition. It is worthwhile to distinguish between two types of Lotka—Volterra models of competition. Consider the Lotka—Volterra model

$$\begin{aligned} \dot{N}_1 &= N_1[b_1 - c_{11}N_1 - c_{12}N_2], \\ \dot{N}_2 &= N_2[b_2 - c_{21}N_1 - c_{22}N_2] \end{aligned} \tag{5.10.12}$$

where $b_1, b_2, c_{11}, c_{22}, c_{12}, c_{21}$ are positive constants. If $c_{11}c_{22} - c_{12}c_{21} > 0$ model (5.10.12) is said to describe a *tolerant* competition. If $c_{11}c_{22} - c_{12}c_{21} < 0$, then (5.10.12) is said to describe a *severe* competition. A Lotka—Volterra model for a tolerant competition is globally stable if it has a positive equilibrium. Similarly, the general Lotka—Volterra model

$$\dot{N}_i = N_i \left[b_i - \sum_{s=1}^{m} c_{is} N_s\right], \quad i = 1, 2, \ldots, m \tag{5.10.13}$$

where $C = (C_{is})$ is a nonnegative matrix, is said to describe a tolerant competition if there exists a positive diagonal matrix D such that $DC + C^T D$ is positive definite. Otherwise, model (5.10.13) is said to describe a severe competition.

Levin (1974, 1976 b) studied a network of *identical* patches which contain severe competition between two species. In this case the Lotka—Volterra equations for a patch in the absence of migration may have two locally sector stable nonnegative equilibria. Thus the outcome of the competition in an isolated patch depends on the initial densities of the species. If the diffusion coefficients between patches are small, an equilibrium of a typical patch will have the same stability behavior as that of a nearby nonnegative equilibrium which exists if there is no migration between patches; this is shown by Levin (1976 b). From this result and computer simulations, we make the conjecture that a positive equilibrium of an interconnected network of *identical* patches which contain species in severe competition could only be locally stable.

Example 5.10.3. Let X_i and Y_i be the populations of two competing species in the ith patch. Suppose we have two identical patches and a model of the patches is

$$\begin{aligned}
\dot{X}_1 &= X_1(3 - X_1 - 2Y_1) + D(X_2 - X_1),\\
\dot{Y}_1 &= Y_1(3 - 2X_1 - Y_1) + D(Y_2 - Y_1),\\
\dot{X}_2 &= X_2(3 - X_2 - 2Y_2) + D(X_1 - X_2),\\
\dot{Y}_2 &= Y_2(3 - 2X_2 - Y_2) + D(Y_1 - Y_2). \qquad (5.10.14)
\end{aligned}$$

For $D = 0$ each patch has locally sector-stable equilibria at (3, 0) and (0, 3) and two unstable equilibria at (0, 0) and (1, 1). The species are in severe competition.

Using the results in Levin's (1974) paper, we deduce that (5.10.14) has two locally stable positive equilibria if $0 < D < 0.3$. If $D = 0.1$ these locally stable equilibria are $A \equiv (2.696, 0.104, 0.104, 2.696)$ and $B \equiv (0.104, 2.696, 2.696, 0.104)$. The equilibrium at A corresponds to the case in which species X is established in the first patch and species Y is established in the second patch and immigration maintains the presence of the other species in each patch. Similar comments apply to the equilibrium at B. For $D = 0.1$, model (5.10.14) has an unstable equilibrium at (1, 1, 1, 1). Using Theorem 5.10.1 we can show that there are two locally sector-stable equilibria at (0, 3, 0, 3) and (3, 0, 3, 0). For $D > 0.3$ there is no stable positive equilibrium.

We conclude that (5.10.14) describes a relatively fragile system. The reasons for this fragility are: (i) the patches are identical; and (ii) the two species are in severe competition with each other.

In the next example we shall demonstrate that spatial variations in the carrying capacity for a species can allow two species in severe competition with each other to coexist in a robust manner.

Example 5.10.4. Let X and Y be the populations of two competing species in an isolated patch. Suppose a model of the competition is

$$\dot{X} = X(b - X - Y),$$
$$\dot{Y} = Y(3 - 2X - Y) \qquad (5.10.15)$$

where b is a positive constant. Suppose the X species is sensitive to an environmental factor which varies spatially; and as a result of this the parameter b varies from patch to patch. Comparing (5.10.12) and (5.10.15), we note that the latter describes a severe competition because $c_{11}c_{22} - c_{12}c_{21} = -1$. Thus the species cannot coexist in an isolated patch.

If $1.5 < b < 3$, model (5.10.15) has an unstable positive equilibrium at $(3 - b, 2b - 3)$. In this case the dynamical behavior of (5.10.15) is similar to that of an isolated patch of (5.10.14). The most interesting case is when $b > 3$ in one patch and $b < 1.5$ in the other patch. If $b > 3$ it can be shown graphically or by computer simulations that $(b, 0)$ is globally sector-stable. This means that every solution of (5.10.15) which begins in the region $\{(X, Y) \mid X > 0, Y \geqslant 0\}$ remains in it and converges to $(b, 0)$. In this case the X species wins in the competition for all admissible initial densities. Similarly for $b < 1.5$, the Y species wins in the competition for all initial densities; $(0, 3)$ is globally sector-stable.

It follows that if there are two isolated patches in a region, the two species in severe competition with each other could persist in the region for a wide range of initial conditions. As expected, the two species could also persist in the region if there is a small diffusion between the patches.

Let X_i, Y_i be the populations of the two competing species in the ith patch where $i = 1, 2$. Consider the model

$$\dot{X}_1 = X_1(4 - X_1 - Y_1) + D(X_2 - X_1),$$
$$\dot{Y}_1 = Y_1(3 - 2X_1 - Y_1) + D(Y_2 - Y_1),$$
$$\dot{X}_2 = X_2(1 - X_2 - Y_2) + D(X_1 - X_2),$$
$$\dot{Y}_2 = Y_2(3 - 2X_2 - Y_2) + D(Y_1 - Y_2). \qquad (5.10.16)$$

If $D = 0$ the Y species is excluded in the first patch and the X species is excluded in the second patch. If $D = 0.1$ model (5.10.16) has a positive equilibrium at $(3.855, 0.051, 0.226, 2.471)$. Computer simulations suggest that this equilibrium is globally stable. We conclude that environmental variations and spatial heterogeneity can allow two species in severe competition with each other to coexist for a wide range of initial conditions.

If D is relatively large the two patches behave like one, and one species is excluded. For example, for $D = 0.5$, computer simulations suggest that the equilibrium $(3.72, 0, 1.64, 0)$ is globally sector-stable. Thus species X wins the competition in the two interconnected patches.

In the next example, we demonstrate that spatial variations in the carrying capacity of a species can allow two species in tolerant competition with each other to coexist in a robust manner even if the diffusion coefficients are relatively high. We shall consider the nontrivial case in which the two species cannot coexist in an isolated and homogenous patch.

Example 5.10.5. Let X and Y be the populations of two competing species in an isolated patch. Suppose a model of the competition is

$$\dot{X} = X(2 - X - Y),$$
$$\dot{Y} = Y(b - X - 2Y) \qquad (5.10.17)$$

where b is a positive constant. We shall assume that b varies from patch to patch because the Y species is sensitive to an environmental factor (e.g. limiting nutrient in the soil for a plant species) which varies spatially. Comparing (5.10.12) with (5.10.17), we deduce that the latter models a tolerant competition because $c_{11}c_{22} - c_{12}c_{21} = 1 > 0$. This does not immediately imply that both species can coexist in an isolated patch because the interaction may not have a positive equilibrium.

If $2 < b < 4$, model (5.10.17) has a globally stable equilibrium at $(4 - b, b - 2)$. In this case both species can coexist in an isolated patch.

Using Theorem 5.8.3 we can show that the equilibrium (2, 0) is globally sector-stable if $b < 2$. Here, the X species wins the competition in an isolated patch. Similarly, it can be shown that the Y species wins the competition in an isolated patch if $b > 4$. In this case $(0, b/2)$ is globally sector-stable.

Let X_i, Y_i be the populations of the two species in the ith patch where $i = 1, 2$. Suppose a model of the coupled patches is

$$\dot{X}_1 = X_1(2 - X_1 - Y_1) + D(X_2 - X_1),$$
$$\dot{Y}_1 = Y_1(5 - X_1 - 2Y_1) + D(Y_2 - Y_1),$$
$$\dot{X}_2 = X_2(2 - X_2 - Y_2) + D(X_1 - X_2),$$
$$\dot{Y}_2 = Y_2(1.5 - X_2 - 2Y_2) + D(Y_1 - Y_2). \qquad (5.10.18)$$

If $D = 0$, the Y species wins in the first patch and the X species wins in the second patch.

For the relatively large value of $D = 0.5$, model (5.10.18) has a positive equilibrium at (0.471, 2.10, 1.01, 1.72). Computer simulations suggest that this equilibrium is globally stable.

This example suggests that an interconnected network of patches, each of which contains a set of tolerant competing species which cannot coexist together in an isolated patch, could allow all the species to coexist in a robust manner even if the diffusion coefficients are relatively high.

From these simple examples, it is clear that the qualitative behavior of the dynamics of a community in a spatially heterogenous environment could be very different from that of the same community in a homogenous environment.

We note that in the study of a community in a heterogenous environment the concept of global sector stability is useful. We expect that an interconnected network of patches, each of which in isolation has a globally sector stable nonnegative equilibrium, would be a globally stable system if the diffusion coefficients between the patches are small.

In the third approach to the study of the role of spatial heterogeneity in population dynamics, it is assumed that the populations are continuously spread out in space. Thus, partial differential equations are used to model the dynamics.

Let R be a simply connected region inside an ocean. Suppose a community in R can be described by the equations

$$\frac{\partial N_i}{\partial t} = N_i F_i(N) + \sum_{s=1}^{3} \frac{\partial}{\partial x_s}\left(D_i \frac{\partial N_i}{\partial x_s}\right), \quad i = 1, 2, \ldots, m \tag{5.10.19}$$

where $D_i(t, x, N)$ are the diffusion coefficients which are nonnegative functions of t, x, N. Let n denote a unit vector which is normal to ∂R, the boundary of R. Let $\nabla = (\partial/\partial x_1, \partial/\partial x_2, \partial/\partial x_3)$ be the gradient operator. We shall assume that there is no net migration across every point of ∂R. Hence, at every point on ∂R we have

$$n \cdot \nabla N_i = 0 \quad \text{for all } i = 1, 2, \ldots, m. \tag{5.10.20}$$

Here $n \cdot \nabla N_i$ is the projection of ∇N_i on n.

We shall establish that (5.10.19) has a spatially independent solution which is globally stable relative to all positive solutions of (5.10.19) if: (i) it has a globally stable solution when $D_i = 0$ for $i = 1, 2, \ldots, m$; and (ii) for the reduced model there exists a Liapunov function

$$V(N) = \sum_{i=1}^{m} c_i V_i(N_i), \quad i = 1, 2, \ldots, m \tag{5.10.21}$$

where $c_1, c_2, \ldots, c_m$ are positive constants such that $\partial^2 V/\partial N_i^2 > 0$ for $i = 1, 2, \ldots, m$. It follows that

$$\dot{V}(N) = \sum_{i=1}^{m} (\partial V/\partial N_i) N_i F_i(N) \tag{5.10.22}$$

is negative semidefinite and $\dot{V}(N)$ does not vanish identically along a non-trivial solution of the reduced model (i.e. $D_i = 0, i = 1, 2, \ldots, m$)

Let $\mathrm{d}v = \mathrm{d}x_1 \mathrm{d}x_2 \mathrm{d}x_3$ and $\mathrm{d}s$ be an element of the surface ∂R. We shall show that

$$E = \int_R V(N)\mathrm{d}v \tag{5.10.23}$$

is a Liapunov functional of (5.10.19). Along a positive solution of (5.10.19), we have

$$\dot{E} = \int_R \sum_{i=1}^{m} \frac{\partial V}{\partial N_i} \frac{\partial N_i}{\partial t} \mathrm{d}v$$

$$= \int_R \sum_{i=1}^{m} \left[\frac{\partial V}{\partial N_i} N_i F_i(N) + \sum_{s=1}^{3} \frac{\partial V}{\partial N_i} \nabla \cdot (D_i \nabla N_i)\right] \mathrm{d}v$$

$$= \int_R \sum_{i=1}^{m} \left[\frac{\partial V}{\partial N_i} N_i F_i(N) + \nabla \cdot \left(\frac{\partial V}{\partial N_i} D_i \nabla N_i \right) - \nabla \left(\frac{\partial V}{\partial N_i} \right) \cdot (D_i \nabla N_i) \right] dv.$$

Using the divergence theorem and condition (5.10.20), we get

$$\dot{E} = \sum_{i=1}^{m} \left[\int_{\partial R} (n \cdot \nabla N_i) \frac{\partial V}{\partial N_i} D_i ds + \int_R \frac{\partial V}{\partial N_i} N_i F_i - \frac{\partial^2 V}{\partial N_i^2} D_i \nabla N_i \cdot \nabla N_i \, dv \right]$$

$$= \sum_{i=1}^{m} \left[\int_R \frac{\partial V}{\partial N_i} N_i F_i dv - \int_R \frac{\partial^2 V}{\partial N_i^2} D_i \, \|\nabla N_i\|^2 \, dv \right]$$

where $\|a\|^2 = a_1^2 + a_2^2 + a_3^2$. As $D_i \geqslant 0$ and $\partial^2 V / \partial N_i^2 > 0$ for $i = 1, 2, \ldots, m$, it follows that $\dot{E}$ is negative semidefinite and does not vanish identically along a nontrivial solution. We conclude that the spatially uniform solution $N^\star$ is globally stable relative to all positive solutions of (5.10.19). This result was first established by Hastings (1978). Special cases of this result were established by Rosen (1975) and Jorne and Carmi (1977). Similar results can also be established in the case where R is a simply connected region of a two-dimensional space.

Hastings (1978) also showed that the spatially homogenous solution of an interconnected network of n identical patches, which is described by (5.10.9) and in which $D_i^{sr} = D_i^{rs} \geqslant 0$, is globally stable for all initial conditions $N_i^s(0) > 0$ for all i, s. This result implies that a system of equations which models a spatially homogenous and well mixed community can also be used to model a collection of identical patches of the same community.

CHAPTER 5: SELECTED REFERENCES

Complexity versus stability

MacArthur (1955); Elton (1958); Gardner and Ashby (1970); May (1974); Roberts (1974); DeAngelis (1975); Gilpin (1975 b); Goh (1975, 1978 d); Goodman (1975); Gilpin (1975); Gilpin and Case (1976); Goh and Jennings (1977); Pimm and Lawton (1977); Mazanov (1978); McNaughton (1978); Siljak (1978); Vincent and Anderson (1979).

Stability of nonlinear multispecies models

Strobeck (1973); May (1974); Freedman and Waltman (1977); Goh (1977 a, 1978 d); May (1977); Hsu (1978 a, c); Siljak (1978); Gard and Hallam (1979).

Sector stability

Gilpin and Case (1976); Leung (1976); Levin (1976 b); Goh (1978 a); Takeuchi et al. (1978); Hsu (1978 b); Case and Casten (1979).

Population models with time delays

Volterra (1931); Wangersky and Cunningham (1957); Scudo (1971); May (1974); MacDonald (1976); Brauer (1977, 1978); Cushing (1977); Wörz-Busekros (1978).

Spatially heterogenous populations

Skellam (1951); Huffaker (1958); Reddingius and Den Boer (1970); Levins and Culver (1971); Othmer and Scriven (1971); Horn and MacArthur (1972); Karlin and McGregor (1972); Comins and Blatt (1974); Hadeler et al. (1974); Levin (1974, 1976 a, b); Slatkin (1974); Steele (1974); Rosen (1975); Christiansen and Fenchel (1977); Gopalsamy (1977); Hastings (1977, 1978); Jorne and Carmi (1977); Leung (1978); Ashkenazi and Othmer (1978).

Chapter 6

MANAGEMENT OF PESTS AND GREENHOUSE CROPS

6.1. INTRODUCTION

Insecticide pollution is recognized as a major health hazard to human beings. This has led to an increased interest in biological control methods for insect and plant pests (DeBach, 1964; Huffaker and Messenger, 1976). Another important reason for the intense interest in biological control is the decreasing effectiveness of chemical pesticides due to the adaptation of the pests to such pesticides. This leads to exponential increases in the cost of spraying chemicals on a pest. For instance, in the Ord River region of the northwestern part of Australia, the spraying of insecticides to suppress cotton pests increased steadily until it had to be applied every other day on the same crop. By that time, it had become too costly to continue spraying chemicals which has become decreasingly effective; consequently, the growing of cotton in the region was abandoned in 1976.

Biological control was practised by Chinese citrus growers since ancient times (Debach, 1964, p.22). They placed nests of predaceous ants, *Oecophylla smaragdina* F., in mandarin orange trees to reduce the numbers of foliage-feeding insects. But the scientific approach to biological control began with the dramatic and successful control of the cottony-cushion scale, *Icerya purchasi* Mask., by the introduction of the vedalis beetle (lady bird), *Rodolia cardinalis* (Muls.), into California in 1888 (DeBach, 1964, Chapter 2). In 1976, Laing and Junji Hamai reported that 157 species of insect pests have been successfully suppressed by means of biological control. However, one third of these successes are only partial.

Usually in the biological control of a pest, a parasite or a predator of the pest is imported and bred in laboratories. When there is an adequate number of the imported natural enemy, they are released with the expectation that a stable prey—predator or host—parasite system will be created. There are a few cases in which additional releases of the natural enemy are made now and then (Hussey and Bravenboer, 1971; Parker, 1971). It is plausible that the manipulation of the rate of release of the natural enemy and the rate of release of the pest itself will increase the number of successes in the biological control of pests.

We can distinguish two uses of mathematical analysis in pest control. Firstly, mathematics is used for *designing* robust and stable prey—predator or host—parasite systems. This is achieved by finding out the characteristics of a natural enemy which will enhance stability in the system consisting of the pest and its natural enemy. It is useful in this type of study to compute the range of parameter values for which the system is stable. Having discovered the best characteristics of a natural enemy, it is then possible to choose the most appropriate natural enemy to control a pest.

The second way in which mathematics is used in pest control is to formulate good *feedback control* policies in the dynamic manipulation of the control variables acting on a system consisting of the pest and its natural enemy. Due to difficulties in applying optimal control theory, this application of mathematics in pest control is limited to a few studies (Goh et al., 1974 a; Conway, 1977; Shoemaker, 1977).

In a design problem in pest control, a good model of the dynamics of the system consisting of the pest and its natural enemy is essential. Otherwise, we will not be justified in making any conclusions on the characteristics of a natural enemy which enhance stability in the system in the real world.

On the other hand, it is not necessary to have a good model of the natural dynamics of a pest—predator or pest—parasite system before we can formulate good feedback control policies. In Section 6.3 we shall reaffirm this assertion by considering a poorly defined prey—predator system which is subjected to a few control variables which could enhance the desirable trends in the natural dynamics so as to drive the state of the system to a desirable target and maintain the state inside the target.

Biological control is not the panacea to all pest problems. In the forseeable future it will be necessary to continue the use of chemical pesticides. A unified approach which uses both biological and chemical methods in a harmonious way to suppress a pest is called integrated control (Geier, 1970; Corbet and Smith, 1976).

In Section 6.4 we shall use optimal control theory to formulate optimal feedback policies in the integrated control of a pest—predator system. The control variables are rates of releases of the pests and the predators, and rates of application of chemical agents. We shall use a Lotka—Volterra model to describe the natural dynamics. As previously noted, the qualitative features of an efficient feedback control policy in this type of situation is not critically dependent on the precise details of a model of the natural dynamics.

Potentially, the use of viruses, fungi and bacteria is one of the most effective methods for controlling pests. One reason for this optimism is that the generation time lapse of microbes is much shorter than that of an insect pest. This advantage is subdued in the case where the transmission is carried out by an insect vector. The control of rabbit pests in Australia (Fenner and Ratcliffe, 1965; Davis et al., 1976) by the virus disease called myxomatosis, provides an extremely spectacular example of a virus controlling a pest. In

Section 6.5 we shall formulate optimal feedback control policies for generating epidemics to control a pest.

Section 6.6 contains an interesting example of managing, without an analytical model, a real system in which the state variable is highly variable and unpredictable. We shall show that it is possible to prescribe in an optimal manner the duration of a tomato crop grown in a greenhouse.

6.2. SOME APPLICATIONS OF STABILITY CONCEPTS IN PEST CONTROL

The most common method of biological control is to introduce a natural enemy which is usually a predator or a parasite (Laing and Junji Hamai, 1976). This method will succeed if the pest—predator or pest—parasite system has a stable equilibrium in which the pest level is relatively low. In order to provide a robust control of the pest the desirable equilibrium should have a large domain of attraction. It follows that an understanding of the factors which enhance stability and which generate a positive equilibrium in a prey—predator or host—parasite system will provide useful guidelines for biological control. This type of analysis has been carried out by Hassell and May (1973), Diamond (1974), Beddington et al. (1976), Freedman (1976), Anderson and May (1978), May and Anderson (1978), Goh and Agnew (1979) and many other authors.

In this section we shall examine a few population models in order to identify some of the factors which stabilize or destabilize a prey—predator system. The first example demonstrates that a prey—predator system is fragile if the predation rate is proportional to the product of the prey and predator populations; and the prey birth rate and the predator death rate are density independent.

Example 6.2.1. A pest population is usually unstable. For simplicity we shall model it by

$$\dot{H} = rH \tag{6.2.1}$$

where r is a positive constant and H denotes the pest population.

Let us suppose the natural enemy is a predator whose population is denoted by P. We shall assume that: (i) the predator consumes the prey (pest) at a rate which is proportional to the product of the prey and predator populations; (ii) the eaten prey biomass is converted into predator biomass with a constant efficiency; and (iii) the per capita prey birth rate and the per capita predator death rate are independent of the prey and predator populations respectively. Under these assumptions a model of the prey—predator system is

$$\dot{H} = H[r - a_{12}P],$$

$$\dot{P} = P[-d + Ea_{12}H] \tag{6.2.2}$$

where r, a_{12}, d and E are positive constants.

This is a well known Lotka—Volterra model. Its solutions are limit cycles and the model is fragile. Suppose the predator is subjected to small but unpredictable disturbances which increase the mortality rate of the population of the predator by $u(t)$ where $0 \leqslant u(t) \leqslant b$. Accordingly, we have

$$\dot{H} = H[r - a_{12}P],$$

$$\dot{P} = P[-d + Ea_{12}H] - u(t)P. \tag{6.2.3}$$

Let $(H^\star, P^\star) = (d/(Ea_{12}), r/a_{12})$ and

$$V(H, P) = E[H - H^\star - H^\star \ln (H/H^\star)] + [P - P^\star - P^\star \ln (P/P^\star)]. \tag{6.2.4}$$

Computing $\dot{V}$ along the solutions of (6.2.3) we get

$$\dot{V} = (P - P^\star)(-u). \tag{6.2.5}$$

If $u(t) \equiv 0$, then $\dot{V} \equiv 0$. Clearly a solution of (6.2.2) is $V(H, P)$ = constant. The worse possible way $u(t)$ can act on (6.2.3) is $u(t) = b$ for $P \geqslant P^\star$ and $u(t) = 0$ for $P < P^\star$. If this happens the solutions of (6.2.3) will approach the coordinate axis planes $H = 0$ and $P = 0$. This leads to a breakdown of the prey—predator system. In other words, model (6.2.3) is vulnerable to small continual disturbances.

In the next example, we shall show that a prey—predator system can be globally stable if the prey or predator sustains density dependent mortalities. This means that the death rate of the prey or predator increases with density.

Example 6.2.2. If the death rate of the prey population increases with density, model (6.2.2) becomes

$$\dot{H} = H[r - a_{11}H - a_{12}P],$$

$$\dot{P} = P[-d + Ea_{12}H], \tag{6.2.6}$$

where r, a_{11}, a_{12}, d and E are positive constants. This model has a nontrivial equilibrium at $(H^\star, P^\star) = (d/(Ea_{12}), [r - a_{11}d/(Ea_{12})]/a_{12})$. The equilibrium is positive if $r > a_{11}d/(Ea_{12})$.

Let $V(H, P)$ be the function in (6.2.4). Computing $\dot{V}(H, P)$ along solutions of (6.2.6), we get

$$\dot{V}(H, P) = -Ea_{11}(H - H^\star)^2. \tag{6.2.7}$$

Clearly, $\dot{V} \leqslant 0$ and $\dot{V}$ does not vanish identically along solutions of (6.2.6). Therefore $(H^\star, P^\star)$ is globally stable if $P^\star > 0$.

On the other hand let us assume that the death rate of the predator population increases with density. In the above case model (6.2.2) becomes

$$\dot{H} = H[r - a_{12}P],$$

$$\dot{P} = P[-d + Ea_{12}H - a_{22}P] \tag{6.2.8}$$

where r, a_{12}, a_{22}, d and E are positive constants. This model has an equilibrium at $(H^\star, P^\star)$ where $H^\star = d/(Ea_{12}) + ra_{22}/(Ea_{12}^2)$ and $P^\star = r/a_{12}$. This equilibrium is positive if all the parameters are positive. Model (6.2.8) is more robust than model (6.2.6). The latter has a positive equilibrium only if $r > a_{11}d/(Ea_{12})$. For practical purposes it is desirable that r is significantly larger than $a_{11}d/(Ea_{12})$.

Let $V(H, P)$ be the function in (6.1.2). Computing $\dot{V}(H, P)$ along solutions of (6.2.6), we get

$$\dot{V}(H, P) = -a_{22}(P - P^\star)^2 \leqslant 0. \tag{6.2.9}$$

Hence the equilibrium $(H^\star, P^\star)$ of (6.2.8) is globally stable.

For pest control, model (6.2.6) has an advantage over model (6.2.8). The equilibrium level of the pest in model (6.2.6) is $d/(Ea_{12})$. This is less than $d/(Ea_{12}) + ra_{22}/(Ea_{12}^2)$, the equilibrium level of the pest in model (6.2.8). In both models it is desirable that d is small and E and a_{12} be relatively large. This means that the per capita death rate of the predator at low densities should be small and the predator should be efficient in converting prey biomass into predator biomass and be efficient also in searching and consuming the pest.

By definition, the functional response of a predator is the rate at which a predator captures and consumes the prey. In the previous example the functional response of the predator was assumed to be proportional to the prey population irrespective of the density of the prey. In practice, the rate at which a predator eats its prey becomes saturated at high prey densities. More realistic functional responses of predators have been proposed by Holling (1959). He classified functional responses into three categories. In type I the response is a piecewise linear function: as the prey density increases the response increases at a constant rate but, beyond a certain prey density it is equal to a constant. In type II the response rises at a decreasing rate towards an asymptotic value as the prey density increases. In type III the response is sigmoid and it tends towards an asymptotic value as the prey density increases. Experimental results on the functional responses of predators can be found in the papers by Hassell et al. (1976 a, 1977).

We shall show that in simple population models type I and type II functional responses cannot contribute to the stability of a prey—predator system (Hassell and May, 1973; Oaten and Murdoch, 1975). If a predator has a type I or a type II functional response and the prey—predator system is stable, the stability is due to other processes affecting the predator or prey.

Let $f(H)$ be the functional response of a predator. Suppose a model of the prey—predator system is

$$\dot{H} = rH - Pf(H),$$

$$\dot{P} = P(-d + Ef(H)) = Pg(H) \tag{6.2.10}$$

where r and E are positive constants. Let $(H^\star, P^\star)$ be an equilibrium which satisfies the equations,

$$f(H^\star) = d/E, \tag{6.2.11}$$

$$P^\star = ErH^\star/d. \tag{6.2.12}$$

For relatively low values of H the functional response $f(H)$ is usually a strictly monotonic increasing function of H as H increases. It follows that if d/E is less than the supremum of $f(H)$ for $H > 0$, then (6.2.11) and (6.2.12) have a unique positive solution. If otherwise, every solution of (6.2.10) has the behavior $H(t) \to \infty$ and $P(t) \to 0$ as $t \to \infty$.

Suppose (6.2.10) has a positive equilibrium at $(H^\star, P^\star)$. Consider the function

$$V(H, P) = \int_{H^\star}^{H} \frac{g(s)}{f(s)}\, ds + \int_{P^\star}^{P} \frac{(s - P^\star)}{s}\, ds. \tag{6.2.13}$$

Computing $\dot{V}(H, P)$ along the solutions of (6.2.10), we get

$$\dot{V}(H, P) = [EHP^\star/f(H)]\ [f(H) - f(H^\star)]\,[f(H^\star)/H^\star - f(H)/H]. \tag{6.2.14}$$

If $f(H)$ is a type I functional response, we have

$$f(H) = \begin{cases} cH & \text{for } H \leqslant H_0, \\ cH_0 & \text{for } H > H_0. \end{cases} \tag{6.2.15}$$

For $H_0 \geqslant H > 0$, the function $\dot{V}(H, P)$ of (6.2.14) is equal to zero; and for $H > H_0$, it is positive. It follows that the solutions of (6.2.10) in the set $\{(H, P) \mid V(H, P) \leqslant V(H_0, P^\star)\}$ are fragile limit cycles; and those in the set $\{(H, P) \mid V(H, P) > V(H_0, P^\star)\}$ are unstable.

Suppose $f(H)$ is a type II functional response. This means that $f(H)$ is strictly monotonic increasing, and the slope of $f(H)$ is strictly monotonic decreasing as H increases. Hence, (i) $f(H^\star) > f(H)$ and $f(H)/H > f(H^\star)/H^\star$ for $H^\star > H > 0$; and (ii) $f(H^\star) < f(H)$ and $f(H)/H < f(H^\star)/H^\star$ for $H > H^\star$. Therefore, $\dot{V}(H, P) > 0$ for all $H > 0$ and $H \neq H^\star$. We conclude that the solutions of (6.2.10) are unstable for type II functional response. The next example illustrates the result obtained above.

Example 6.2.3. Let the functional response be

$$f(H) = aH/(1 + wH) \tag{6.2.16}$$

where a and w are positive constants. We have

$$f(H) - f(H^\star) = a(H - H^\star)/[(1 + wH)(1 + wH^\star)], \tag{6.2.17}$$

$$f(H^\star)/H^\star - f(H)/H = aW(H - H^\star)/[(1 + wH)(1 + wH^\star)]. \tag{6.2.18}$$

It follows that for this particular functional response the function $\dot{V}(H, P)$ is positive for all $H > 0$ and $H \neq H^\star$. Hence the solutions of (6.2.1) (6.2.10) with this particular $f(H)$ function are unstable.

Suppose $f(H)$ is a type III functional response. By definition this is a sigmoid function. Let the function $f(H)/H$ have a maximum at H_m; the function $f(H)/H$ being the slope of the line joining the points (0, 0) and $(H, f(H))$. For

$H > H_m$, $f(H)$ behaves locally like a type II functional response. Hence, if $H^\star > H_m$ the equilibrium $(H^\star, P^\star)$ is unstable.

If $H^\star$ satisfies the inequality $H_m > H^\star > 0$, the graph of the function $(f(H^\star)/H^\star)H$ intersects that of the function $f(H)$ at two points. One of these points is $H^\star$; let H_u denote the other point. For $H^\star > H > 0$, we have $f(H) < f(H^\star)$ and $f(H^\star)/H^\star > f(H)/H$; thus $\dot{V}(H, P) < 0$. For $H_u > H > H^\star$, we have $f(H) > f(H^\star)$ and $f(H^\star)/H^\star < f(H)/H$; thus $\dot{V}(H, P) < 0$. It follows that $\dot{V}(H, P) < 0$ for all H such that $H_u > H > 0$ and $H \neq H^\star$. Therefore $\{(H, P) \mid V(H, P) < V(H_u, P^\star)\}$ is a region of attraction of $(H^\star, P^\star)$.

The next example will illustrate these results.

Example 6.2.4. Consider the prey—predator model

$$\dot{H} = 0.25H - PH^2/(0.64 + H^2),$$

$$\dot{P} = P[-d + 0.1H^2/(0.64 + H^2)]. \tag{6.2.19}$$

For this model the function $f(H)/H = H/(0.64 + H^2)$ has a maximum at $H_m = 0.8$. Hence if $0.8 > H^\star > 0$, the positive equilibrium $(H^\star, P^\star)$ of (6.2.19) is locally stable. If $H^\star > 0.8$, then $(H^\star, P^\star)$ is unstable. Using (6.2.11) we deduce that (6.2.19) has a locally stable equilibrium if $d < 0.05$, and no stable equilibrium if $d > 0.05$.

If $d = 0.02$, the equilibrium $(H^\star, P^\star) = (0.5, 0.89)$. In this case, the graph of the function $(f(H^\star)/H^\star)H$ intersects that of the function $f(H)$ at the points, $H^\star = 0.5$ and $H_u = 1.28$. Hence, the region $\{(H, P) \mid V(H, P) < V(1.28, 0{,}89)\}$ is a domain of attraction of the equilibrium where the function $V(H, P)$ is given in (6.2.13).

In a well known example, Doutt and Nakata (1965) (see Hagen et al., 1971, p.275) discovered that the presence of an alternate host which infests blackberries was crucial in the biological control of the grape leaf hopper, *Erythroneura elegantula*, by the egg parasite, *Anagrus epos.* Hedgerows can be important refuges for the natural enemies of a wide variety of pests. Thus spatial heterogeneity can play an important role in biological control.

We shall demonstrate that spatial heterogeneity can stabilize a prey—predator system in biological control (Murdoch, 1977).

Let us assume that the predator has two subpopulations: one is with the pest population in the cultivated area; the other is outside the cultivated area. At time t, let these subpopulations be $P_1(t)$ and $P_2(t)$ respectively. Let the P_2 population be under stress, with the result that its mortality rate increases with density. Let the diffusion coefficient between the two subpopulations be small. A simplified model of this type of prey—predator system is

$$\dot{H} = H(r - a_{12}P_1),$$

$$\dot{P}_1 = P_1(-d + Ea_{12}H) + D(P_2 - P_1),$$

$$\dot{P}_2 = P_2(b - a_{33}P_2) + D(P_1 - P_2) \tag{6.2.20}$$

where r, a_{12}, a_{33}, d, E and D are positive constants.

Model (6.2.20) has an equilibrium at $(H^\star, P_1^\star, P_2^\star)$ where

$$P_1^\star = r/a_{12},$$

$$P_2^\star = \{(b-D) + [(b-D)^2 + 4a_{33}DP_1^\star]^{1/2}\}/(2a_{33}),$$

$$H^\star = [d + D(1 - P_2^\star/P_1^\star)]/(Ea_{12}). \tag{6.2.21}$$

By assumption, P_2 is an environment which is harsher than that of P_1. We can then expect that $P_2^\star < P_1^\star$. It follows that it is likely that $(H^\star, P_1^\star, P_2^\star)$ is a positive equilibrium.

Model (6.2.20) may be rewritten in the form $\dot{N}_i = N_iF_i(N_1, N_2, N_3)$, $i = 1, 2, 3$, where $(N_1, N_2, N_3) = (H, P_1, P_2)$ and

$$F_1 = r - a_{12}N_2,$$

$$F_2 = -d + Ea_{12}N_1 - D + D(N_3/N_2),$$

$$F_3 = b - D - a_{33}N_3 + D(N_2/N_3). \tag{6.2.22}$$

The Jacobian $(\partial F_i/\partial N_j)$ is

$$\begin{pmatrix} 0 & -a_{12} & 0 \\ Ea_{12} & -D(N_3/N_2^2) & D/N_2 \\ 0 & D/N_3 & -a_{33} - D(N_2/N_3^2) \end{pmatrix}. \tag{6.2.23}$$

At the equilibrium $N^\star = (H^\star, P_1^\star, P_2^\star)$, let $A = (A_{ij}) = (\partial F_i/\partial N_j)$. If $c_1 = E$, $c_2 = 1$, $c_3 = N_3^\star/N_2^\star$ and $C = \text{diag}(c_1, c_2, c_3)$, then the matrix $CA + A^TC$ is semidefinite. Hence $(H^\star, P_1^\star, P_2^\star)$ is locally stable. The function

$$V(N) = \sum_{i=1}^{3} c_i[N_i - N_i^\star - N_i^\star \ln(N_i/N_i^\star)] \tag{6.2.24}$$

is a Liapunov function and may be used to estimate the region of attraction of $N^\star = (H\ , P_1^\star, P_2^\star)$.

If $b > 0$, then the P_2 predator subpopulation lives on an alternative prey. If $b < 0$, there may be no alternative prey in the region that the P_2 population occupies. In either case the existence of the second subpopulation stabilizes the prey—predator system. From (6.2.2) we know that in the absence of the P_2 population, the prey—predator system is fragile.

We consider next the case in which the pest has a refuge (e.g. a hedgerow). In the refuge the pest subpopulation is under stress and its death rate increases with density. Let H_1 denote the pest subpopulation in the cultivated area and let H_2 be the pest subpopulation in the refuge. Suppose the diffusion coefficient between the two subpopulations is small. A simplified model of this type of prey—predator system is

$$\dot{H}_1 = H_1(r_1 - a_{12}P) + D(H_2 - H_1),$$

$$\dot{H}_2 = H_2(r_2 - a_{22}H_2) + D(H_1 - H_2),$$

$$\dot{P} = P(-d + Ea_{12}H_1) \qquad (6.2.25)$$

where r_1, a_{12}, a_{22}, D, E and d are positive constants. The parameter r_2 can be positive or negative.

It is reasonable to assume that $r_1 > D$. If this assumption is satisfied, then (6.2.25) has a positive equilibrium at $(H_1^\star, H_2^\star, P^\star)$ where

$$H_1^\star = d/(Ea_{12}),$$

$$H_2^\star = \{(r_2 - D) + [(r_2 - D)^2 + 4a_{22}DH_1^\star]^{1/2}\}/(2a_{22}),$$

$$P^\star = (r_1 - D)/a_{12} + DH_2/(a_{12}H_1). \qquad (6.2.26)$$

Model (6.2.26) may be rewritten in the form $\dot{N}_i = N_iF_i(N_1, N_2, N_3)$ where $(N_1, N_2, N_3) = (H_1, H_2, P)$. Let $C = \text{diag}\,(1, H_2^\star/H_1^\star, 1/E)$ and $A = (\partial F_i/\partial N_j)$ at $N^\star = (H_1^\star, H_2^\star, P^\star)$. We find that $CA + A^TC$ is negative semidefinite. It follows that $(H_1^\star, H_2^\star, P^\star)$ is locally stable and the function of the form given in (6.2.24) can be used as a Liapunov function to estimate the region of attraction of $(H_1^\star, H_2^\star, P^\star)$.

We note that (6.2.5) has a stable positive equilibrium at $(H_1^\star, H_2^\star, P^\star)$ even if r_2 is negative. In this case the H_2 subpopulation is in a refuge that is in a contiguous area which is not cultivated or which contains another crop. This suggests that monoculture in a large area is undesirable for pest control.

The stability of the equilibrium $(H_1^\star, H_2^\star, P^\star)$ depends critically on the parameter a_{22}. If $a_{22} = 0$, we can show that the solutions of (6.2.25) in a neighbourhood of $(H_1^\star, H_2^\star, P^\star)$ are fragile limit cycles.

6.3. GRAPHICAL ANALYSIS OF A CONTROL TWO-SPECIES SYSTEM

In this section we shall establish four simple rules for determining the stability or instability of an equilibrium in the graphical analysis of a two-species system. We shall demonstrate that the graphical method can be used for developing efficient feedback control policies in the management of a pest—predator system.

Let X and Y be the populations of two species whose interaction can be described by the model,

$$\dot{X} = G_1(X, Y)F_1(X, Y),$$

$$\dot{Y} = G_2(X, Y)F_2(X, Y). \qquad (6.3.1)$$

Let the functions $G_1(X, Y)$ and $G_2(X, Y)$ be positive in the positive quadrant. By definition the curve $G_1(X, Y)F_1(X, Y) = 0$ is called the X isocline. Strictly speaking, it should be called the X *zero* isocline. As $G_1(X, Y) > 0$ the X isocline is given by $F_1(X, Y) = 0$. Similarly $F_2(X, Y) = 0$ is the Y isocline.

Before the advent of modern computers the method of isoclines was often used to analyse the behavior of a given nonlinear model with one of two variables (Cunningham, 1958). Rosenzweig and MacArthur (1963) popularized this method in population dynamics and showed that it is a very flexible tool for studying two-species interactions. Strebel and Goel (1973) have re-examined this method and have pointed out several difficulties in applying this method to data from an experimental two-species interaction.

In the case where the predator isocline is vertical at a positive equilibrium and the prey isocline has a single hump, Rosenzweig and MacArthur (1963) and Freedman (1976) showed that: (i) the equilibrium is locally stable if the isoclines intersect to the right of the maximum of the prey isocline; and (ii) the equilibrium is unstable if the isoclines intersect to the left of the maximum of the hump. We shall generalize and clarify these graphical rules for determining the local stability of a positive equilibrium.

Consider the linear equations

$$\dot{x} = k_1(a_{11}x + a_{12}y),$$

$$\dot{y} = k_2(a_{21}x + a_{22}y) \tag{6.3.2}$$

where k_1, k_2 are positive constants. The characteristic equation for this system is

$$\lambda^2 - (k_1a_{11} + k_2a_{22})\lambda + k_1k_2 \det A = 0. \tag{6.3.3}$$

The Routh—Hurwitz necessary and sufficient conditions for (6.3.2) to be stable are

$$k_1a_{11} + k_2a_{22} < 0 \tag{6.3.4}$$

and

$$\det A > 0. \tag{6.3.5}$$

Hence (6.3.2) is stable if (i) $a_{11} < 0$, $a_{22} \leqslant 0$ and $\det A > 0$; or (ii) $a_{11} \leqslant 0$, $a_{22} < 0$ and $\det A > 0$. It is unstable if (i) $\det A < 0$; or (ii) $a_{11} \geqslant 0$ and $a_{22} > 0$; or (iii) $a_{11} > 0$ and $a_{22} \geqslant 0$.

Let model (6.3.1) have a positive equilibrium at $(X^\star, Y^\star)$. Linearizing the equations of (6.3.1) we get $k_1 = G_1(X^\star, Y^\star)$, $k_2 = G_2(X^\star, Y^\star)$, $a_{11} = \partial F_1(X^\star, Y^\star)/\partial X^\star$, $a_{12} = \partial F_1/\partial Y$, $a_{21} = \partial F_2/\partial X$, and $a_{22} = \partial F_2/\partial Y$.

The gradient of the X isocline at $(X^\star, Y^\star)$ is $\nabla F_1 = (a_{11}, a_{12})^T$ and it is normal to the X isocline at $(X^\star, Y^\star)$. Similarly $\nabla F_2 = (a_{21}, a_{22})^T$ is normal to the Y isocline at $(X^\star, Y^\star)$. Thus the signs of a_{11} and a_{22} can be determined graphically be examining, at the point $(X^\star, Y^\star)$, the sign of the first component of the normal to the X isocline, and the sign of the second component of the normal to the Y isocline.

Let u be a unit vector perpendicular to ∇F_1 and ∇F_2. The vector cross product,

$$\nabla F_1 \times \nabla F_2 = (\det A)u, \tag{6.3.6}$$

$$= \sin\theta \, \| \nabla F_1 \| \, \| \nabla F_2 \| u, \tag{6.3.7}$$

where θ is the angle between the vectors ∇F_1 and ∇F_2 in the anticlockwise direction from ∇F_1 to ∇F_2 and $\|\nabla F_1\|$ is the magnitude of ∇F_1. It follows that $\det A > 0$ if $\pi > \theta > 0$, and $\det A < 0$ if $2\pi > \theta > \pi$. Therefore the sign of $\det A$ can be determined graphically by examining the angle in the anticlockwise direction between the normals to the X and Y isoclines.

From this analysis we deduce four graphical rules for determining the local asymptotic stability of an equilibrium. Note that in an application of these rules we have some freedom in choosing the G_1 and G_2 functions. The only constraint on the G_1 and G_2 functions is that they must be positive at the equilibrium point.

Rule I. At the equilibrium $(X^\star, Y^\star)$ let θ be the angle in the anticlockwise direction between the normals of the X and Y isoclines respectively. If $2\pi > \theta > \pi$, then $(X^\star, Y^\star)$ is unstable.

Rule II. The equilibrium $(X^\star, Y^\star)$ is unstable if, at the point $(X^\star, Y^\star)$, the first component of the normal to the X isocline is positive (nonnegative) *and* the second component of the normal to the Y isocline is nonnegative (positive).

Rule III. The equilibrium $(X^\star, Y^\star)$ is locally stable if: (i) the first component of the normal to the X isocline is negative (nonpositive); (ii) the second component of the normal to the Y isocline is nonpositive (negative); and (iii) the angle θ in the anticlockwise direction between the normals of the X and Y isoclines satisfies the inequality $\pi > \theta > 0$.

Rule IV. The equilibrium $(X^\star, Y^\star)$ of (6.3.1) is locally asymptotically stable (unstable) if, at the point $(X^\star, Y^\star)$, $G_1 \partial F_1/\partial X + G_2 \partial F_2/\partial Y$ is negative (positive) and $\pi > \theta > 0$.

The next two examples demonstrate that the above rules can be used to determine quickly the instability or local stability of an equilibrium.

Example 6.3.1. Consider the competition model,

$$\dot{X} = X(3 - 2X - Y),$$

$$\dot{Y} = Y(2 - X - Y). \tag{6.3.8}$$

It has four equilibria at (1, 1), (0, 2), (3/2, 0) and (0, 0).

At (1, 1) the X and Y isoclines are $3 - 2X - Y = 0$ and $2 - X - Y = 0$ respectively. The components of their normals have the signs $(-, -)$, $(-, -)$. Thus the first component of the normal to the X isocline and the second component of the normal to the Y isocline are both negative. The angle θ between these normals satisfies the inequality $\pi/2 > \theta > 0$. By Rule III this equilibrium is locally stable (see Fig.6.3.1).

At the equilibrium (0, 2) let $G_1 = 3 - 2X - Y$. We have $G_1(0, 2) = 1$. The X and Y isoclines are $X = 0$ and $2 - X - Y = 0$ respectively. Their normals have the signs $(+, 0)$ and $(-, -)$. The angle θ between them satisfies the

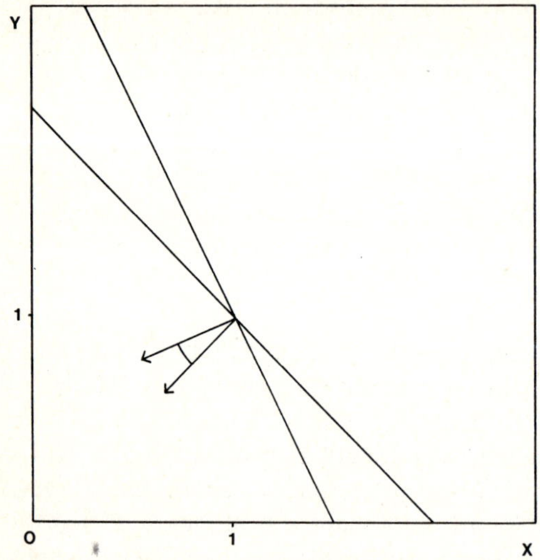

Fig.6.3.1. The angle in the anticlockwise direction between the normals of the X and Y zero isoclines of model (6.3.8).

inequality $3\pi/2 > \theta > \pi$. By Rule I this equilibrium is unstable. Examination of the vector field (XF_1, YF_2) in a nonnegative neighbourhood of (0, 2) shows that the equilibrium is not sector stable.

At the equilibrium (3/2, 0) let $G_2(X, Y) = 2 - X - Y$. We have $G_2(3/2, 0) = 1/2$. The X and Y isoclines are $3 - 2X - Y = 0$ and $Y = 0$ respectively. Their normals have the signs (—, —) and (0, +) and the angle θ between them satisfies the inequality $3\pi/2 > \theta > \pi$. By Rule I this equilibrium is unstable.

At the equilibrium (0, 0) let $G_1 = 3 - 2X - Y$ and $G_2 = 2 - X - Y$. We have $G_1(0, 0) = 3$, $G_2(0, 0) = 2$. The X and Y isoclines are $X = 0$ and $Y = 0$ respectively. Their normals have the signs (+, 0) and (0, +). By Rule II this equilibrium is unstable.

Example 6.3.2. Suppose a predator has the sigmoid functional response $f(X) = X^2/(0.64 + X^2)$. Let a model of the prey—predator interaction be

$$\dot{X} = X[0.2 - XY/(0.64 + X^2)], \tag{6.3.9}$$

$$\dot{Y} = Y[-d + 0.1X^2/(0.64 + X^2)] \tag{6.3.10}$$

where d is a positive constant. Eq. (6.3.9) may be rewritten in the form

$$\dot{X} = 0.2[X^2/(0.64 + X^2)](X + 0.64/X - 5Y). \tag{6.3.11}$$

At a positive equilibrium $(X^\star, Y^\star)$ the X and Y isoclines are $X + 0.64/X - 5Y = 0$ and $0.1X^2/(0.64 + X^2) - d = 0$. For $X > 0$ the curve $Y = (0.64/X + X)/5$ is convex and it has a global minimum at $X = 0.8$. At (X, Y) the normals to the X and Y isoclines are $(1 - 0.64/X^2, -5)$ and $(0.128X/(X^2 + 0.64)^2, 0)$.

If $d > 1/20$, then $X^\star > 0.8$. In this case the components of the normals to the X and Y isoclines at the positive equilibrium $(X^\star, Y^\star)$ have the signs $(+, -)$ and $(+, 0)$ respectively. By Rule II the equilibrium is unstable (Fig.6.3.2). If $1/20 > d > 0$ then $0.8 > X^\star > 0$. In this case the components of the normals to the X and Y isoclines at the positive equilibrium $(X^\star, Y^\star)$ have the signs $(-, -)$ and $(+, 0)$ respectively. Furthermore the angle θ between the two normals satisfies the inequality $\pi > \theta > \pi/2$. By Rule III the equilibrium is locally stable.

It is worthwhile to distinguish three types of control policies. The first type uses constant control variables; the second type uses control variables which are continuous functions of the populations (state variables); and the third type uses discontinuous feedback control functions. We shall examine the effects of the three types of control polciies by applying them to simple prey—predator models.

Let $u(X, Y)$ and $v(X, Y)$ be control variables acting on a prey—predator system; and let a model of the system be

$$\dot{X} = G_1(X, Y)F_1(X, Y) - u(X, Y),$$
$$\dot{Y} = G_2(X, Y)F_2(X, Y) - v(X, Y). \qquad (6.3.12)$$

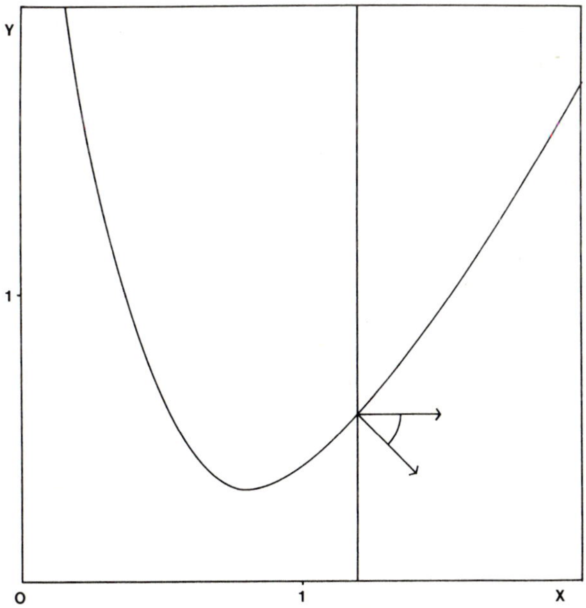

Fig.6.3.2. The angle in the anticlockwise direction between the normals of the X and Y zero isoclines of model (6.3.9) and (6.3.10).

Each choice of the control functions $u(X, Y)$ and $v(X, Y)$ from an admissible set gives a set of X and Y isoclines. Thus a way to design a desirable control pest—predator system is to choose $u(X, Y)$ and $v(X, Y)$ which will reshape and shift the isoclines so as to create an economically desirable and locally stable equilibrium. Using Liapunov functions, or otherwise, we could check that the equilibrium has a large region of attraction.

Example 6.3.3. At relatively moderate levels, an insecticide tends to kill a pest and its predator in a density-independent (proportional) manner. Let us assume that a nonselective insecticide is applied at a constant rate to a pest—predator system. If b, a_{12}, a_{22}, d, e, E and c are positive constants, let a model of the system be

$$\dot{X} = X(b - a_{12}Y) - EX,$$

$$\dot{Y} = Y(-d + ea_{12}X - a_{22}Y) - cEY. \tag{6.3.13}$$

This model has an equilibrium at $(X^\star, Y^\star)$ where $Y^\star = (b - E)/(a_{12})$ and $X^\star = (d + cE + a_{22}Y^\star)/(ea_{12})$.

In this case the X and Y isoclines are shifted when E is varied. If $b > E$ the predator equilibrium level is decreased and the pest equilibrium level is increased as the control parameter E is increased. The normals to the isoclines remain unchanged. Therefore the stability of $(X^\star, Y^\star)$ remains the same for all nonnegative values of E, provided that $Y^\star$ remains positive. Hence $(X^\star, Y^\star)$ is globally stable.

By lowering the predator equilibrium level, the application of the insecticide makes the system more vulnerable to perturbations.

Example 6.3.4. Let us assume that a nonselective insecticide is applied at a rate which is proportional to the pest density. This gives an example of a continuous feedback control policy. Let a model of the control prey—predator system be

$$\dot{X} = X[0.2 - XY/(0.64 + X^2)] - 1.5cX^2,$$

$$\dot{Y} = Y[-0.0559 + 0.1X^2/(0.64 + X^2)] - scXY \tag{6.3.14}$$

where c and s are nonnegative constants.

If the insecticide is not used, we have $c = 0$, and model (6.3.14) has a unique positive equilibrium at (0.9, 0.322). The prey isocline is convex and it has a global minimum at $X = 0.8$. The normals of the prey and predator isoclines at this equilibrium have the signs (+, —) and (+, 0). By Rule II it is unstable.

Let $c = 0.0508$ and $s = 0.1$. Here the prey isocline

$$Y = 0.128/X + 0.2X - 0.96c - 1.5cX^2 \tag{6.3.15}$$

is monotonic decreasing and intersects the X axis at $X = 2.62$. Model (6.3.14) has a unique positive equilibrium at (1, 0.203). The normals to the prey and predator isoclines at this equilibrium have the signs (—, —) and (+, 0). Hence this equilibrium is locally stable. For $1 > X > 0$ the prey isocline is above the line $Y = 0.203$; and for $1 < X < 2.62$ the prey isocline is below the line $Y = 0.203$. Hence (1, 0.203) is globally stable (Hsu, 1978 a).

If $c = 0.0508$ and $s = 0.298$, there are two predator isoclines, namely, $X = 1.618$ and $X = 2$. The prey isocline is given in (6.3.15). The two positive equilibria are (1.62, 0.154) and (2, 0.11).

At the equilibrium (1.62, 0.154) the normals to the prey and predator isoclines have the signs (—, —) and (+, 0) respectively. By Rule III this equilibrium is locally stable.

At the equilibrium (2, 0.11) the normals to the prey and predator isoclines have the signs (—, —) and (—, 0). The angle θ between the normals to the prey and predator isoclines satisfies the inequality $2\pi > \theta > \pi$. By Rule I this equilibrium is unstable. It is interesting to note that in this case the equilibrium is unstable even though the slope of the prey isocline is negative and the predator isocline is vertical. This should be a warning that the stability rule given by Rosenzweig and MacArthur (1963) ought to be used with some care.

Example 6.3.5. An increasingly popular method of biological control is to release predators which are bred in laboratories. If v is the rate of release of the predators, let a model of the system be

$$\dot{X} = X(1 - Y),$$

$$\dot{Y} = Y(-1 + X) + v. \tag{6.3.16}$$

Suppose the pest level at the equilibrium (1, 1) is above the economic threshold for pest damage. It is desirable to reduce the pest level below $X = 0.8$. Tentatively we shall examine whether or not it is possible to create a stable equilibrium at (0.5, 1) if predators are released. This requires the

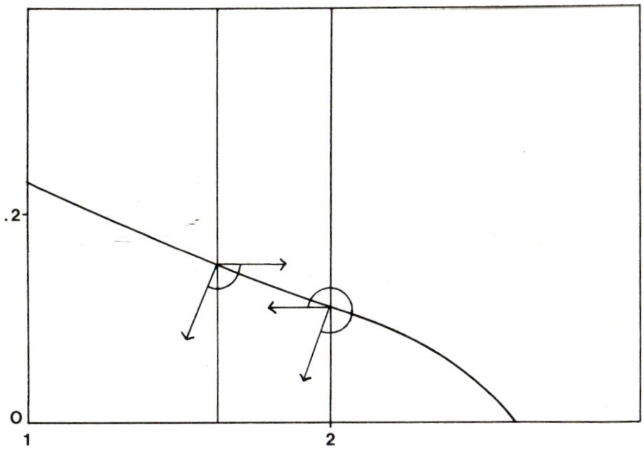

Fig.6.3.3. Angles between the normals of the zero isoclines of model (6.3.14) for $c = 0.0508$ and $s = 0.298$.

predators to be released at the rate $v = 0.5$.

In this case the prey and predator isoclines are $0 = 1 - Y$ and $Y(X-1) + 0.5 = 0$ respectively. At (0.5, 1), their normals have the signs (0, —) and (+, —). The angle θ between these normals satisfies the inequality $\pi > \theta > 0$. By Rule III the equilibrium (0.5, 1) is locally stable. Using the Liapunov function

$$V(X, Y) = X - 0.5 - 0.5 \ln (X/0.5) + Y - 1 - \ln (Y), \qquad (6.3.17)$$

we find that $\dot{V}(X, Y) = -0.5(Y-1)^2/Y \leqslant 0$. Hence the equilibrium at (0.5, 1) is globally stable.

In (6.3.16) the prey—predator model for the natural dynamics is very fragile. But the conclusion that the equilibrium (0.5, 1) is locally stable is not critically dependent on the model of the natural dynamics of the prey—predator system. At the equilibrium the slopes of the isoclines vary with changes in the model of the natural dynamics. The angle θ between the normals of the isoclines of the control prey—predator system is arctan(2) and it satisfies the inequality $\pi > \theta > 0$. Thus small changes in the slopes of the isoclines will not make θ violate the inequality $\pi > \theta > 0$.

We shall now examine the use of discontinuous feedback control functions to manage a prey—predator system (Goh et al., 1974 b) by analysing a few simple models. If there is a sufficient number of control variables it is possible to formulate good and robust management policies for a prey—predator system without an accurate model of the natural dynamics of the system. This is because the control variables can be used to dominate the system and to drive the system to wherever we want. As an illustration, consider a daily life situation. The natural dynamics of a car is unknown to the average driver; yet he can manoeuvre it satisfactorily because he has enough control variables to dominate the natural dynamics of the car.

In the management of the dynamics of a pest—predator system we must specify a target to which we will drive the system (Vincent et al., 1974, 1977). We need also an index for measuring the performance of the control policies. In most cases it is impossible to eliminate a pest completely. The next best target is an equilibrium of the natural dynamics provided the corresponding level of the pest is below the economic threshold for pest damage. Otherwise, we have to create a new equilibrium by using one or more of the control variables in a continuous manner as in Example 6.3.5.

Example 6.3.6. Let $u(t)$ be the rate at which an insecticide is applied. Clearly $u(t)$ is a control variable. Suppose a model of the control prey—predator system is

$$\dot{X} = X(1 - Y) - u(t)X,$$
$$\dot{Y} = Y(X - 1). \qquad (6.3.18)$$

The point (1, 1) is an equilibrium of the natural dynamics. Suppose $X = 1$

is an acceptable pest level. Due to limitations of facilities and manpower, let the control variable be bounded from above, i.e.,

$$0 \leqslant u(t) \leqslant b \tag{6.3.19}$$

where b is a positive constant. The control problem is to choose u as a function of X and Y so as to drive the system from any positive state at time $t = 0$ to the equilibrium (1, 1) with a minimum use of the control variable.

Consider the Liapunov function

$$V(X, Y) = X - 1 - \ln(X) + Y - 1 - \ln(Y). \tag{6.3.20}$$

Computing $\dot{V}(X, Y)$ along solutions of (6.3.18) we get

$$\dot{V}(X, Y) = -(X-1)u. \tag{6.3.21}$$

To make $V(X, Y)$ as negative as possible we choose $u = b$ for $X \geqslant 1$. With this feedback control policy we can drive the system from any positive initial state to (1, 1). But it is not clear that this control policy provides an efficient use of the control variable.

To develop an efficient control policy we examine in each region of the state space the directions of the components of the velocity vector of the natural dynamics and the effects of the control variable on the dynamics of the control prey—predator system. In the region $\{(X, Y) \mid X > 1, Y < 1\}$ we have

$$\begin{pmatrix} \dot{X} \\ \dot{Y} \end{pmatrix} = \begin{pmatrix} + \\ + \end{pmatrix} + \begin{pmatrix} -uX \\ 0 \end{pmatrix}. \tag{6.3.22}$$

The first term in (6.3.22) gives the signs of the components of the velocity vector of the natural dynamics. If $u = b$ there is wastage in the use of the control variable for it acts against the first component of the velocity vector of the natural dynamics. This suggests that we should let $u = 0$.

In the region $\{(X, Y) \mid X > 1, Y > 1\}$ we have

$$\begin{pmatrix} \dot{X} \\ \dot{Y} \end{pmatrix} = \begin{pmatrix} - \\ + \end{pmatrix} + \begin{pmatrix} -uX \\ 0 \end{pmatrix}. \tag{6.3.23}$$

Therefore if we let $u = b$, the control variable will *enhance* the natural dynamics. This means that the control variable is used in an efficient manner. Thus an efficient control policy for model (6.2.18) is $u = b$ at all points (X, Y) where $X \geqslant 1$ and $Y \geqslant 1$; and $u = 0$ elsewhere.

However, the above policy may be unsatisfactory in terms of other performance index. For example it would be useless if it is absolutely essential to keep the pest level below $X = 1$ for all values of t.

Example 6.3.7. Let $u(t)$ be the rate of application of a nonselective insecticide. Suppose a model of the control pest—predator system is

$$\dot{X} = X(1 - Y) - uX,$$

$$\dot{Y} = Y(X - 1) - suY \qquad (6.3.24)$$

where s is a positive constant and represents the relative toxicity of the chemical to the prey and predator.

If V is the Liapunov function in (6.3.20) the derivative $\dot{V}$ along solutions of (6.3.24) is

$$\dot{V}(X, Y) = -[(X - 1) + s(Y - 1)]u. \qquad (6.3.25)$$

Hence $u = b$ is stabilizing if it is applied at the maximum level in the region $\{(X, Y) \mid (X - 1) + s(Y - 1) > 0, X > 0, Y > 0\}$ and $u = 0$ elsewhere.

On examining the signs of the components of the velocity vector of the natural dynamics and the effects of the control variable, we deduce, as in Example 6.3.6, that the control variable is used in an efficient manner if it is applied at the maximum level in the region $\{(X, Y) \mid (X - 1) + s(Y - 1) > 0, 1 > X > 0\}$. If s is small this represents a relatively small region in the state space. A compromise policy is to let $u = b$ for all (X, Y) in the region $\{(X, Y) \mid (X - 1) + s(Y - 1) > 0, (Y - 1) - s(X - 1) > 0, X > 0\}$ and $u = 0$ elsewhere. As expected this region reduces to $\{(X, Y) \mid X > 1, Y > 1\}$ as $s \to 0$. When $s = 0$ the insecticide acts only on the pest and model (6.3.24) reduces to (6.3.18).

Example 6.3.8. Let u_1 and u_2 be the rates of release of pests and predators which are bred in laboratories. Let v_1 and v_2 be the rates of application of chemical agents which kill the prey and predator selectively. Let $F(X, Y)$ and $G(X, Y)$ be the growth rates of the pest and predator in the absence of the control variables. A model of the control prey—predator system is

$$\begin{pmatrix} \dot{X} \\ \dot{Y} \end{pmatrix} = \begin{pmatrix} F(X, Y) \\ G(X, Y) \end{pmatrix} + \begin{pmatrix} u_1 \\ u_2 \end{pmatrix} + \begin{pmatrix} -v_1 X \\ -v_2 Y \end{pmatrix}. \qquad (6.3.26)$$

Let $u_1(t)$, $u_2(t)$, $v_1(t)$ and $v_2(t)$ be piecewise continuous functions of t; and let them be subject to the constraints, $0 \leqslant u_1 \leqslant a_1$, $0 \leqslant u_2 \leqslant a_2$, $0 \leqslant v_1 \leqslant b_1$ and $0 \leqslant v_2 \leqslant b_2$ where a_1, a_2, b_1 and b_2 are positive constants.

Suppose the prey isocline $F(X, Y) = 0$ and $G(X, Y) = 0$ intersect at a positive point $(X^\star, Y^\star)$. Let this equilibrium be an acceptable target for driving the system to. Let these isoclines divide the positive quadrant into four subregions which are denoted by $R(+, +)$, $R(-, +)$, $R(-, -)$ and $R(+, -)$. By definition, $R(+, +) = \{(x, y) \mid F(X, Y) > 0, G(X, Y) > 0\}$. The other regions are defined in a similar manner.

From the analyses of a number of explicit prey—predator models, we find that an efficient control policy is as follows: (i) in $R(+, +)$ the predators which are bred in laboratories are released at the maximum rate; (ii) in $R(-, +)$ the chemical agent to kill the pest is used at the maximum rate; (iii) in

$R(-,-)$ the chemical agent to kill the predators is used at the maximum rate; and (iv) in $R(+,-)$ the pests (preys) which are bred in laboratories are released at the maximum rate. In each case the other control variables are set equal to zero.

For (X, Y) in $R(+,-)$, the rationale for releasing the pests is to feed the predators. This control policy drives the state rapidly into the region $R(+,+)$ where the predator population is increasing.

The most important conclusion to be drawn from this analysis is that it is possible to formulate an efficient feedback control policy which is insensitive to the natural dynamics of a prey—predator system. The reason for this is that if there are enough control variables it would be possible to enhance the desirable trends in the natural dynamics so as to drive the state of the system to a desirable target in an efficient manner.

6.4. OPTIMAL CONTROL OF A PREY—PREDATOR SYSTEM

We shall examine the optimal control of a prey—predator system. An object of the analysis is to illustrate the use of optimal control for formulating optimal policies in the dynamic management of a prey—predator system. For simplicity we shall use the classical Lotka—Volterra model for the natural dynamics of the system. The solutions of the model are limit cycles and its behavior is fragile to small continual perturbations (see model (6.2.3)). But from Example 6.3.8 we note that the qualitative features of an efficient feedback control policy is not critically dependent on the precise details of a model for the natural dynamics of a prey—predator system if there is a sufficient number of control variables to enhance the desirable trends in the natural dynamics so as to drive the system to a desirable target.

We shall formulate five optimal control problems and consider their optimal feedback control policies (Goh et al., 1974 a); and examine the mathematical details of the optimal control analysis of two of these problems.

In the first three problems the control variable is the rate of application of a chemical which (i) kills the prey selectively; (ii) kills the predators selectively; and (iii) kills the preys and predators. In the fourth problem the control variable is the rate of release of preys (pests) which are bred in laboratories or imported. In the last problem the control variable is the rate of release of predators which are bred in laboratories or imported. The object of the optimal control problem is to drive a prey—predator system to a desirable target and to do so with minimum cost, expressed as a weighted sum of the pest damage and the cost in using the control variable.

Using nondimensional variables (see Vincent et al., 1974) the classical Lotka—Volterra model for the natural dynamics of a prey—predator system is

$$\dot{X} = X(1 - Y),$$
$$\dot{Y} = Y(X - K) \qquad (6.4.1)$$

where K is a positive constant. The solutions are limit cycles which satisfy the equation

$$X - K - K \ln(X/K) + Y - 1 - \ln(Y) = C. \tag{6.4.2}$$

The parameter C in (6.4.2) is a constant of integration and depends on the initial values of X and Y.

Let the control variable $u(t)$ be the rate of application of a chemical which kills the pests and the predators. At moderate levels of the prey population and at moderate values of $u(t)$ the rate at which the prey is killed is proportional to the product of u and the population of the prey; the predator is killed at a similar rate. Suppose the chemical leaves no residue. The dynamics of the prey—predator system subject to this control variable is

$$\dot{X} = X(1 - Y) - e_1 u X,$$

$$\dot{Y} = Y(X - K) - e_2 u Y \tag{6.4.3}$$

where e_1 and e_2 are nonnegative constants.

At the initial time $t = 0$ let

$$X(0) = X_0, \qquad Y(0) = Y_0. \tag{6.4.4}$$

This means that the population levels can be measured accurately.

The point $(K, 1)$ is an equilibrium of the natural dynamics of the prey—predator system. Thus in a neighbourhood of $(K, 1)$ we can expect that the components of the natural velocity vector will be small. If the state of the system is in a neighbourhood of this equilibrium it would be relatively cheap to use control variables to maintain the state in the neighbourhood. This is one reason why it may be desirable to drive the state of a system to an equilibrium of its natural dynamics. Assuming that the pest level at the equilibrium $(K, 1)$ is acceptable we shall drive the system to $(K, 1)$. Hence we have the target (terminal) conditions

$$X(T) = K, \qquad Y(T) = 1 \tag{6.4.5}$$

where T, the terminal time, is unconstrained. If $(K, 1)$ is unacceptable, we have to use some of the control variables to create an artificial equilibrium and use it as a target. Example 6.3.5 shows how an artificial equilibrium may be created.

Because of limitations in facilities and manpower the control variable will be bounded from above. For simplicity we assume that

$$0 \leqslant u(t) \leqslant b \tag{6.4.6}$$

where b is a positive constant. In practice $u(t)$ is a piecewise continuous function.

We need a criterion to measure the performance of the control program. One eminently reasonable criterion is a cost function which is a weighted combination of the total cost of using the chemical and the total cost of pest damage to crops. We assume that the crop damage is negligible when the system is at the equilibrium $(K, 1)$. The final choice of the performance index may take into account other considerations e.g. health. In this study let the performance index for the control problem be

$$J = \int_0^T (cX + u)\,\mathrm{d}t \tag{6.4.7}$$

where c is a nonnegative constant.

The optimal control is to choose an admissible control program which will drive the system from the initial state (X_0, Y_0) to the target $(K, 1)$; and such that the performance index is minimized.

Example 6.4.1. Consider the case in which the chemical kills only the preys. In brief the optimal control problem is as follows:

System:	$\dot{X} = X(1 - Y) - uX, \qquad \dot{Y} = Y(X - 1).$	(6.4.8)
Initially:	$X(0) = X_0, \qquad Y(0) = Y_0,$	(6.4.9)
Constraint:	$0 \leqslant u(t) \leqslant b.$	(6.4.10)
Terminally:	$X(T) = 1, \qquad Y(T) = 1.$	(6.4.11)
Objective:	$\min \int_0^T (cX + u)\,\mathrm{d}t.$	(6.4.12)

For a given initial state the optimal control program is a sequence of $u(t) = 0$ arcs and $u(t) = b$ arcs. The exact sequence has to be determined numerically. In Fig.6.4.1 the optimal control program is given for all positive initial states in a region. The control variables given in this manner make up an *optimal feedback control policy*.

The curves AP and BP in Fig.6.4.1 are switching curves. Let the typical initial state of the system be at G_1. The system is allowed to evolve under null control until its solution intersects the switching curve BP. At the point of intersection, the control variable is switched on and applied at maximum level until the state of the system reaches the target P at (1, 1). If the initial state is at G_2, the control variable is immediately applied at maximum level until the solution intersects the switching curve AP. At that point the control variable is switched off, and the system evolves under its own natural dynamics until its solution intersects the switching curve BP. There the control variable switched on again at maximum rate and it drives the state of the system to the target at P.

From Fig.6.4.1 we observe that the spraying of the chemical must stop as soon as the state of the system reaches the target at P. Otherwise, the system

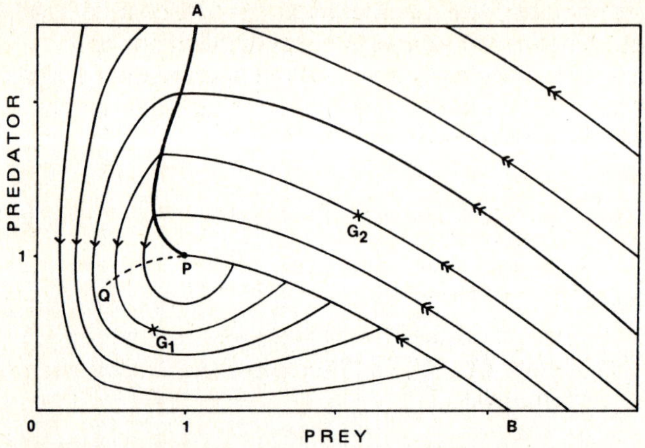

Fig.6.4.1. Optimal control of a prey-predator system using an ideal pesticide. ↠ denotes a u_{max} trajectory, and → denotes a $u = 0$ trajectory.

may be driven to the point Q via the broken curve PQ. If the spraying is then stopped at Q, the state of the system will move along a trajectory of the natural dynamics which is some distance from the target P. Ultimately this leads to a pest outbreak.

Example 6.4.2. We shall examine the mathematical details of the optimal control problem which is considered in the previous example.

The Hamiltonian function for the problem in (6.4.8) to (6.4.12) is

$$H(t, X, Y, u, p) = p_0(cX + u) + p_1[X(1 - Y) - uX] + p_2 Y(X - 1). \quad (6.4.13)$$

Let $X^\star, Y^\star, u^\star, p^\star$ denote an optimal set.

The optimal set $X^\star, Y^\star, u^\star, p^\star$ must satisfy the costate equations

$$\dot{p}_1 = -\partial H/\partial X = -cp_0 - p_1(1 - Y - u) - p_2 Y, \quad (6.4.14)$$

$$\dot{p}_2 = -\partial H/\partial Y = p_1 X - p_2(X - 1), \quad (6.4.15)$$

and p_0 = constant $\geqslant 0$. Along an optimal set we have

$$H(t, X^\star, Y^\star, u^\star, p^\star) = \text{constant} \quad (6.4.16)$$

because H does not contain t explicitly. This follows from the condition $\mathrm{d}H/\mathrm{d}t = \partial H/\partial t$, and the condition that H is continuous.

The transversality condition gives

$$H(T, X^\star, Y^\star, u^\star, p^\star) = 0 \quad (6.4.17)$$

at $t = T$ because T is unconstrained. Therefore along an optimal trajectory

$$H(t, X^\star, Y^\star, u^\star, p^\star) = 0. \quad (6.4.18)$$

The parameters $p_1(T)$ and $p_2(T)$ are constants of integration.

Along an optimal trajectory the optimal control $u^\star(t)$ must minimize $H(X^\star, Y^\star, u, p^\star)$ with respect to all admissible controls. It follows that

$$u^\star(t) = 0 \quad \text{only if } \partial H(X^\star, Y^\star, u^\star, p^\star)/\partial u^\star > 0, \tag{6.4.19}$$

$$u^\star(t) = b \quad \text{only if } \partial H(X^\star, Y^\star, u^\star, p^\star)/\partial u^\star < 0. \tag{6.4.20}$$

If during the subinterval $[t_1, t_2]$ of $[0, T]$,

$$\partial H(X^\star, Y^\star, u^\star, p^\star)/\partial u^\star = p_0 - p_1 X = 0, \tag{6.4.21}$$

then $u^\star(t)$ is a singular control. This is because u appears linearly in the Hamiltonian H.

For clarity and convenience, let the total derivative with respect to t be denoted by D or a dot. When eq. (6.4.21) is differentiated totally with respect to t, we get

$$D(\partial H/\partial u) = -p_1 \dot{X} - \dot{p}_1 X. \tag{6.4.22}$$

Employing (6.4.8), (6.4.14) and (6.4.22), we get

$$D(\partial H/\partial u) = X(cp_0 + p_2 Y) = 0. \tag{6.4.23}$$

As $X > 0$ we deduce that

$$cp_0 + p_2 Y = 0. \tag{6.4.24}$$

Similarly, we get

$$D^2(\partial H/\partial u) = p_2 XY(X - 1) = 0. \tag{6.4.25}$$

But $X > 0$, $Y > 0$ and $X \neq 1$. It follows that $p_2 = 0$. Using (6.2.24) we deduce that $p_0 = 0$. Employing (6.4.21) it follows that $p_1 = 0$. Thus $(p_0, p_1, p_2) = 0$. But the vector (p_0, p_1, p_2) must not be equal to a zero vector. Hence there is no optimal singular control.

The precise switching sequence is determined numerically. This is done by integrating the equations in (6.4.8), (6.4.14) and (6.4.15) in reverse time; and using the optimality conditions (6.4.19) and (6.4.20); and setting $p_0 = 1$. The "initial" conditions in this reverse time integration are $X(T) = 1$, $Y(T) = 1$. Furthermore $p_1(T), p_2(T)$, and T are constants of integration. Each choice of the set of constants of integration, $\{T, p_1(T), p_2(T)\}$, generates an initial state (X_0, Y_0) at $t = 0$.

Strictly speaking, we cannot conclude that the solutions given in Fig.6.4.1 are optimal because we have not used sufficiency theorems to establish that the solutions are indeed optimal. We are only justified in saying that these solutions satisfy all the necessary conditions for optimality. *Thus the usage of the phrase "optimal control" in this section should be understood in this limited sense.*

Example 6.4.3. Let $u(t)$ be the rate of application of a nonresidual chemical which kills the prey and the predators. Let $K = 1$ in (6.4.3). In brief, the optimal control problem is as follows:

System: $\dot{X} = X(1 - Y) - e_1 uX, \qquad \dot{Y} = Y(X - 1) - e_2 uY.$ (6.4.26)

Initially: $X(0) = X_0, \qquad Y(0) = Y_0.$ (6.4.27)

Constraint: $0 \leqslant u \leqslant b.$ (6.4.28)

Terminally: $X(T) = 1, \qquad Y(T) = 1.$ (6.4.29)

Objective: $\min \int_0^T (cX + u)\, dt.$ (6.4.30)

If $e_1 = 1$ and $e_2 = 0$, the problem reduces to that in Example 6.4.1. If $e_1 = 0$ and $e_2 = 1$, the chemical kills only the predators. Using arguments similar to those in Example 6.4.2 we can show that there is no optimal singular control. Thus the optimal program for a given initial state is a combination of $u = 0$ arcs and $u = b$ arcs. This has to be determined numerically.

The optimal feedback control policy for this problem is displayed in Fig.6.4.2. The curves AP, PS and SB are switching curves. Typically, if the initial state is at G_1 the system is allowed to move with null control until the solution intersects the switching curve PS. At the point of intersection, the control is switched on and maintained at the maximum rate until the state of the system reaches P. If the initial state is at G_2 the control variable is immediately applied at the maximum level until the solution intersects the switching curve AP. There the control is switched off until the solution intersects the switching curve PS. The control variable is then applied again at maximum

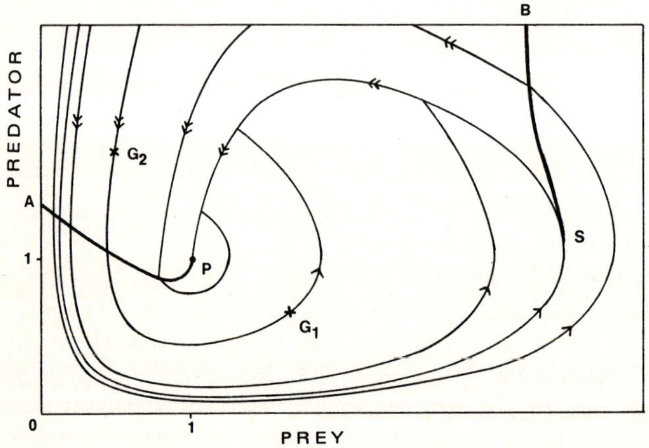

Fig.6.4.2. Optimal control of a prey—predator system using a chemical which kills the predator. ↠ denotes a u_{max} trajectory, and → denotes a $u = 0$ trajectory.

until the state of the system reaches P. At P the control variable is switched off.

Suppose the chemical kills the prey and the predator. In this case e_1 and e_2 are positive. If $e_1 > e_2$ the is a locally optimal singular extremal which satisfies the equation

$$cX(e_1^2 + e_1e_2 + e_2^2Y - e_1e_2Y) + e_1X(1 - Y) + e_2Y(X - 1) = 0. \tag{6.4.31}$$

This singular extremal passes through the origin (0, 0). If $c = 0$ it also passes through the target (1, 1); and the corresponding singular control is

$$u^\star(t) = X^\star(t)/e_2 - Y^\star(t)/e_1. \tag{6.4.32}$$

The full significance of this singular remains to be unravelled. For simplicity we ignore the singular control and use only the optimality conditions for $u = 0$ and $u = b$. From the practical point of view, it is easier to use a combination of $u = 0$ and $u = b$ than a combination of $u = 0$, $u = b$ and singular control.

In Fig.6.4.3 the optimal feedback control policy is displayed for the parameters $e_1 = e_2 = 1$, $b = 1$. In this case there is no optimal singular control. The arcs AP and BP are switching curves. Typically, if the state of the system is at G_1 the system is allowed to evolve under null control until the solution intersects the switching curve PB. At the point of intersection the maximum control is switched on until the state of the system reaches the target P. If the initial state of the system is at G_2 the maximum control is switched on until the solution intersects the switching curve AP where it is switched off until the solution intersects the switching curve PB. It is then driven by $u = b$ until it reaches the target P at (1, 1).

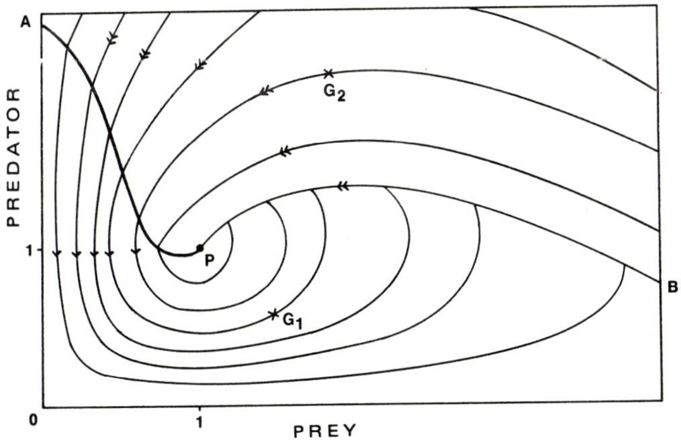

Fig.6.4.3. Optimal control of a prey—predator system using a chemical which kills prey and predator. ↠ denotes a u_{max} trajectory, and → denotes a $u = 0$ trajectory.

Example 6.4.4. Let the control variable $u(t)$ be the rate of release of preys (pests) which are bred in laboratories or are imported. Suppose a model of the control prey—predator system is

$$\dot{X} = X(1 - Y) + u,$$

$$\dot{Y} = Y(X - 1). \tag{6.4.33}$$

At any time t, let us assume that the populations can be measured accurately. This gives the initial conditions,

$$X(0) = X_0, \qquad Y(0) = Y_0, \tag{6.4.34}$$

where X_0 and Y_0 are measurements of the populations at time $t = 0$.

Because of limitations of facilities and manpower, let $u(t)$ satisfy the constraint

$$0 \leqslant u(t) \leqslant b \tag{6.4.35}$$

where b is a constant.

Let the target of the optimal control problem be P = (1, 1). Thus at the terminal time,

$$X(T) = 1, \qquad Y(T) = 1. \tag{6.4.36}$$

We assume that T is unconstrained.

The problem is to find an admissible control function $u^\star(t)$ which drives the system from (X_0, Y_0) to the target at P and which minimizes the function

$$J = \int_0^T (cX + u)\,\mathrm{d}t. \tag{6.4.37}$$

This function is a weighted sum of a measure of pest damage and the cost of using the control variable.

It can be shown that the optimal control program is some combination of $u = 0$ and $u = b$. The exact sequence of $u = 0$ and $u = b$ has to be determined numerically.

The optimal feedback control is displayed in Fig.6.4.4. The arcs AP, PS and SB are switching curves. Typically, if G_1 is the initial state of the system the control is switched on at maximum rate until the solution intersects AP. At the point of intersection the control is switched off until the solution intersects the switching curve PS. There the control variable is switched on at maximum rate until the state of the system reaches the target P. If the initial state is at G_2 the system is allowed to evolve under its own natural dynamics until the solution intersects the switching curve PS. There the control $u = b$ is switched on until the state of the system reaches the target P. At P the control variable is switched off and theoretically it will remain there because P is an equilibrium of the natural dynamics. In practice, the state of the system

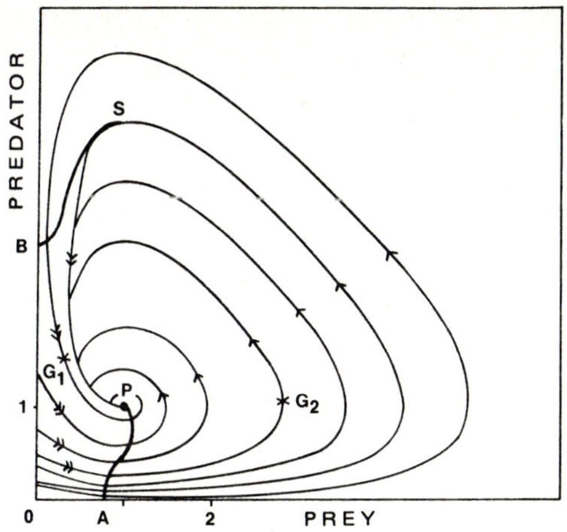

Fig.6.4.4. Optimal control of a prey—predator system by releasing pests. ↠ denotes a u_{max} trajectory, and → denotes a $u = 0$ trajectory.

has to be maintained at the target by using small amounts of control variables. It is best to consider this question of stabilizing the system at a target as a separate problem (Vincent et al., 1977; Vincent and Skowronski, 1979).

Initially, the idea that it may be desirable to release some pests at the right time in order to maintain the pest population at a low level may sound absurd. However, the reason for doing so is to feed the predators during a period when their numbers are decreasing. By doing this the predator population is prevented from falling too low. Subsequently, when the pest population begins to build up there will be enough predators around to suppress it.

Example 6.4.5. Let $u(t)$ be the rate of release of predators which are bred in laboratories or are imported. Suppose a model of the control pest—predator system is

$$\dot{X} = X(1 - Y),$$

$$\dot{Y} = Y(X - 1) + u. \tag{6.4.38}$$

At time $t = 0$, let

$$X(0) = X_0, \qquad Y(0) = Y_0; \tag{6.4.39}$$

and let $u(t)$ satisfy the constraint

$$0 \leqslant u(t) \leqslant b \tag{6.4.40}$$

where b is a constant.

Let T be unconstrained; and let the target of the control problem be

$X(T) = 1, \qquad Y(T) = 1.$ (6.4.41)

Suppose it is desirable (i) to drive the system quickly to the target; (ii) to have a low pest damage; and (iii) to have a low cost in using the control variable. To take these objectives into account we use the weighted performance index

$$J = T + \int_0^T (c_1 X + c_2 u)\mathrm{d}t \qquad (6.4.42)$$

where c_1 and c_2 are nonnegative constants. The problem is to choose an admissible $u^{\star}(t)$ which will drive the system from (X_0, Y_0) to P, and which minimizes the performance index.

The optimal feedback control policy for this problem when b is relatively large ($b = 0.5$) is given in Fig.6.4.5. The curves BP and PA are switching curves. Typically, if G_1 is the initial state the system moves under the control $u = b$ until the solution intersects the switching curve PA. There the control is switched off until the solution intersects the switching curve BP. It is then driven to P along BP by using $u = b$. At P the control is switched off.

Biologically this policy is as expected. The predators are released when the pest population begins to increase. This reduces the pest immediately and also prevents subsequent large increases in the pest population.

To illustrate the application of optimal control theory we shall examine some of the mathematical details in the solution of the optimal control problem given in (6.4.38) to (6.4.42).

By definition, the Hamiltonian function is

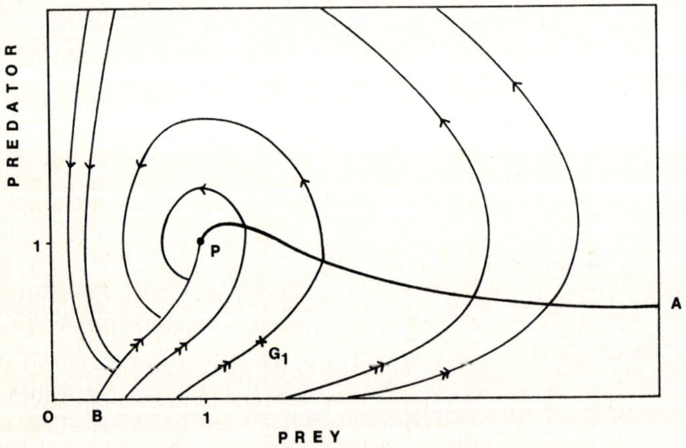

Fig.6.4.5. Optimal control of a prey—predator system by releasing predators. ↠ denotes a u_{max} trajectory, and → denotes a $u = 0$ trajectory.

$$H(t, X, Y, u, p) = p_0(c_1 X + c_2 u) + p_1 X(1 - Y) + p_2[X(X-1) + u] \quad (6.4.43)$$

where p_0 = constant $\geqslant 0$.

The costate equations are

$$\dot{p}_1 = -\partial H/\partial X = -p_0 c_1 - p_1(1-Y) - p_2 Y, \quad (6.4.44)$$

$$\dot{p}_2 = -\partial H/\partial Y = p_1 X - p_2(X-1). \quad (6.4.45)$$

Let $(X^\star, Y^\star, u^\star, p^\star)$ be an optimal solution. The Hamiltonian function in (6.4.43) does not contain t explicitly. Using the condition $dH/dt = \partial H/\partial t$ we get

$$H(X^\star, Y^\star, u^\star, p^\star) = \text{constant} \quad (6.4.46)$$

for all values of t in $[0, T]$. The transversality condition at $t = T$ gives

$$H[X^\star(T), Y^\star(T), u^\star(T), p^\star(T)] + 1 = 0. \quad (6.4.47)$$

Hence the constant in (6.4.46) is equal to -1. The values of $p_1(T)$ and $p_2(T)$ are constants of integration. These and the value of T are chosen so that the optimal solution in reverse time passes through the initial state (X_0, Y_0) at time $t = 0$.

Along an optimal trajectory the optimal control $u^\star(t)$ must minimize $H(X^\star, Y^\star, u, p^\star)$ with respect to all admissible controls. It follows that

$$u^\star(t) = 0 \quad \text{only if } \partial H(X^\star, Y^\star, u^\star, p^\star)/\partial u^\star > 0, \quad (6.4.48)$$

$$u^\star(t) = b \quad \text{only if } \partial H(X^\star, Y^\star, u^\star, p^\star)/\partial u^\star < 0. \quad (6.4.49)$$

If on a subinterval $[t_1, t_2]$ of $[0, T]$,

$$\partial H(X^\star, Y^\star, u^\star, p^\star)/\partial u^\star = p_0 c + p_2 = 0, \quad (6.4.50)$$

then $u^\star(t)$ is a singular control. This is because the control variable appears linearly in the Hamiltonian H.

We shall denote the total derivative with respect to t by D or by a dot. Differentiate (6.4.50) totally with respect to t; we get

$$D(\partial H/\partial u) = \dot{p}_2 = p_1 X - p_2(X-1) = 0, \quad (6.4.51)$$

$$D^2(\partial H/\partial u) = \dot{p}_1 X + p_1 \dot{X} - \dot{p}_2(X-1) - p_2 \dot{X} = (c_2 - c_1) p_0 X = 0. \quad (6.4.52)$$

If $p_0 > 1$ this is satisfied only if $c_1 = c_2$. This is a very special case. For $c_1 \neq c_2$ there is no optimal singular control; in particular, if $c_1 = 0$ and $c_2 = 2$, there is no optimal singular control. Hence the optimal program is a combination of $u = 0$ and $u = b$; this has to be determined numerically.

6.5. OPTIMAL CONTROL OF EPIDEMICS FOR PEST CONTROL

The use of bacteria, fungi and viruses is potentially one of the most important methods in pest control. For example, *Bacillus thuringiensis*, which is available in commercial preparations, is used in the control of a large number of pests (Falcon, 1971; Burges and Hussey, 1971). An advantage of using insect pathogens is that they are safe to man and are usually safe to beneficial insects.

There is a vast amount of literature on the applications of microbial diseases to suppress pests (Tanada, 1964; Fenner and Ratcliffe, 1965; Burges and Hussey, 1971; Davis et al., 1976; Falcon, 1976). But there are only a few papers on mathematical models of the dynamics of microbial diseases in pest control (Anderson and May, 1978; Goh et al., 1980). On the other hand there are many mathematical models of infectious diseases of man (Dietz, 1967; Ludwig, 1974; Waltman, 1974; Bailey, 1975; Wickwire, 1977).

At present, insect pathogens are used in two ways. In the first method, a small amount of pathogen is introduced into a pest population with the expectation that it will generate an epidemic and that it will subsequently be endemic. The success of this approach depends on the survival of the microbes which in turn depends on environmental factors, such as temperature, humidity and crop conditions. In the second method, an insect pathogen is used like a nonresidual chemical insecticide. In this case, it is applied whenever a pest population is at an economically significant level for pest damage and there is no expectation that the pathogen will survive for an appreciable period.

We shall examine the use of pathogens in a more flexible manner. We are especially interested in the situation when environmental conditions do not allow a significant epidemic to be generated if only a small amount of pathogen is introduced into a pest population on a once-and-for-all basis. Until more realistic models are available, we shall use the simplest model of an epidemic.

At time t, let $S(t)$ be the number of susceptible insects, and $I(t)$ be the number of infective insects. A model of an epidemic

$$\dot{S} = -rSI,$$

$$\dot{I} = rSI - wI, \qquad (6.5.1)$$

where r and w are positive constants. In practice, the parameters are functions of the humidity and the physiological time which is some measure of the period in which the temperature is above a certain threshold (Clark et al., 1967; Gilbert et al., 1976). Insects are cold blooded animals; below a certain species specific temperature their metabolism slows down considerably. The parameter w is the death rate of the insect population, and r is the infection rate.

Let $S(t)$ denote the number of insects in the larvae and adult stages. In each generation there is a burst of new larvae and adults. This resets the initial value $S(0)$ for each generation in an insect population with nonoverlapping generations. Thus the time variable t is on a fast time scale. It is appropriate for describing processes within a generation.

If eq. (6.5.1) is written in the form,

$$\dot{S} = -rSI,$$
$$\dot{I} = rI(S - w/r), \qquad (6.5.2)$$

we deduce immediately that S and I decrease monotonically if $0 < S < w/r$. Therefore the infection dies out rapidly if the initial state $(S(0), I(0))$ satisfies the conditions $I(0) > 0$ and $0 < S(0) < w/r$. This is a well known threshold phenomenon of an epidemic. If $(S(0), I(0))$ satisfies the conditions $I(0) > 0$ and $S(0) > w/r$, then $I(t)$ will increase for an initial period until $S(t) = w/r$.

The solutions of model (6.5.1) for various initial values are shown in Fig.6.5.1. This figure could be used to predict the final outcome of an epidemic if only a small amount of pathogen is released into the pest population. For example, if $S(0) = 2.56(w/r)$ and $I(0) = 0.0001$, then $S(T) \to 0.256(w/r)$ and $I(T) \to 0$ as $T \to \infty$. Again, if $S(0) = 4(w/r)$ and $I(0) = 0.0001$, then $S(T) \to 0.08(w/r)$ and $I(T) \to 0$ as $T \to \infty$.

Suppose pests which are bred in laboratories or which have been captured from the field are infected by a pathogen and are then released during the brief period $[-\Delta t, 0]$ where Δt is a small positive number. This means that we can manipulate the initial value $I(0)$. Thus $I(0)$ is a control parameter. Given a pest population $S(0)$, we can use Fig.6.5.1 to compute the least value of $I(0)$ which is needed to generate an epidemic so that the pest population at the end of an epidemic will be below a desirable level. In other words, we can use Fig.6.5.1 to compute the least value of $I(0)$ so that the limit of $S(T)$ as $T \to \infty$ will be below a desirable level.

For example, let $S(\infty)$ denote the limit of $S(T)$ as $T \to \infty$. Suppose it is

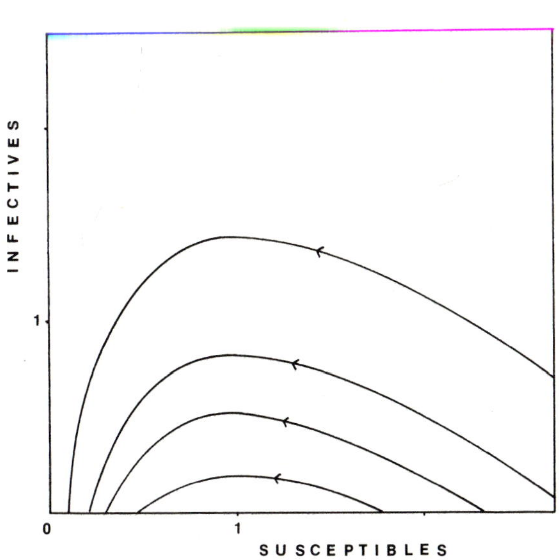

Fig.6.5.1. Solutions of a simple epidemic.

desirable that $S(\infty)$ should be below $0.1(w/r)$. If $S(0) = 3.26(w/r)$, then $I(0)$ must be greater than $0.325(w/r)$ in order to generate an epidemic which will reduce the pest population to below $0.1(w/r)$.

We shall examine two optimal control problems in the management of an epidemic to control a pest population. In the first problem the control variable is the rate of release of infected pests; and the control problem is to drive the pest population below a certain level with a minimum use of the control variable. In the second problem the pathogen is used in the same way as a non-residual chemical pesticide. In this case the control variable is the rate of release of the pathogen. The control problem is to drive the pest population below a given pest level and to do so in a manner which minimizes a weighted sum of the cost of using the control and the time it takes to drive the system to the target.

Example 6.5.1. Let $u(t)$ be the rate of pests infected in a laboratory. Suppose a model of the epidemic under this control variable is

$$\dot{S} = -rSI,$$

$$\dot{I} = rSI - wI + u. \tag{6.5.3}$$

At time $t = 0$, let S_0 and I_0 be the measured values of S and I. Thus we have

$$S(0) = S_0, \qquad I(0) = I_0. \tag{6.5.4}$$

Let us assume that the control variable is bounded from above by a constant. This limitation on the control variable is due to a lack of facilities and manpower. Hence we have

$$0 \leqslant u(t) \leqslant b \tag{6.5.5}$$

where b is a constant. Let $u(t)$ be piecewise continuous.

Let the control program be over the period $[0, T]$ where T is unconstrained. In order that the pest population is below a certain level at the end of an epidemic, we shall impose the conditions,

$$S(T) = d, \qquad I(T) \geqslant a, \tag{6.5.6}$$

where a is a positive constant. The value of d is specified so that the damage due to the pests is negligible at this level.

The object is to choose an admissible function $u(t)$ which drives the system from (S_0, I_0) to the target specified in (6.5.6); and which minimizes the objective function

$$J = \int_0^T u \, dt. \tag{6.5.7}$$

Let ABC be the $u = 0$ solution of (6.5.3) which passes through the point

$(S, I) = (d, a)$. Below ABC the cost of driving the system from a given point to another is independent of the path of integration. Therefore an optimal feedback policy is $u = b$ at a point below ABC; and is $u = 0$ at a point on ABC, or at a point above ABC.

We can show that the generalized Legendre condition is satisfied in an indeterminate manner by singular extremals below ABC. This suggests that the cost of control below ABC is independent of the path of integration. We shall establish that this is indeed the case. Let (S_1, I_1) and (S_2, I_2) be any two given points below ABC; and let t_1 and t_2 be the corresponding values of t. Using (6.5.3) and (6.5.7), we get

$$\int_{t_1}^{t_2} u \, \mathrm{d}t = \int_{S_1}^{S_2} \mathrm{d}I + \left(1 - \frac{w}{rS}\right) \mathrm{d}S. \tag{6.5.8}$$

The integral on the right hand side of (6.5.8) is independent of the path of integration.

Example 6.5.2. Let $u(t)$ be the rate of application of a pathogen. We shall assume that at moderate levels of $u(t)$ the rate at which a pest is infected by the pathogen is uS. Thus a model of an epidemic which is modified by the application of the pathogen is

$$\dot{S} = -rSI - uS,$$

$$\dot{I} = rSI - wI + uS. \tag{6.5.9}$$

For reasons similar to those given in Example 6.5.1, the constraints in the optimal control problem are

$$S(0) = S_0, \qquad I(0) = I_0, \tag{6.5.10}$$

$$S(T) = d, \qquad I(T) \geqslant a, \tag{6.5.11}$$

$$0 \leqslant u(t) \leqslant b. \tag{6.5.12}$$

The problem is to choose an admissible control program $u(t)$ which will drive the system from the initial state (S_0, I_0) to the target (6.5.11) so as to minimize the function

$$J = c_1 T + \int_0^T c_2 u \, \mathrm{d}t \tag{6.5.13}$$

where c_1 and c_2 are nonnegative constants. Eq. (6.5.13) may be rewritten as

$$J = \int_0^T (c_1 + c_2 u) \, \mathrm{d}t. \tag{6.5.14}$$

If $c_1 = 0$ and $c_2 = 1$, then the optimal feedback policy is similar to that in Example 6.5.1. Let ABC be the $u = 0$ trajectory of (6.5.9) which passes through the point $(S, I) = (d, a)$. Below ABC the cost of the control program between any two given points is independent of the path of integration. Therefore, for an initial state below ABC, the optimal program could be any admissible control program which drives the state of the system to ABC. On and above ABC the optimal program is $u = 0$.

We shall establish that the cost of control is independent of the path of integration if $c_1 = 0$. Let $y = S + I$. Eq. (6.5.9) is equivalent to

$$\dot{y} = -wI,$$

$$\dot{I} = r(y - I)I - wI + u(y - I). \tag{6.5.15}$$

Using (6.5.15) we get

$$\int u\,\mathrm{d}t = \int \frac{\mathrm{d}I}{(y - I)} + \left[\frac{r}{w} - \frac{1}{(y - I)}\right]\mathrm{d}y. \tag{6.5.16}$$

The integral on the right hand side of (6.5.16) is independent of the path of integration.

If c_1 and c_2 are both positive, then the optimal control trajectory is a sequence of $u = 0$ arcs and $u = b$ arcs. The exact sequence has to be computed numerically. Fig.6.5.2 gives the optimal feedback control policy for the case in which $c_1 = 0.05$ and $c_2 = 1.0$.

We shall establish that there is no optimal singular control if c_1 and c_2 are both positive.

By definition, the Hamiltonian function is

$$H = p_0(c_1 + c_2 u) - p_1(rSI + uS) + p_2(rSI - wI + uS). \tag{6.5.17}$$

The costate equations are

$$\dot{p}_1 = -\partial H/\partial S = (p_1 - p_2)(rI + u), \tag{6.5.18}$$

$$\dot{p}_2 = -\partial H/\partial I = p_1 rS + p_2(w - rS) \tag{6.5.19}$$

The terminal conditions (6.5.11) can be decomposed into two cases: (i) $I(T) > a$; and (ii) $I(T) = a$. In the first case the transversality conditions at $t = T$ are $p_2(T) = 0$ and $H = 0$. In the second case we have $H = 0$ at $t = T$; and $p_2(T)$ is a constant of integration. Given $p_2(T)$, the condition $H = 0$ gives

$$p_1(T) = [p_0(c_1 + c_2 u) + p_2(rSI - wI + uS)]/(rSI + uS). \tag{6.5.20}$$

The time variable t does not appear explicitly in H. Using the equation $\mathrm{d}H/\mathrm{d}t = \partial H/\partial t$, and the continuity of H, we get

$$H[S, I, p_1, p_2, u] = 0 \tag{6.5.21}$$

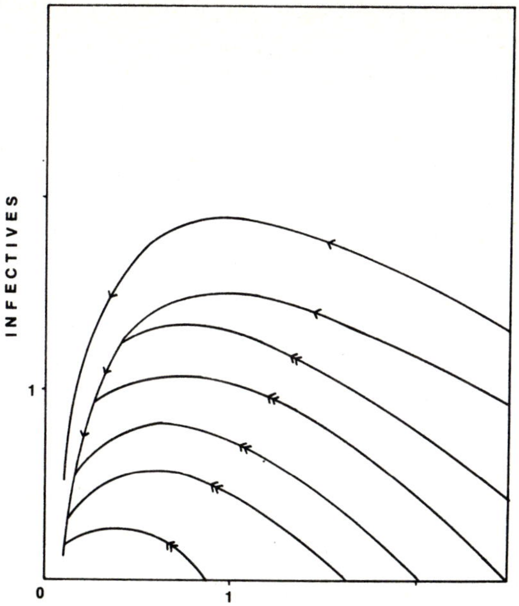

Fig.6.5.2. Optimal control of an epidemic for pest control. ↠ denotes a u_{max} trajectory, and → denotes a $u = 0$ trajectory.

for all values of t in the period $[0, T]$.

Along a singular extremal we have

$$\partial H/\partial u = c_2 p_0 + (p_2 - p_1)S = 0. \tag{6.5.22}$$

For the sake of clarity the total derivative with respect to time is denoted by D or by a dot. Using (6.5.18) and (6.5.19), we find that along an optimal singular extremal, we have

$$D(\partial H/\partial u) = [p_1 rS + p_2(w - rS)]S = 0. \tag{6.5.23}$$

As $S \neq 0$ we have

$$p_1 rS + p_2(w - rS) = 0. \tag{6.5.24}$$

Using (6.5.21) and (6.5.22), we get

$$H = p_0 c_1 - [p_1 rS + p_2(w - rS)]I = 0. \tag{6.5.25}$$

Eqs. (6.5.24) and (6.5.25) and the assumption that c_1 is positive give $p_0 = 0$. From (6.5.22) it follows that $p_1 = p_2$. Substituting $p_1 = p_2$ into (6.5.24), we get $p_2 = 0$. Therefore $p_0 = 0$, $p_1 = 0$ and $p_2 = 0$. But (p_0, p_1, p_2) must not be a zero vector. Hence there is no optimal singular control. We conclude that the optimal trajectory from a given initial state is a sequence of $u = 0$ arcs and $u = b$ arcs.

6.6. OPTIMAL MANAGEMENT OF GREENHOUSE CROPS

In desert areas the major limiting factor for crops is water. To conserve it, vegetable crops are grown in greenhouses. The temperature in this type of greenhouse is lowered at night so that food made by the plants during the day is not entirely lost during the night. Associated with the greenhouse is a nursery used to produce seedlings which are ready for transplanting after a fixed time interval. Since it is relatively cheap to grow the seedlings, the costs in doing so are ignored in the subsequent analysis. After transplanting the seedlings grow to maturity and produce fruit.

The quantity of fruit produced per week varies considerably with the lapse of time. It declines after a number of weeks. Thus the manager has to decide when to terminate the current crop and to plant the next one. If the crop is terminated too soon, too much time will have been devoted to unproductive vegetative plant growth. On the other hand, if the crop is not terminated soon enough, too much time will have been devoted to perpetuating old plants which have falling productivity rates (see Goh et al., 1975).

The determination of an optimal crop length falls in the class of problems known as the "replacement problem". Similar problems occur in the production of milk (Smith, 1973), eggs and timber (Gaffney, 1960; Pease, 1967; Clark, 1976 a; Dillon, 1977). Some of the papers on the replacement problem make use of dynamic programming (Faris, 1960; Burt and Allison, 1963); Schreuder, 1968; Smith, 1973).

A common assumption in the analysis of the replacement problem is that a predictive and deterministic model of the value function or the cumulative production function of a crop is available. This assumption is not satisfied by the tomato crop in a greenhouse. Over a fixed period of 26 weeks, and under normal conditions, the cumulative production of a tomato crop in a greenhouse in Arizona can vary between 80 000 lbs and 170 000 lbs per acre. Because of this variability we shall develop an algorithm based on past and current production data and current production trends. This extends the Faustmann algorithm which is based on a deterministic model of the cumulative production function.

For simplicity let us consider the case in which (i) the price for the produce is constant; (ii) the market is large and will accept all the production of the crop; and (iii) there is a single greenhouse. After the initial analysis we shall discuss how these assumptions can be relaxed.

Firstly it is necessary to build a production model of an average crop. Suppose that production data from N crops of tomatoes in a greenhouse are available. Let t be the number of weeks from the date the seedlings are planted. At t weeks from the transplant date, let $Y_n(t)$ be the weekly production per acre from the nth crop. By definition, the weekly production per unit area for the average crop at t weeks from the transplant date is

$$Y(t) = \frac{1}{N} \sum_{n=1}^{N} Y_n(t). \qquad (6.6.1)$$

Let $S(t)$ be the cumulative yield per acre for the average crop at t weeks from the transplant date. By definition,

$$S(t) = \sum_{i=0}^{t} Y(i). \qquad (6.6.2)$$

The average production per week from a greenhouse at time t is

$$R(t) = S(t)/t. \qquad (6.6.3)$$

This is the slope of the line which joins the origin to a point $(t, S(t))$ on the graph of $S(t)$. In most replacement problems the function $S(t)$ is a sigmoid function such that $R(t)$ has a unique maximum at $t = t^\star$. According to the Faustmann algorithm, $t^\star$ is the optimal crop length (Gaffney, 1960; Clark, 1976 a).

Instead of being weekly production data, $Y(t)$ could denote the net weekly revenue per acre from an average crop at t weeks from the transplant date. This would be one way of handling the variation in the costs of maintaining and harvesting the crop as t varies. In this case $S(t)$ would denote the net cumulative revenue per acre from the crop at t weeks from the transplant date (Dillon, 1977, p.72).

We shall now examine some ways to overcome the large and unpredictable variations in the total (cumulative) yield from different crops. The main

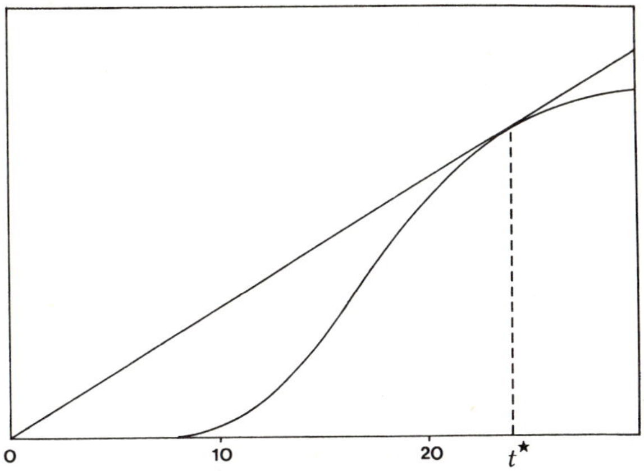

Fig.6.6.1. A graphical method for computing the optimal length of an average crop.

thrust of our approach is to make decisions based on the latest available information on the current crop.

Let $Y^c(t)$ denote the weekly production of the current crop at t weeks from the transplant date. Let $S^c(t)$ be the cumulative yield per acre of the current crop at t weeks from the transplant date. The average yield per acre from the current crop at t weeks from the transplant date is denoted by $R^c(t) = S^c(t)/t$.

By examining the unripened fruits on the plants it is possible to predict a few weeks ahead of time the value of $S^c(t)$. Thus we can estimate a few weeks ahead of time the week t'_c which will maximize the average yield per acre for the current crop. With this information the following scheduling algorithm may be used: (i) if $R^c(t'_c)$ is significantly smaller than $R(t^\star)$ the current crop should be terminated immediately; (ii) if $R^c(t')$ is approximately equal to $R(t^\star)$ the current crop should be terminated at $t = t^\star$; (iii) if $R^c(t'_c)$ is significantly greater than $R(t^\star)$, the crop should be prolonged beyond t' until t^e_c where $Y^c(t^e_c)$ is approximately equal to $R(t^\star)$, or until t''_c where $R^c(t''_c) = R(t^\star)$.

The rationale behind this feedback algorithm is as follows. In case (i) we are comparing the current crop with an average crop. If the current crop is projected to be bad, i.e. $R^c(t'_c) \ll R(t^\star)$, we are wasting our time with it. If a new crop is started there is a good chance that the average yield per week, including the weeks required for a new crop to mature, will be greater. Further extending the crop until $t = t'_c$ gives an immediate short-term advantage, but it lowers the total production in the long run.

In case (ii) the current crop is approximately average. Since $t^\star$ is the best time to terminate an average crop we should terminate the current crop accordingly, i.e. at $t = t^\star$.

In case (iii) the current crop is projected to be better than the average crop as $R^c(t'_c) > R(t^\star)$. Thus the chances are that if we terminate the crop at t'_c, the maximum average production per week of the next crop $R(t^\star)$ will not be as large as $R^c(t'_c)$. It is desirable to prolong the current crop. In the case of tomatoes, the weekly production function $Y^c(t)$ is very erratic as t varies. It depends to some extent on the skill of the labour that harvests the fruit. Thus it is not always a good idea to terminate a crop when the weekly production $Y^c(t^e_c)$ for the current crop falls below $R(t^\star)$. Here, the manager must use his own judgement; from observations of the unripened fruit on the plants he should estimate t''_c where $R^c(t''_c) = R(t^\star)$. If t^e_c is close to t''_c and $Y^c(t) \simeq R(t^\star)$, then it is desirable to terminate the crop.

We shall now relax some of the assumptions of the above analysis. In the real world the price of tomatoes varies considerably during the year. For the southern states of U.S.A. the price of tomatoes is depressed in May and from the second half of July to the end of October because of an influx into the market of (i) tomatoes which are locally grown in open fields; and (ii) tomatoes which are imported from Mexico. In practice it is desirable to sell a significant portion of the produce by contract. This requires that the weekly production throughout the year must be greater than a minimum level. On the other hand, an upper limit should be arbitrarily imposed on the total weekly production so as not to saturate the market and cause the price of tomatoes to fall below economically viable levels.

These complications in the real world can be successfully accomodated by partitioning a greenhouse into several sections so that different stages of maturity of the crop can be achieved simultaneously. If the previously stated algorithm is used in each section independently, then the total production in the *long run* is maximized; this is independent of the number of partitions made. We can make as many partitions as we like without significantly altering the long-term maximum production. Naturally the more partitions there are the "smoother" the total weekly production curve for a group of greenhouses will be.

Let $G(s)$ denote the total weekly production from a group of greenhouses at time s. The ability to partition a greenhouse without significantly altering the maximum long-term total production can be used to shape the function $G(s)$ so that it satisfies the maximum and minimum production constraints, and has "peaks" which coincide with the peaks in the market price for the produce. This part of the analysis can be carried out very simply by simulations on a computer.

The results of this analysis have been successfully applied to the management of the production of tomatoes from 10 acres of greenhouses in Tucson, Ariz. (Goh et al., 1975).

An important lesson from this real life problem is that it is worthwhile to reduce a problem to its barest essentials in the initial analysis. The results of the initial analysis can provide considerable insight into the solution of the problem in the real world. These insights can then be used in a computer simulation model which incorporates all those complications which are thought to be important. If we begin with a complex computer simulation model it is difficult to know how to extract useful information from the model.

CHAPTER 6: SELECTED REFERENCES

Pest management

DeBach (1964); Becker (1970); Geier (1970); Hussey and Bravenboer (1971); Jacquette (1972); Goh et al. (1974 a); Giese et al. (1975); Vincent (1975); Corbet and Smith (1976); Huffaker and Messenger (1976); Ruesink (1976); Tummala et al. (1976); Cohen (1977); Conway (1977); Shoemaker (1977); Tummala and Haynes (1977); Ludwig et al. (1978).

Stability concepts in pest control

Hassell and May (1973); Diamond (1974); Oaten and Murdoch (1975); Beddington et al.

(1976); Hassell et al. (1976 a, 1977); Freedman (1976); Murdoch (1977); Anderson and May (1978); May and Anderson (1978).

Graphical analysis

Cunningham (1958); Rosenzweig and MacArthur (1963); Strebel and Goel (1973); Goh et al. (1974 b); Noy-Meir (1975, 1978); Freedman (1976); Goh (1979 a).

Optimal control of pests

Becker (1970); Mann (1971); Jacquette (1972); Shoemaker (1973, 1977); Goh et al. (1974 a, b); Conway (1977).

REFERENCES

Aggarwal, J.K., 1972. Notes on Nonlinear Systems. Van Nostrand Rheinhold, New York, N.Y., 279pp.
Agnew, T.T., 1979. Optimal exploitation of a fishery employing a non-linear harvesting function. Ecol. Modelling, 6:47—57.
Albrecht, F., Gatzke, H., Haddad, A. and Wax, N., 1976. On control of certain interacting populations. J. Math. Anal. Appl., 53: 578—603.
Allen, K.R., 1953. A method for computing the optimum size limit for a fishery. Nature, 172: 210.
Allen, K.R., 1973. Analysis of stock-recruitment relations in Antarctic fin whales. Rapp. P.V. Reun. Cons. Int. Explor. Mer, 164: 132—137.
Allen, K.R. and Kirwood, G.P., 1977. Were 1946 Sperm Whale Stocks at the Unexploited Level? CSIRO Div. Fish. and Oceanogr., Rep. No. 87.
Anderson, L.G., 1977. The Economics of Fisheries Management. The Johns Hopkins Univ. Press, Baltimore, Md., 214pp.
Anderson, R.M. and May, R.M., 1978. Regulation and stability of host—parasite population interactions. I. Regulatory Processes. J. Anim. Ecol., 47: 219—247.
Annual Reports of the International Whaling Commission, Cambridge, U.K.
Araki, M. and Kondo, B., 1972. Stability and transient behavior of composite nonlinear systems. IEEE Trans. Autom. Control, 17: 537—541.
Arrow, K.J. and Hahn, F.H., 1971. General Competitive Analysis. Holden-Day, San Francisco, Calif., 452pp.
Arrow, K.J., Block, H.D. and Hurwicz, L., 1959. On the stability of the competitive equilibrium, II. Econometrica, 27: 82—109.
Ashkenazi, M. and Othmer, H.G., 1978. Spatial patterns in coupled biochemical oscillators. J. Math. Biol., 5: 305—350.
Astrom, K.J., 1970. Introduction to Stochastic Control Theory. Academic Press, New York, N.Y., 299pp.
Auslander, D., Oster, G. and Huffaker, C.B., 1974. Dynamics of interacting populations. J. Franklin Inst., 297: 345—376.
Bailey, N.T.J., 1975. The Mathematical Theory of Infectious Diseases and its Applications. Griffin, London, 413pp.
Banks, H.T., 1975. Modelling and control in the biomedical sciences. In: S.A. Levin (Editor), Lecture Notes in Biomathematics, Vol. 6. Springer-Verlag, Berlin, pp. 1—114.
Barbashin, E.A. and Krasovskii, N.N., 1952. On the stability of a motion in the large. Dokl. Akad. Nauk. SSSR, 86: 453—456.
Barnett, S., 1975. Introduction to Mathematical Control Theory. Clarendon Press, Oxford, 264pp.
Becker, N.G., 1970. Control of a pest population. Biometrics, 26: 365—375.
Beddington, J.R., 1978. On the dynamics of Sei whales under exploitation. Annu. Rep. Int. Whaling Comm., 28: 169—172.

Beddington, J.R. and May, R.M., 1977. Harvesting natural populations in a randomly fluctuating environment. Science, 197: 463—465.
Beddington, J.R. and Taylor, D.B., 1973. Optimum age specific harvesting of a population. Biometrics, 29: 801—809.
Beddington, J.R., Free, C.A. and Lawton, J.H., 1975. Dynamic complexity in predator—prey models framed in difference equations. Nature, 255: 58—60.
Beddington, J.R., Free, C.A. and Lawton, J.H., 1976. Concepts of stability and resilience in predator—prey models. J. Anim. Ecol., 45: 791—816.
Bell, D.J. and Jacobson, D.H., 1975. Singular Control Problems. Academic Press, New York, N.Y., 190pp.
Bellman, R.E., 1957. Dynamic Programming. Princeton University Press, Princeton, N.J., 342pp.
Berkovitz, L.A., 1961. Variational methods of control and programming. J. Math. Anal. Appl., 3: 145—169.
Beverton, R.J.H. and Holt, S.J., 1957. On the dynamics of exploited fish populations. Fish. Invest. Minist. Agric. Fish. Food G.B. Ser. II Salmon Freshwater Fish., 19: 1—533.
Bliss, G.A., 1946. Lectures on the Calculus of Variations. University of Chicago Press, Chicago, Ill., 296pp.
Booth, D.E., 1972. A model for optimal salmon management. Fish. Bull., 70: 497—507.
Botkin, D.B. and Sobel, M.J., 1974. The complexity of ecosystem stability. In: S.A. Levin (Editor), Ecosystem Analysis and Prediction. SIAM, Philadelphia, Pa., pp. 144—150.
Brauer, F., 1977. Stability of some population models with delay. Math. Biosci., 33: 345—358.
Brauer, F., 1979. Asymptotic behavior of a class of differential-difference equations. To appear.
Brauer, F. and Sanchez, D.A., 1975. Constant rate population harvesting: equilibrium and stability. Theor. Popul. Biol., 8: 12—30.
Brauer, F., Soudack, A.C. and Jarosch, H.S., 1976. Stabilization and de-stabilization of predator—prey systems under harvesting and nutrient enrichment. Int. J. Control, 23: 553—573.
Bryson, A.E. and Ho, Y.C., 1969. Applied Optimal Control. Blaisdell, Waltham, Mass., 481 pp.
Bulmer, M.G., 1976. The theory of prey—predator oscillations. Theor. Popul. Biol., 9: 137—150.
Burges, H.D. and Hussey, N.W., 1971. Microbial Control of Insects and Mites. Academic Press, New York, N.Y., 861pp.
Burt, O.R. and Allison, J.R., 1963. Farm management decisions with dynamic programming. J. Farm Econ., 45: 121—136.
Canon, M.D., Cullum, C.D. and Polak, E., 1970. Theory of Optimal Control and Mathematical Programming. McGraw-Hill Book Company, New York, N.Y., 285pp.
Case, T.D. and Casten, R., 1979. Global stability and multiple domains of attraction in ecological systems. Am. Nat., in press.
Caswell, H., 1978. Predator-mediated coexistence: a nonequilibrium model. Am. Nat., 112: 127—154.
Christiansen, F.B. and Fenchel. T.M., 1977. Theories of Populations in Biological Populations. Springer-Verlag, New York, N.Y., 144pp.
Clark, C.W., 1971. Economically optimal policies for the utilization of biologically renewable resources. Math. Biosci., 12: 245—260.
Clark, C.W., 1973 a. The economics of over-exploitation. Science, 181: 630—634.
Clark, C.W., 1973 b. Profit maximization and the extinction of animal species. J. Polit. Econ., 81: 950—961.
Clark, C.W., 1976 a. Mathematical Bioeconomics: The optimal Management of Renewable Resources. Wiley-Interscience, New York, N.Y., 352pp.
Clark, C.W., 1976 b. Delayed-recruitment model of population dynamics with an appli-

cation to baleen whale population. J. Math. Biol., 3: 381—392.
Clark, C.W., 1978. Restricted entry to common-property fishery resources: a game-theoretic analysis. Inst. Appl. Math. and Statist., Tech. Rep. No. 78-9, Univ. of British Columbia, pp. 1—31.
Clark, C.W., 1979. Mathematical models in the economics of renewable resources. SIAM Rev., 21: 81—99.
Clark, C.W. and Mangel, M., 1979. Aggregation and fishery dynamics: a theoretical study of schooling and the purse-seine tuna fisheries. Fish. Bull., in press.
Clark, C.W. and Munro, G.R., 1978. Renewable resource management and extinction. J. Environ. Econ. Manage., 5: 198—205.
Clark, C.W., Clarke, F.H. and Munro, G.R., 1979. The optimal exploitation of renewable resource stocks: problems of irreversible investment. Econometrica, 47: 25—47.
Clark, L.R., Geier, P.W., Hughes, R.D. and Morris, R.F., 1967. The Ecology of Insect Populations. Methuen, London, 232pp.
Cliff, E.M. and Vincent, T.L., 1973. An optimal policy for a fish harvest. J. Optimization Theory Appl., 12: 485—496.
Cohen, J.E., 1977. Mathematical models of schistosomiasis. Annu. Rev. Ecol. Syst., 8: 209—233.
Comins, H.N. and Blatt, W.E., 1974. Prey—predator models in spatially heterogeneous environments. J. Theor. Biol., 48: 75—83.
Conway, G.R., 1977. Mathematical models in applied ecology. Nature, 269: 291—297.
Corbet, P.S. and Smith, R.F., 1976. Is integrated control realistic? In: C.B. Huffaker and P.S. Messenger (Editors). Theory and Practice of Biological Control. Academic Press, New York, N.Y., pp. 661—682.
Cross, G.W., 1978. Three types of matrix stability. Linear Algebra Appl., 20: 253—263.
Cunningham, W.J., 1958. Introduction to Nonlinear Analysis. McGraw-Hill, New York, N.Y., 349pp.
Cushing, D.H., 1975. Fisheries Resources of the Sea and their Management. Oxford Univ. Press, Oxford, 87pp.
Cushing, J.M., 1977. Integrodifferential equations and delay models in population dynamics. In: S.A. Levin (Editor), Lecture Notes in Biomathematics, Vol. 20. Springer-Verlag, Berlin, pp. 1—196.
Cushing, D.H. and Harris, J.G.K., 1973. Stock and recruitment and the problem of density-dependence. Rapp. P.V. Reun. Cons. Int. Explor. Mer, 164: 142—155.
Dantzig, G.B., 1963. Linear Programming and Extensions. Princeton University Press, Princeton, N.Y., 621pp.
Davis, P.E., Myers, K. and Hoy, J.B., 1976. Biological control among vertebrates. In: C.B. Huffaker and P.S. Messenger (Editors), Theory and Practice of Biological Control. Plenum Press, New York, N.Y., pp. 501—519.
DeAngelis, D.L., 1975. Stability and connectance in food web models. Ecology, 56: 238—243.
DeBach, P. (Editor), 1964. Biological Control of Insect Pests and Weeds. Chapman and Hall, London, 844pp.
Diamond, P., 1974. The stability of the interaction between entomophagous parasites and their hosts. Math. Biosci., 19: 121—129.
Diamond, P., 1976 a. Domains of stability and resilience for biological populations obeying difference equations. J. Theor. Biol., 61: 287—306.
Diamond, P., 1976 b. Chaotic behaviour of systems of difference equations. Int. J. Syst. Sci., 7: 953—956.
Dietz, K., 1967. Epidemics and rumours: a survey. J. R. Statist. Soc. A, 130: 505—527.
Dillon, J.L., 1977. The Analysis of Response in Crop and Livestock Production. Pergamon Press, Oxford, 216pp.

Dixon, L.C.W. and Szego, G.P., 1975. Towards Global Optimization. North-Holland, Amsterdam, 472pp.

Doutt, R.L. and Nakata, J., 1965. Parasites for control of grape leafhopper. Calif. Agric., 19: 3.

Elton, C.S., 1958. The Ecology of Invasion by Animals and Plants. Methuen, London, 181pp.

Falcon, L.A., 1971. Use of bacteria for microbial control of insects. In: H.D. Burges and N.W. Hussey (Editors), Microbial Control of Insects and Mites. Academic Press, New York, N.Y., pp. 67—95.

Falcon, L.A., 1976. Problems associated with the use of arthropod viruses in pest control. Annu. Rev. Entomol. 21: 305—324.

Faris, J.E., 1960. Analytical techniques used in determining the optimum replacement Pattern. J. Farm Econ., 42: 755—766.

Feldman, M.W. and Roughgarden, J., 1975. A population's stationary distribution and chance of extinction in a stochastic environment with remarks on the theory of species packing. Theor. Popul. Biol., 7: 197—207.

Fenner, F. and Ratcliffe, F.N., 1965. Myxomatosis. Cambridge Univ. Press, Cambridge, 379pp.

Fiacco, A.V. and McCormick, G.P., 1968. Nonlinear Programming: Sequential Unconstrained Minimization Techniques. Wiley, New York, N.Y., 210pp.

Fisher, M.E. and Goh, B.S., 1977. Stability in a class of discrete time models of interaction populations. J. Math. Biol., 4(1977) 265—274.

Fisher, M.E. and Kloeden, P.E., 1978. Asymptotic behaviour of a class of discontinuous difference equations. J. Aust. Math. Soc., 20 (Series B): 370—374.

Fisher, M.E., Goh, B.S. and Vincent, T.L., 1979. Some stability conditions for discrete time single species models. Bull. Math. Biol., in press.

Fleming, W.H. and Rishel, R.W., 1975. Deterministic and Stochastic Optimal Control. Springer-Verlag, Berlin, 222pp.

Fletcher, R., 1975. An ideal penalty method for constrained optimization. J. Inst. Math. Appl., 15: 319—342.

Freedman, H.I., 1976. Graphical stability, enrichment, and pest control by a natural enemy. Math. Biosci., 31: 207—225.

Freedman, H.I. and Waltman, P., 1977. Mathematical analysis of some three species food-chain models. Math. Biosci., 33: 257—276.

Gaffney, M.M., 1960. Concepts of Financial Maturity in Timber. North Carolina State College, Dep. of Agric. Econ. Ser. No. 62.

Gambell, R., 1977. Whale conservation. Role of the International Whaling Commission. Mar. Policy, 1: 301—310.

Gard, T.C. and Hallam, T.G., 1979. Persistence in food webs: I. Lotka—Volterra food chains. Bull. Math. Biol., in press.

Gardner, M.R. and Ashby, W.R., 1970. Connectance of large dynamical (cybernetic) systems: critical values for stability. Nature, 228: 784.

Gatto, M. and Rinaldi, S., 1977. Stability analysis of predator—prey models via the Liapunov method. Bull. Math. Biol., 39: 339—347.

Gause, G.F., 1934. The Struggle for Existence. Williams and Wilkins, Baltimore, Md., 163pp.

Geier, P.W., 1970. Temporary suppression, population management, or eradication: how and when to choose. In: R.L. Rabb and F.E. Guthrie (Editors), Concepts of Pest Management. North Carolina State University, Raleigh, N.C., pp. 170—189.

Getz, W.M., 1977. On harvesting two competing populations. CSIR Tech. Rep., TWISK 7, NRIMS, CSIR, Pretoria, pp. 1—29.

Giese, R.L., Peart, R.M. and Huber, R.T., 1975. Pest management. Science, 187: 1045—1052.

Gilbert, N., Gutierrez, A.P., Frazer, B.D. and Jones, R.E., 1976. Ecological Relationships. W.H. Freeman and Co., San Francisco, Calif.

Gilpin, M.E., 1972. Enriched predator—prey systems: theoretical stability. Science, 177: 902—904.

Gilpin, M.E., 1974. A Liapunov function for competition communities. J. Theor. Biol., 44: 35—48.

Gilpin, M.E., 1975 a. Group Selection in predator—prey Communities. Princeton Univ. Press, Princeton, N.J., 108pp.

Gilpin, M.E., 1975 b. Stability of feasible predator—prey systems. Nature 254: 137—139.

Gilpin, M.E. and Ayala, F.J., 1973. Global models of growth and competition. Proc. Nat. Acad. Sci. (U.S.A.), 70: 3590—3593.

Gilpin, M.E. and Case, T.J., 1976. Multiple domains of attraction in competition communities. Nature, 261: 40—42.

Goh, B.S., 1966 a. The second variation of the singular Bolza problem. SIAM J. Control, 4: 309—325.

Goh, B.S., 1966 b. Necessary conditions for singular extremals involving multiple control variables. SIAM J. Control, 4:716—731.

Goh, B.S., 1966 c. Necessary Conditions for Singular Extremals in the Calculus of Variations. Ph.D. Thesis, Univ. of Canterbury, Canterbury, N.Z., 109pp. mimeographed.

Goh, B.S., 1969/70. Optimal control of a fish resource. Malay. Sci., 5: 65—70.

Goh, B.S., 1971. The potential utility of control theory to pest management. Proc. Ecol. Soc. Aust., 6: 84—89.

Goh, B.S., 1973 a. Optimal control of renewable resources and pest populations. Proc. 6th Hawaii Int. Conf. on System Sciences, pp. 26—28.

Goh, B.S., 1973 b. Compact forms of the generalized Legendre conditions and the derivation of singular extremals. Proc. 6th Hawaii Int. Conf. on System Sciences, pp. 115—117.

Goh, B.S., 1974. Stability and diversity in ecosystems. Proc. 7th Hawaii Int. Conf. on System Sciences, 241—243.

Goh, B.S., 1975. Stability, vulnerability and persistence of complex ecosystems. Ecol. Modelling, 1: 105—116.

Goh, B.S., 1976 a. Nonvulnerability of ecosystems in unpredictable environments. Theor. Popul. Biol., 10: 83—95.

Goh, B.S., 1976 b. Global stability in two species interactions. J. Math. Biol., 3: 313—318.

Goh, B.S., 1977 a. Global stability in many species systems. Am. Natr., 111: 135—143.

Goh, B.S., 1977 b. Optimum size limit for a fishery with a limited fishing season. Ecol. Modelling, 3: 3—15.

Goh, B.S., 1977 c. Stability in a stock recruitment model of an exploited fishery. Math. Biol. Sci., 33: 359—372.

Goh, B.S., 1978 a. Sector-stability of complex ecosystem models. Math. Biosci., 40: 157—166.

Goh, B.S., 1978 b. Global stability in a class of prey—predator models. Bull. Math. Biol., 40: 525—533.

Goh, B.S., 1978 c. The usefulness of optimal control theory to ecological problems. In: E. Halfon (Editor), Theoretical Systems Ecology. Academic Press, New York, N.Y.

Goh, B.S., 1978 d. Robust stability concepts for ecosystem models. In: E. Halfon (Editor), Theoretical Systems Ecology. Academic Press, New York, N.Y., pp.467—487.

Goh, B.S., 1979. Stability in models of mutualism. Am. Nat., 113: 261—275.

Goh, B.S., 1980. Stability of a scalar nonlinear difference equation. To appear.

Goh, B.S. and Agnew, T.T., 1977. Stability in Gilpin and Ayala's models of competition. J. Math. Biol., 4: 275—279.

Goh, B.S. and Agnew, T.T., 1978. Stability in a harvested population with delayed recruitment. Math. Biosci., 42: 187—197.

Goh, B.S. and Agnew, T.T., 1980. Global and finite stability of a population with delayed recruitment. Ecol. Modelling, in press.

Goh, B.S. and Jennings, L.S., 1977. Feasibility and stability in randomly assembled Lotka—Volterra models. Ecol. Modelling, 3: 63—71.

Goh, B.S., Leitmann, G. and Vincent, T.L., 1974 a. The optimal control of a prey—predator system. Math. Biosci., 19: 263—286.

Goh, B.S., Wilson, D.J. and Vincent, T.L., 1974 b. Suboptimal control of prey-predator system by graphical analysis. Proc. 7th Hawaii Int. Conf. on System Sciences, pp. 196—199.

Goh, B.S., Peng, W.Y., Vincent, T.L. and Riley, J.J., 1975. Optimal management of green house crops. Hortscience, 10 (1): 7—11.

Goh, B.S., Vincent, T.L. and Leitmann, G., 1980. Optimal epidemic programs for pest control. To appear.

Goodman, D., 1975. The theory of diversity—stability relationships in ecology. Q. Rev. Biol., 50: 237—266.

Gopalsamy, K., 1977. Competition and coexistence in spatially heterogenous environments. Math. Biosci., 36; 229—242.

Gordon, H.S., 1954. Economic theory of a common-property resource: the fishery. J. Polit. Econ., 62: 124—142.

Guckenheimer, J., Oster, G. and Ipaktchi, A., 1977. The dynamics of density dependent population models. J. Math. Biol., 4: 101—147.

Gulland, J.A., 1974. The Management of Marine Fisheries. Scientechnica, Bristol, 198pp.

Gulland, J.A., 1977. World fisheries and fish stocks. Mar. Policy, 1: 179—189.

Gulland, J.A., 1978. Fishery management: new strategies for new conditions. Trans. Am. Fish. Soc., 107: 1—11.

Gurel, O. and Lapidus, L., 1968. Stability via Liapunov's second method. Ind. Eng. Chem., 60(6): 1—26.

Gurel, O. and Lapidus, L., 1969. A guide to the generation of Liapunov functions. Ind. Eng. Chem., 61(3): 30—41.

Hadeler, K.P., Van der Heiden, U. and Rothe, F., 1974. Nonhomogeneous spatial distributions of populations. J. Math. Biol., 1: 165—176.

Hagen, K.S., Van den Bosch, R. and Dahlsten, D.L., 1971. The importance of naturally-occurring biological control in the Western United States. In: C.B. Huffaker (Editor), Biological Control. Plenum Press, New York, N.Y., pp. 253—293.

Hahn, W., 1967. Stability of Motion. Springer-Verlag, Berlin, 446pp.

Harrison, G.W., 1979 a. Persistent sets via Liapunov functions. Nonlinear Anal., 3: 73—80.

Harrison, G.W., 1979 b. Global stability of predator—prey interactions. J. Math. Biol., in press.

Hassell, M.P., 1975. Density dependence in single-species populations. J. Anim. Ecol., 44: 283—295.

Hassell, M.P., 1976. The Dynamics of Competition and Predation. Edward Arnold, London, 68pp.

Hassell, M.P., 1978. The Dynamics of Arthropod Predator—Prey System. Princeton University Press, Princeton, N.J., 237pp.

Hassell, M.P. and Comins, H.N., 1976. Discrete time models for two species competition. Theor. Popul. Biol., 9: 202—221.

Hassell, M.P. and May, R.M., 1973. Stability in insect host—parasite models. J. Anim. Ecol., 42: 693—726.

Hassell, M.P., Lawton, J.H. and Beddington, J.R., 1976 a. The components of arthropod predation. I. The prey death rate. J. Anim. Ecol., 45: 135—164.

Hassell, M.P., Lawton, J.H. and May, R.M., 1976 b. Patterns of dynamical behavior in

single-species populations. J. Anim. Ecol., 45: 471—486.
Hassell, M.P., Lawton, J.H. and Beddington, J.R., 1977. Sigmoid functional responses by invertebrate predators and parasitoids. J. Anim. Ecol., 46: 249—262.
Hastings, A., 1977. Spatial heterogeneity and the stability of predator—prey systems. Theor. Popul. Biol., 12: 37—48.
Hastings, A., 1978. Global stability in Lotka—Volterra systems with diffusion. J. Math. Biol., 6: 163—168.
Holling, C.S., 1959. The components of predation as revealed by a study of small mammal predation of the European pine sawfly. Can. Entomol., 91: 293—320.
Holling, C.S., 1966. The functional response of predators to prey density and its role in mimicry and population regulation. Mem. Entomol. Soc. Can., 48: 1—86.
Holling, C.S., 1973. Resilience and stability of ecological systems. Annu. Rev. Ecol. Syst., 4: 1—23.
Holling, C.S. (Editor), 1978. Adaptive Environmental Assessments and Management. Wiley, New York, N.Y., 377pp.
Horn, H.S. and MacArthur, R.H., 1972. Competitive among fugitive species in a harlequin environment. Ecology, 53: 749—752.
Hsu, S.B., 1978 a. On global stability of a predator—prey system. Math. Biosci., 39: 1—10.
Hsu, S.B., 1978 b. Limiting behavior for competing species. SIAM J. Appl. Math., 34: 760—763.
Hsu, S.B., 1978 c. The application of the Poincare transform to the Lotka—Volterra model. J. Math. Biol., 6: 67—73.
Huffaker, C.B., 1958. Experimental studies on predation: Dispersion factors and predator—prey oscillations. Hilgardia, 27: 343—383.
Huffaker, C.B. and Messenger, P.S., 1976. Theory and Practice of Biological Control. Academic Press, New York, N.Y., 788pp.
Hussey, N.W. and Bravenboer, L., 1971. Control of pests in glasshouse culture by the introduction of natural enemies. In: C.B. Huffaker (Editor), Biological Control. Plenum Press, New York, N.Y., pp. 195—216.
Innis, G., 1974 a. Dynamics analysis in "soft science" studies: in defence of difference equations. In: P. van den Driessche (Editor), Mathematical Problems in Biology. Lecture Notes in Biomath., Vol. 2. Springer-Verlag, Berlin, pp. 102—122.
Innis, G., 1974 b. Stability, sensitivity, resilience, persistence. What is of interest? In: S.A. Levin (Editor), Ecosystem Analysis and Prediction. SIAM, Philadelphia, Pa., pp. 131—140.
Intrilligator, M.D., 1971. Mathematical Optimization and Economic Theory. Prentice-Hall, Englewood Cliffs, N.J., 508pp.
Jaquette, D.L., 1972. Mathematical models for controlling growing populations: a survey. Oper. Res., 20: 1142—1151.
Johnson, C.R., 1974. Sufficient conditions for D-stability. J. Econ. Theory, 9: 53—62.
Jorne, J. and Carmi, S., 1977. Liapunov stability of the diffusive Lotka—Volterra equations. Math. Biosci., 37: 51—61.
Jury, E.I., 1974. Inners and Stability of Dynamic Systems. Wiley, New York, N.Y., 308pp.
Kalman, R.E., 1956. Nonlinear aspects of sampled-data control systems. Proc. 2nd Symp. on Nonlinear Circuit Theory, Poly. Inst. of Brooklyn, pp. 273—313.
Kalman, R.E. and Bertram, J.E., 1960 a. Control system analysis and design via the second method of Liapunov. I. Continuous-time systems. ASME J. Basic Eng., 82: 371—393.
Kalman, R.E. and Bertram, J.E., 1960 b. Control system analysis and design via the second method of Liapunov. II. Discrete time systems. ASME J. Basic Eng., 82: 394—400.
Karlin, S. and McGregor, J., 1972. Polymorphism for genetic and ecological systems with

weak coupling. Theor. Popul. Biol., 3: 210—238.
Kazarinoff, N.D. and Van den Driessche, P., 1978 A model predator—prey system with functional response. Math. Biosci., 39: 125—134.
Kelley, H.J., Kopp, R.E. and Moyer, H.G., 1967. Singular extremals. In: G. Leitmann (Editor), Topics in Optimization. Academic Press, New York, N.Y., pp. 63—101.
Kilmer, W.L., 1972. On some realistic constraints in prey—predator systems. J. Theor. Biol., 36: 9—22.
Kolmogoroff, A.N., 1936. Sulla teoria di Volterra della lotta per l'esistenza. G. Inst. Ital. Attuari, 7: 74—80.
Kuester, J.L. and Mize, J.H., 1973. Optimization Techniques with FORTRAN. McGraw-Hill, New York, N.Y., 500pp.
Kushner, H.J., 1967. Stochastic Stability and Control. Academic Press, New York, N.Y., 161pp.
Laing, J.E. and Junji Hamai, 1976. Biological control of insect pests and weeds by imported parasites, predators and pathogens. In: C.B. Huffaker and P.S. Messenger (Editors), Theory and Practice of Biological Control. Academic Press, New York, N.Y., pp. 686—743.
Larkin, P.A., 1963. Interspecific competition and exploitation. J. Fish. Res. Board Can., 20: 647—678.
Larkin, P.A., 1966. Exploitation on a type of predator—prey relationship. J. Fish. Res. Board Can., 23: 349—356.
Larkin, P.A., 1977. An epitaph for the concept of maximum sustained yield. Trans. Am. Fish. Soc., 106: 1—11.
LaSalle, J.P., 1976. The Stability of Dynamical Systems. SIAM, Philadelphia, Pa., 76pp.
LaSalle, J.P., 1980. Stability and Control of Discrete Time Systems. Springer-Verlag, Berlin, to appear.
LaSalle, J. and Lefschetz, S., 1961. Stability by Liapunov's Direct Method. Academic Press, New York, N.Y., 134pp.
Lashiher, A.M. and Storey, C., 1972. Final stability with some applications. J. Inst. Math. Appl., 9: 397—410.
Lawden, D.F., 1963. Optimal Trajectories for Space Navigation. Butterworth, London, 126pp.
Lee, E.B. and Markus, L., 1967. Foundations of Optimal Control Theory, John Wiley, New York, N.Y., 576pp.
Lefkovitch, L.P., 1965. The study of population growth in organisms grouped by stages. Biometrics., 21: 1—18.
Leitmann, G., 1966. An Introduction to Optimal Control. McGraw-Hill, New York, N.Y., 163pp.
Leitmann, G. and Marzollo, A., 1975. Multicriteria Decision Making. Springer-Verlag, Wien, 386pp.
Leslie, P.H., 1945. The use of matrices in certain population mathematics. Biometrika, 33: 183—212.
Leslie, P.H., 1948. Some further notes on the use of matrices in population mathematics. Biometrika, 35: 213—245.
Leung, A., 1976. Limiting behavior for several interacting populations. Math. Biosci., 29: 85—98.
Leung, A., 1978. Limiting behaviour for a prey—predator model with diffusion and crowding effects. J. Math. Biol., 6: 87—93.
Levin, S.A., 1974. Dispersion and population interactions. Am. Nat., 108: 207—228.
Levin, S.A., 1976 a. Population dynamic models in heterogenous environments. Annu. Rev. Ecol. Syst., 7: 287—310.
Levin, S.A., 1976 b. Spatial patterning and the structure of ecological communities.

In: S.A. Levin (Editor), Some Mathematical Questions in Biology, Vol. 7. Am. Math. Soc., Providence, R.I., pp.1—36.
Levins, R., 1966. The strategy of model building. Am. Sci., 54: 421—431.
Levins, R., 1968. Ecological engineering: theory and practice. Q. Rev. Biol., 43: 301—305.
Levins, R. and Culver, D., 1971. Regional coexistence of species and competition between rare species. Proc. Nat. Acad. Sci. (U.S.A.), 68: 1246—1248.
Lewontin, R.C., 1969. The meaning of stability. In: Diversity and Stability in Ecological Systems. Brookhaven Symp. Biol., 22: 13—24.
Li, T.Y. and Yorke, J.A., 1975. Period three implies chaos. Am. Math. Monthly, 82: 985—992.
Liapunov, A.M., 1966. Stability of Motion. Academic Press, New York, N.Y., 203pp.
Lotka, A.J., 1925. Elements of Physical Biology. Williams and Wilkins, Baltimore, Md., 465pp.
Ludwig, D., 1974. Stochastic Population Theories. Springer-Verlag, Berlin, 108pp.
Ludwig, D., 1975. Persistence of dynamical systems under random perturbations. SIAM Rev., 17: 605—640.
Ludwig, D., Jones, D.D. and Holling, C.S., 1978. Qualitative analysis of insect outbreak systems: the spruce budworm and forest. J. Anim. Ecol., 47: 315—332.
MacArthur, R.H., 1955. Fluctuations of animal populations and a measure of community stability. Ecology, 36: 533—536.
MacDonald, N., 1976. Time delay in prey—predator models. Math. Biosci., 28: 321—330.
Mancill, J.D., 1950. Identically non-regular problems in the calculus of variations. Rev. Mat. Y. Fis. Teo., Ser. A, 7: 131—139.
Mann, S.H., 1971. Mathematical models for the control of pest populations. Biometrics, 27: 357—368.
Marotto, F.R., 1978. Snap-back repellers imply chaos in R^n. J. Math. Anal. Appl., 63: 199—223.
May, R.M., 1972. Limit cycles in predator—prey communities. Science, 177: 900—902.
May, R.M., 1974. Stability and Complexity in Model Ecosystems. Princeton University Press, Princeton, N.J., 2nd edn, 265pp.
May, R.M., 1976 a. Simple mathematical models with very complicated dynamics. Nature 261: 459—467.
May, R.M., 1976 b. Theoretical Ecology: Principles and Applications. Blackwell, Oxford, 317pp.
May, R.M., 1977. Thresholds and breakpoints in ecosystems with a multiplicity of stable states. Nature, 269: 471—477.
May, R.M. and Anderson, R.M., 1978. Regulation and stability of host-parasite population interactions. II. Destabilizing processes. J. Anim. Ecol., 47: 249—267.
May, R.M. and Oster, G.F., 1976. Bifurcations and dynamic complexity in simple ecological models. Am. Nat., 110: 573—599.
Maynard Smith, J., 1974. Models in Ecology. Cambridge Univ. Press, Cambridge, 146 pp.
Mazanov, A., 1978. Acceptor control in model ecosystems. J. Theor. Biol., 71: 21—38.
McNaughton, S.J., 1978. Stability and diversity of ecological communities. Nature, 274: 251—253.
Miele, A., 1962. Extremization of linear integrals by Green's theorem. In: G. Leitmann (Editor), Optimization Techniques. Academic Press, New York, N.Y., pp. 69—98.
Moran, P.A.P., 1950. Some remarks on animal dynamics. Biometrics, 6: 251—258.
Murdoch, W.W., 1973. The functional response of predators. J. Appl. Ecol., 10: 335—342.
Murdoch, W.W., 1977. Stabilizing effects of spatial heterogeneity in predator—prey systems. Theor. Popul. Biol., 11: 252—273.
Noy-Meir, I., 1975. Stability of grazing systems: an application of predator—prey graphs. J. Ecology, 63: 459—481.

Noy-Meir, I., 1978. Stability in simple grazing models; effects of explicit functions. J. Theor. Biol., 71: 347—380.
Oaten, A. and Murdoch, W.W., 1975. Functional response and stability in predator—prey systems. Am. Nat., 109: 289—298.
Odum, E.P., 1971. Fundamentals of Ecology. W.B. Saunders, Philadelphia, Pa., 3rd edn, 574pp.
Orians, G.H., 1974. Diversity, stability and maturity in natural ecosystems. Proc. 1st Int. Congr. Ecol., The Hague, pp. 62—65.
Othmer, H.G. and Scriven, L.E., 1971. Instability and dynamic pattern in cellular networks. J. Theor. Biol., 32: 507—537.
Paine, R.T., 1966. Food web complexity and species diversity. Am. Nat., 100: 65—75.
Palm, W.J., 1975. Fishery regulation via optimal control theory. Fish. Bull., 73: 830—837.
Palm, W.J., 1977. Stochastic control problems in fishery management. In: E.O. Roxin, P.T. Liu and R.L. Sternberg (Editors), Differential Games and Control Theory II. Marcel Dekker, New York, N.Y., pp. 65—81.
Parker, F.D., 1971. Management of pest populations by manipulating densities of both hosts and parasites through periodic releases. In: C.B. Huffaker (Editor), Biological Control. Plenum Press, New York, N.Y., pp. 365—376.
Patten, B.C. (Editor), 1971. System Analysis and Simulation in Ecology, Vol. 3. Academic Press, New York, N.Y., 601pp.
Pease, P., 1967. The optimal forest rotation. For. Chron., 43; 178—195.
Pella, J.J. and Tomlinson, P.K., 1969. A generalized stock production model. Bull. Inter-Am. Trop. Tuna Comm., 13: 421—496.
Pianka, E.R., 1972. r and K-selection or b and d selection? Am. Nat., 106: 581—588.
Pianka, E.R., 1974. Evolutionary Ecology. Harper and Row Publ., New York, N.Y., 356pp.
Pielou, E.C., 1977. Mathematical Ecology. Wiley, New York, N.Y., 385pp.
Pimm, S.L. and Lawton, J.H., 1977. Number of trophic levels in ecological communities. Nature, 268: 329—331.
Pontryagin, L.S., Boltyanskii, V.G., Gamkrelidze, R.V. and Mischenko, E.F., 1962. The Mathematical Theory of Optimal Processes. Interscience Publishers, Inc., New York, N.Y., 360pp.
Quirk, J. and Saposnik, R., 1968. Introduction to General Equilibrium Theory and Welfare Economics. McGraw-Hill, New York, N.Y., 221pp.
Reed, W.J., 1974. A stochastic model for the economic management of renewable resources. Math. Biosci., 22: 313—337.
Reed, W.J., 1979. Optimal escapement levels in stochastic and deterministic harvesting models. To appear.
Reddingius, J. and Den Boer, P.J., 1970. Simulation experiments illustrating stabilization of animal numbers by spreading of risk. Oecologia, 5: 240—248.
Rescigno, A. and Richardson, I.W., 1973. The deterministic theory of population dynamics. In: R. Rosen (Editor), Foundations of Mathematical Biology, Vol. 3. Academic Press, New York, N.Y., pp. 283—360.
Ricker, W.E., 1954. Stock and recruitment. J. Fish. Res. Board Can., 11: 559—623.
Ricker, W.E., 1975. Computation and interpretation of biological statistics of fish populations. Bull. Fish. Res. Board Can. 191: 1—382.
Roberts, A., 1974. The stability of feasible random system. Nature, 250: 607—608.
Rorres, C. and Fair, W., 1975. Optimal harvesting policy for an age-specific population. Math. Biosci., 24: 31—47.
Rosen, G., 1975. Solutions to systems of nonlinear reaction—diffusion equations. Bull. Math. Biol., 37: 277—289.
Rosenzweig, M.L., 1969. Why the prey curve has a hump. Am. Nat., 103: 81—87.
Rosenzweig, M.L., 1971. Paradox of enrichment: destabilization of exploitation eco-

systems in ecological time. Science, 171: 385—387.
Rosenzweig, M.L. and MacArthur, R., 1963. Graphical representation and stability conditions of predator—prey interactions. Am. Nat., 97: 209—223.
Ruesink, W.G., 1976. Status of systems approach to pest management. Annu. Rev. Entomol., 21: 27—44.
Schaefer, M.B., 1954. Some aspects of the dynamics of populations important to the management of commercial marine fisheries. Bull. Inter-Am. Trop. Tuna Comm., 1: 26—58.
Schevill, W.E. (Editor), 1974. The Whale Problem: a Status Report. Harvard University Press, Cambridge, Mass., 419pp.
Schnute, J., 1977. Improved estimates from the Schaefer production model: theoretical considerations. J. Fish. Res. Board Can., 34: 583—603.
Schoener, T.W., 1973. Population growth regulated by intraspecific competition for energy or time: some simple representations. Theor. Popul. Biol., 4: 56—84.
Schoener, T. W., 1974. Competition and the form of habitat shift. Theor. Popul. Biol., 6: 265—307.
Schreuder, G.F., 1968. Optimal forest investment decisions through dynamic programming. Yale Univ. School, Forestry Bull., No. 72: 1—70.
Schultz, D.G., 1965. The generation of Liapunov functions. In: C.T. Leondes (Editor), Advances in Control Systems, Vol. 2. Academic Press, New York, N.Y., pp. 1—64.
Scudo, F.M., 1971. Vito Volterra and theoretical ecology. Theor. Popul. Biol., 2: 1—23.
Shoemaker, C., 1973. Optimization of agricultural pest management III: results and extensions of a model. Math. Biosci., 18: 1—22.
Shoemaker, C.A., 1977. Pest management models of crop ecosystems. In: C.A.S. Hall and J.W. Day (Editors), Ecosystem Modelling in Theory and Practice. John Wiley and Sons. New York, N.Y., pp. 546—574.
Siljak, D.D., 1978. Large-Scale Dynamic Systems. North-Holland Publ., Amsterdam, 416pp.
Silliman, R.P., 1971. Advantages and limitations of simple fishery models in light of laboratory experiments. J. Fish. Res. Board Can., 28: 1211—1214.
Silvert, W., 1978. The price of knowledge: fisheries management as a research tool. J. Fish. Res. Board Can. 35: 208—212.
Silvert, W. and Smith, W.R., 1977. Optimal exploitation of a multi-species community. Math. Biosci., 33: 121—134.
Simberloff, D., 1976. Experimental zoogeography of islands: effects of island size. Ecology, 57: 629—648.
Skellam, J.G., 1951. Random dispersal in theoretical populations. Biometrika, 38: 196—218.
Slatkin, M., 1974. Competition and regional coexistence. Ecology, 55: 128—134.
Slobodkin, L.B., 1966. Growth and Regulation of Animal Populations. Holt, Rinehart and Winston, New York, N.Y., 184pp.
Smith, B.J., 1973. Dynamic programming of the dairy cow replacement problem. Am. J. Agric. Econ., 55: 100—104.
Smith, V.L., 1969. On models of commercial fishing. J. Polit. Econ., 77: 181—198.
Starr, A.W. and Ho, Y.C., 1969. Nonzero-sum differential games. J. Optimization Theory Appl., 3: 184—206.
Steele, J.H., 1974. The Structure of Marine Ecosystems. Harvard Univ. Press, Cambridge, Mass., 128pp.
Strebel, D.E. and Goel, N.S., 1973. On the isocline methods for analyzing prey—predator interactions. J. Theor. Biol., 39: 211—239.
Strobeck, C., 1973. N species competition. Ecology, 54: 650—654.
Swann, G. and Vincent, T.L., 1977. Optimal control analysis in the chemotherapy of IgG multiple myeloma. Bull. Math. Biol., 39: 317—338.

Tabak, D. and Kuo, B.C., 1971. Optimal Control by Mathematical Programming. Prentice-Inc., Englewood Cliffs, N.J., 237pp.

Takayama, A., 1974. Mathematical Economics. The Dryden Press, Hinsdale, Ill., 744pp.

Takeuchi, Y., Adachi, N. and Tokumaru, H., 1978. The stability of generalized Volterra equations. J. Math. Anal. Appl., 62: 453—473.

Tanada, Y., 1964. Epizootiology of insect diseases. In: P. DeBach (Editor), Biological Control of Insect Pests and Weeds. Chapman and Hall, London, pp. 548—578.

Tuckwell, H.C., 1974. A study of some diffusion models of population growth. Theor. Popul. Biol., 5: 345—357.

Tuljapurkar, S.D. and Semura, J.S., 1979. Stochastic instability and Liapunov stability. J. Math. Biol., 8: in press.

Tummala, R.L. and Haynes, D.L., 1977. On-line pest management. Environ. Entomol., 6: 339—349.

Tummala, R.L., Haynes, D.L. and Croft, B.A. (Editors), 1976. Modeling for Pest Management: Concepts, Techniques and Applications. Michigan State University, East Lansing Mich., 247pp.

Turelli, M., 1977. Random environments and stochastic calculus. Theor. Popul. Biol., 12: 140—178.

Turelli, M., 1978. A reexamination of stability in randomly varying versus deterministic environments with comments on the stochastic theory of limiting similarity. Theor. Popul. Biol., 13: 244—267.

Usher, M.B., 1976. Extensions to models used in renewable resource management which incorporate an arbitrary structure. J. Environ. Manage., 4: 123—140.

Van der Vaart, H.R., 1973. A comparative investigation of certain difference equations and related differential equations: implications for model building. Bull. Math. Biol., 35: 195—211.

Vincent, T.L., 1975. Pest management programs via optimal control theory. Biometrics, 31: 1—10.

Vincent, T.L. and Anderson, L.R., 1979. The vulnerability of long food chains. Theor. Popul. Biol., 15: 213—231.

Vincent, T.L. and Goh, B.S., 1972. Terminally, normality and transversality conditions. J. Optimization Theory Appl., 9: 32—50.

Vincent, T.L. and Leitmann, G., 1970. Control space properties of cooperative games. J. Optimization Theory Appl. 6: 91—113.

Vincent, T.L. and Skowronski, J.M., 1979. Controllability with capture. J. Optimization Theory Appl., in press.

Vincent, T.L., Cliff, E.M. and Goh, B.S., 1974. Optimal direct control programs for a prey—predator system. ASME J. Dynamic Systems, Measurement and Control, 96: 71—76.

Vincent, T.L., Everett, L.G., Lee, C.S. and Pulliam, H.R., 1975. Applications of optimal control to the modelling and management of ecosystems. Simulation, 24 (3): 65—72.

Vincent, T.L. Lee, C.S. and Goh, B.S., 1977. Control targets for the management of biological systems. Ecol. Modelling, 3: 285—300.

Volterra, V., 1931. Lecons sur la Theorie Mathematique de la Lutte pour la Vie. Gauthier-Villars, Paris, 214pp.

Volterra, V., 1938. Population growth, equilibria and extinction under specified breeding conditions: a development and extension of the theory of the logistic equation. Hum. Biol., 10: 1—11.

Walters, C.J., 1975. Optimal harvest strategies for salmon in relation to environmental variability and uncertain production parameters. J. Fish. Res. Board Can., 32: 1777—1785.

Walters, C.J. and Hilborn, R., 1976. Adaptive control of fishing systems. J. Fish. Res. Board Can., 33: 145—159.

Walters, C.J. and Hilborn, R., 1978. Ecological optimization and adaptive management Annu. Rev. Ecol. Syst., 9: 157—188.
Waltman, P., 1974. Deterministic threshold models in the theory of epidemics. In: S.A. Levin (Editor), Lecture Notes in Biomathematics. Springer-Verlag, Berlin, pp. 1—101.
Wangersky, P.J. and Cunningham, W.J., 1957. Time lag in prey—predator population models. Ecology, 38: 136—139.
Watt, K.E.F., 1968. Ecology and Resource Management: A Quantitative Approach. McGraw-Hill, New York, N.Y., 450pp.
Wickwire, K., 1977. Mathematical models for the control of pests and infectious diseases; a survey. Theor. Popul. Biol., 11: 182—238.
Wiegert, R.G., 1975. Simulation models of ecosystems. Annu. Rev. Ecol. Syst., 6: 311—338.
Willems, J.L., 1970. Stability Theory of Dynamical Systems. Nelson, London, 201pp.
Willems, J.L., 1971. Optimum Lyapunov functions and stability regions for multimachine power systems. Proc. IEE, 118: 1631—1632.
Williamson, M.H., 1972. The Analysis of Biological Populations. Edward Arnold, London, 180pp.
Wörz-Busekros, A., 1978. Global stability in ecological systems with continuous time delay. SIAM J. Appl. Math., 35: 123—134.
Wu, L.S.Y., 1974. On the stability of ecosystems. In: S.A. Levin (Editor), Ecosystem Analysis and Prediction. SIAM, Philadelphia, Pa., pp. 155—165.
Yoshizawa, T., 1966. Stability Theory by Liapunov's Second Method. Math. Soc. Japan, Tokyo.

AUTHOR INDEX

The number(s) directly after the initial(s) of the author is (are) the page number(s) on which the author (or his work) is mentioned in the text. Numbers in italics give the page on which the complete reference is listed.

SUBJECT INDEX